D1686103

035371

Growing Food

Growing Food
A Guide to Food Production

By

Tony Winch
Hereford, UK

 Springer

A C.I.P. Catalogue record for this book is available from the Library of Congress.

ISBN-10 1-4020-4827-0 (HB)
ISBN-13 978-1-4020-4827-2 (HB)
ISBN-10 1-4020-4975-7 (e-book)
ISBN-13 978-1-4020-4975-0 (e-book)

Published by Springer,
P.O. Box 17, 3300 AA Dordrecht, The Netherlands.

www.springer.com

Front Cover:
Top left - Beans for seed - Vincent Johnson
Top right - Long grain rice - Keith Weller, ARS/USDA
Lower left - Cambodian vegetable gardener - Vincent Johnson
Lower right - Maize cobs from South America - Keith Weller, ARS/USDA

Back Cover:
Potato seedling in the hand - Scott Bauer, ARS/USDA

Printed on acid-free paper

TABLE OF CONTENTS

SECTION 1
THE PRINCIPLES AND PRACTICES USED IN AGRICULTURE AND HORTICULTURE

SECTION 2
DESCRIPTION AND CHARACTERISTICS
OF THE MAIN FOOD CROPS

SECTION 3

SECTION 1

THE PRINCIPLES AND PRACTICES USED IN AGRICULTURE AND HORTICULTURE

1A. PLANT GROWTH

Green plants need to have an adequate supply of water, warmth, light, air and nutrients in order to grow properly and produce healthy crops. Food growers have some control over all of these factors:

- **Water and Warmth**—the time of planting can be chosen so that plants at each stage of their life are growing when the temperature and rainfall are as near to optimum as possible. Irrigation can sometimes also be used—see pages 93–100.
- **Light and Air**—farmers and other food growers can select the planting date and plant spacing so that adequate light and air is available to the plants–discussed on page 4 **"Plant Population"** and page 42 **"Day length / Photoperiodism"**.
- **Nutrients**—see pages 11–29 **"Soil"** and pages 29–36 **"Fertilizer"**.

The following subsection on plant growth discusses five topics that are of particular interest to food growers:
plant propagation, plant population, leaf area index, the root system and *the Nitrogen Cycle.*

1Aa. Plant Propagation

Plants reproduce themselves, or "propagate", either sexually by means of seeds, or asexually by various processes of vegetative reproduction, described on pages 40–41.

Some plant species, such as onions, can propagate themselves by both methods, using either seed or bulbs ("sets").

Seed, or "grain", provides the biggest proportion of the world's food, especially in poorer countries and, as a result, seed is discussed at greater length in this book than other plant food sources such as root and oilseed crops, fruit and vegetables.

Sexual reproduction in plants, by which they produce seed, is represented in Figure 1. <u>Key</u> to the terms used in the drawing of a flower, Figure 1 overleaf:

Pollen—contains the male gametophytes, and is produced in the anthers. When mature, pollen is released into the air and comes into contact with the stigma of either the same flower, or flower of the same plant, in a process known as self-pollination. When the pollen comes into contact with the stigma in a flower of another plant, this process is known as cross-pollination.

Pollen Tube—a long hollow tube that provides a passageway for the male pollen, containing two male nuclei, to the female ovule, entering via the micropyle.

Pistil—the female part of the flower, consisting of the ovary, stigma and style, around which the other flower parts are arranged.

1

Ovule—after fertilisation the ovule develops into the seed. The egg and the polar nuclei are fertilised by the two male nuclei.

Ovary—develops into a fruit when stimulated by fertilisation, or when auxin or other growth hormones are artificially applied, or in *apomixes*—the production of seed without a male gamete.

Polar Nuclei—the two nuclei near the centre, they develop into the seed endosperm.

Egg Nuclei—develop into the seed embryo.

Integuments—develop into the seed coat, or "testa".

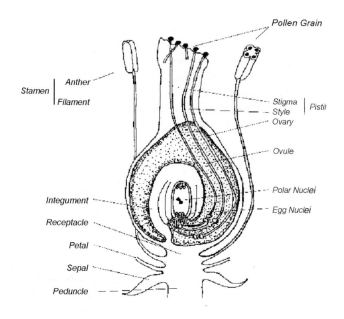

Figure 1. Schematised longisection of a flower

The organisation of the male (*staminate*) and female (*pistillate* or *ovule-producing*) flower parts is different for every plant species, but there are two main types of plants:

Monoecious plants have both male and female flowers on the same plant, eg maize, cucumber, beans and virtually all other food crops.

Dioecious plants have separate male and female flowers, and on different plants, eg asparagus, date palms and most papaya varieties, but very few other food crops.

Flowers are either **self-pollinated** (*autogamy*, producing other "selfer" plants) if pollen is transferred to it from any flower of the same plant, or **cross-pollinated** (*heterogamy*, producing other "out crosser" or "out breeder" plants) if the pollen comes from a flower on a different plant.

About half of the more important food crops are naturally cross-pollinated, and their reproductive systems include various clever mechanisms to encourage cross-pollination and discourage self-pollination, including:

- **Dichogamy, in two forms: Protandry**, the more usual form, often found in insect-pollinated species, where the pollen is shed before the ovules are mature, eg carrot and walnut; and **Protogyny**, where the ovules mature before the pollen is released, eg Arum lilies and many wind-pollinated plants such as grasses; some species such as avocados have both protogynous and protandrous varieties, often grown together to promote cross-fertilisation.
- **Heterostyly**, a structural mechanism in which the plants have variable length *styles* (neck of the *pistil*), eg common primrose, wood sorrel and flax.
- **Dioecy**, where the stamens and pistils are borne on different (dioecious) plants, eg asparagus, date palm, most papaya varieties and hops; and
- **Genetically determined self-incompatibility**, which is activated by chemicals produced by the plant, disables the pollen from growing on the stigma of the same plant, eg cabbage, white clover and many other species.

Cross-pollination has major advantages over self-pollination, since it gives rise to offspring that have much greater genetic variability, providing the potential for the species to adapt. Self-pollination (*"selfing"*) may even lead to the eventual extinction of certain species that are not able to adapt to new circumstances. On the other hand self-pollination may have certain other evolutionary advantages, for example where animal pollinators are temporarily absent, or when individual plants are widely scattered. Self pollination also successfully maintains genotypes that would otherwise be lost in the gene shuffle of sexual reproduction.

Many of the major food crop species are in fact predominantly self-pollinating; examples include wheat, barley, oats, rice, beans, peas and tomatoes. There are relatively few mechanisms that promote self-pollination, the most positive of which is known as **cleistogamy**, where the flowers fail to open, as in certain violets.

Self-pollination is also ensured in various processes known collectively as **apomixis**, which includes *parthenocarpy (parthenogenesis)* and *apogamy* where the ovule develops into a seed without any kind of fertilisation. This development of **seedless fruit** is discussed in Section **2F. "Fruits"**, pages 243–266.

In barley, lettuce and wheat the pollen is shed before or just as the flowers open. With tomatoes, pollination follows opening of the flower, but the stamens form a cone around the stigma. There is always the risk of unwanted cross-pollination occurring with these species.

A cross-pollinated plant, which has two parents, each of which is likely to differ in many genes, produces a diverse population of plants that are hybrid (heterozygous) for many characteristics.

A self-pollinated plant, which has only one parent, produces a more uniform population of plants that are pure breeding (homozygous) for many characteristics.

Thus, in contrast to outbreeders, inbreeders ("self-breeders") are likely to be highly homozygous, and will therefore breed true for certain characteristics.

1Ab. Plant Population

If plants are growing *too close together* (ie they are "too closely spaced") they compete with each other for light, water, nutrients and air, and produce small plants of low quality which are more susceptible to attack by pests and diseases.

If plants are growing *too far apart*, the yield per unit area is reduced and also the plants may become too large and/or woody for consumption or sale. Weeds are also allowed to develop more aggressively in the open spaces between crop plants.

The plant population, or *plant density* or *spacing*, is more critical for some crops than for others. Wheat, for example, can be planted at very different spacings without a big effect on the yield per hectare, because wheat plants can *compensate* for different plant populations. Compensation in plants is the ability to grow large or to remain small in response to the amount of space available to them. Other plants such as maize compensate very poorly and so must be planted at much more precise spacings.

Figures 2 and 3 illustrate the effect of plant population on the yield per hectare and the cob weight of maize:-

Figure 2 shows how maize should be planted more closely together—ie at a higher seed rate - in fertile or highly fertilised soils than in infertile soils or soils with a low or zero fertiliser input.

Figure 3 shows how the cobs of maize become smaller and smaller as the plant population increases, even though the total yield per hectare continues to increase, up to a certain plant population.

The effect of correct spacing of plants is also discussed on page 52 **"Seed Rate"**.

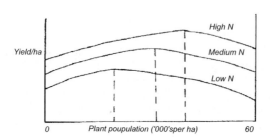

Figure 2. Plant population/Yield per Hectare

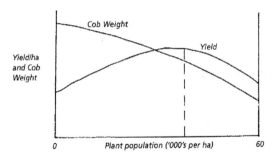

Figure 3. Plant population/Yield per Hectare & Cob Weight

The recommended plant spacings for about 60 of the most common food crops are described in the second paragraph of the Crop Descriptions, in **2A-G** pages 104-287.

1Ac. Leaf Area Index

For optimum plant growth the leaves of crop plants should cover the ground area as soon as possible after planting. By doing this, the plants utilise the sun's energy more efficiently, and they shade out weeds more rapidly; soil moisture loss is also reduced.

The relationship between the leaf area and soil surface area is known as the Leaf Area Index (LAI), and is calculated by dividing the leaf area by the soil surface area. If the LAI at any period in the growing season is less than one, then some of the sun's energy is wasted because some falls onto either bare soil or weeds.

The optimum LAI is different for each crop species. For Irish potatoes (*Solanum tuberosum*) it is about 3, for sugar beet 4–5, and for grasses and most cereals 7–8. If there is not enough leaf—ie if the LAI is too low—then yields will be reduced. If there is too much leaf the lower leaves become too shaded, which also reduces yields because losses due to plant respiration begin to cancel out gains from photosynthesis (**1Ed. "Photosynthesis/Respiration"**, page 39).

Correct plant spacing ensures that the LAI is optimum for plant growth. In temperate climates, where sunlight can be an important limiting factor, the optimum LAI should ideally be reached before the season of maximum light intensity ie during the longest summer days.

From the above, it can be seen that the correct timing of planting (the planting date) can be just as important as the spacing of plants to produce healthy crops.

1Ad. The Root System

The function of the plant roots is to absorb water and nutrients from the soil, to anchor the aerial (above ground) plant parts, and sometimes also to store food.

The young root that bursts out from the seed is called the *radicle*. Depending on the species, this can either persist and become a deep growing *primary root* or *tap root*, or it can be replaced by a more fibrous root system of *secondary roots.*

Adventitious roots are neither primary nor secondary roots, nor do they arise from them, but are roots which develop in an abnormal position from stems or leaves.

Root hairs on the younger roots absorb water by osmosis and nutrients by active selective absorption. This second process requires energy provided by root respiration, which requires oxygen. If the soil is waterlogged, oxygen is unavailable and the roots cannot respire, and so nutrients cannot be absorbed.

1Ae. The Nitrogen Cycle

Plants need nitrogen (N) to make proteins and nucleic acids (DNA and RNA). However, Nitrogen can only be used by plants when it is taken up by their roots, and since Nitrogen is a stable and insoluble gas it has to be changed into soluble Nitrogen compounds before it can be used by plants.

The nitrogen that is found naturally in the soil is mainly in the form of humus and organic matter. Although almost 80% of the air is made up of nitrogen, plants cannot use it (the nitrogen is said to be "unavailable" to them) until it has been broken down by certain specialised soil bacteria, described below, or by lightning, Rhizobia etc. Nitrogen is combined with other atoms to form molecules or ions, when it is said to be "fixed" ie converted from a gas to a solid:

Organic Nitrogen (N_2) ➜ Ammonium (NH_4^+) ➜
Nitrites (NO_2^-)➜ Nitrates (NO_3^-) (& other soluble compounds)

This process is represented in Figure 4 on the next page:

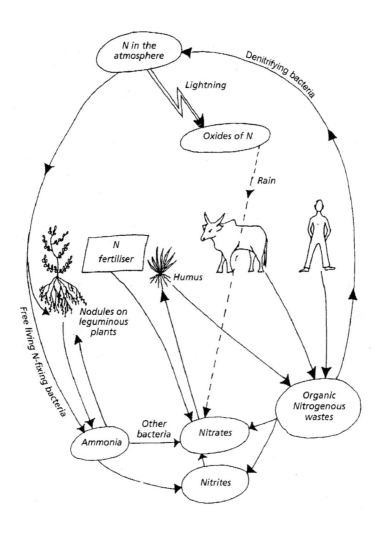

Figure 4. The Nitrogen Cycle

Three kinds of soil bacteria are involved in the Nitrogen Cycle:

1. *Nitrifying bacteria*, such as *Nitrosomonas* (which convert ammonium ions to nitrite ions) and *Nitrobacter* (which oxidise nitrite ions to nitrate ions). They feed on humus and animal excreta to produce soluble compounds that are available to plants.

2. *Nitrogen-fixing bacteria*, which are found in nodules on the roots of leguminous
   plants. They take in Nitrogen gas from the air in the soil and pass it on to the plant,
   discussed in **1Fe. Inoculation/Nitrogen Fixation**, page 54.
3. *De-nitrifying bacteria*, which break down humus and reduce nitrate ions to Nitrogen
   gas, which returns to the atmosphere. Because this reduction requires anaerobic
   conditions, these bacteria are most active in waterlogged soils.

Bacteria are most active, and multiply most rapidly, when the soil is warm, moist and
aerated; their activity is reduced as the soil dries out during the dry season, or if the soil
becomes cooler or less well aerated.

Fixation of nitrogen also occurs during lightning flashes; the huge energy release
breaks apart the nitrogen molecules and allows them to react with oxygen in the air to
form nitrogen oxides. These oxides dissolve in raindrops and fall to the ground, where
they form nitrites and nitrates. Fixation of nitrogen by leguminous plants is discussed in
1Fe, page 54.

1B. ARID REGIONS

1Ba. Plant / Soil / Water Relationships

Plants are said to suffer from water stress, or moisture stress, when their growth is
slower than normal due to insufficient water in the soil. An indication of the
interrelationship between the soil, plants, and water is shown in Figure 5.

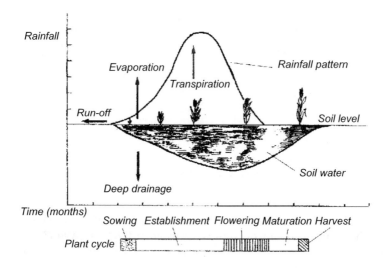

Figure 5. Growth of a cereal crop in relation to moisture availability

The effect of water stress on plant growth and development, and the utilisation of water by plants, is discussed below and also in the section on Irrigation, **1N**, pages 93–100.

1. Soil Water. Soils lose water by drainage, by evaporation from the soil surface and by transpiration by plants.

Plants take up moisture through their roots from the store of *available water* in the soil—this water is the difference between the volume of water held at *field (moisture) capacity* and that held at the *permanent wilting point. Field capacity* is the percentage of moisture held in the soil 2–3 days after being saturated and after free drainage has ceased. The *permanent wilting point* is the moisture content of the soil at which plants fail to recover (to regain full *turgidity*) when water is added again to the soil.

The amount of available water in the soil depends mainly on the soil *texture* and *profile*:

2. Soil Texture. Sandy soils have a lower wilting point and field capacity than clay soils. In other words sandy soils hold less water in reserve for plant growth than clay soils, and so sandy soils need more frequent rainfall or irrigation to support a crop.

3. Soil Profile. Many soils have an impervious layer (a *soil pan* or *hardpan*) a short distance below the surface through which neither water nor roots can easily pass. Thus water is held near to the soil surface, and in addition roots cannot penetrate the soil deeply. Plants growing in these soils rapidly use up the available water, then suffer from water stress and fail to reach their full potential.

1Bb. Plant Reaction to Stress

Plants have a fixed capacity to utilise water from the soil. This water is transpired through very small holes, or pores, called *stomata*, which are found mainly on the underside of leaves. Stomata allow the plant to take in carbon dioxide (CO_2) for the production of carbohydrate, and give out oxygen as a waste product—discussed in **1Ed. "Photosynthesis/Respiration"**, page 39.

However, as the temperature rises above about 32°C, and when there are strong winds, the root system cannot replace the lost water fast enough for the uptake of CO_2 to continue, and the stomata close. As a result, water movement within the plant ceases and the plant wilts and stops making sugars—see paragraph on wilting **1Ef**, page 39.

If this process is repeated for a few hours every day, the plant begins to draw on the moisture within its own plant cells after a few days, and the flowers and young fruits begin to fall to the ground.

Plants are damaged by heat stress at high temperatures even if there is adequate soil moisture. The artificial cooling of crops is discussed in **1N. "Irrigation"**, page 93.

Some crops are more successful at surviving drought than others. For example, fast growing crops such as the millets, grass pea and sorghum often avoid drought because their life cycle is very short—drought resistance via drought avoidance.

The correct choice of variety, or *cultivar*, is also important. Most crops have varieties that have been specifically selected for their drought resistance, such as Kalahari maize. Of course, these drought resistant varieties normally yield less than varieties that need more water.

The growth habit of plants is also relevant to the survival of plants in arid conditions. Cereals, for example, can totally fail to produce grain if there are high temperatures and moisture stress at flowering, even for quite short periods.

Other crops such as legumes may only lose a small part of their overall production in similar stressful conditions because they flower over a much longer period. This so-called *indeterminate* growth habit, as opposed to the *determinate* habit of cereals, is discussed in **1Ei**, page 42.

1Bc. Crop Management in Arid Regions

Food growers and farmers can help to some extent to reduce water stress of plants:

1. Conservation of Water. There are a vast number of techniques to conserve or "hold" water, some of which prevent the available rainwater from running off the field by the construction of ridges or *bunds*. These are raised rows of earth and stones constructed along the contours on sloping ground; or on level ground where they are built around small groups of plants, or even individual plants or trees.

Some other water conservation techniques are discussed later on, such as mulching, green manure, use of shadow, crop rotation, catch crops, cover crops etc.

2. Crop Management. Farmers who make wise decisions about which crops and varieties to grow, when to plant them and how to take care of them can produce crops where their neighbours lose everything. Sensible farming practices such as mixed cropping, staggered planting times, mulching and crop rotations, together with soil and water conservation techniques, can all be used to produce healthy crops even in very arid conditions.

3. Fallow. If level land with a good soil depth of loam or clay is *bare fallowed* or *clean fallowed* for one or more rainy seasons, this makes stored moisture available for the next crop. Bare (or "clean") fallowing is the practice of leaving a field unplanted, with no crop growing—the field should be kept free from weeds and preferably covered with a mulch. Fallowing is not possible in many areas where land is in short supply, but it can sometimes ensure that at least some yield is produced on some parts of the farm where continuous cropping could result in crop failure all over the farm.

4. Weed Control. In arid regions, and elsewhere also, a "clean" (weed-free) field can produce a bumper harvest where a field infested with weeds produces nothing, a subject discussed in **1L. "Weeds"**, pages 88–92.

5. Fertiliser. Application of fertiliser is not always successful in arid regions since plants can only use fertiliser if it is *in moist soil* and if it is *near to the plant roots*. Low rainfall often results in the fertiliser not reaching the root zone, because it remains too near to the soil surface to be used by the plant.

6. Windbreaks. Trees planted around field borders (or hedges/slatted fences) reduce the speed of the wind, reducing the water loss from both plants and soil. Appropriate tree species must be chosen; ideally they should also produce something useful such as fruit or timber as well as providing shelter and shade. The trees should also have a modest water requirement and not be planted too close to crops, as they would remove too much water and nutrients from the soil.

1C. SOIL

Soil consists mainly of particles of sand, silt and clay in close association with organic matter. The relative proportions of each of these soil components determines the *soil type*, such as "sandy", "clay", "loam", etc. The mineral content of soils is derived from the erosion and weathering of rocks. Water percolates down through the soil, which depletes the surface layers of soluble (and fine insoluble) substances in a process known as *leaching*, or *eluviation*. This process affects the soil's pH, discussed below in **1Cc**.

Soil is an extremely complex medium; in addition to the constituents mentioned above it also contains various amounts of bacteria, algae, fungi, viruses and invertebrates such as worms and insects. The inter-relationships between these are enormously complicated. Soil is thus filled with various forms of life from which it produces other forms of life.

The Soil Profile
A vertical section, or *"profile"*, through a soil normally consists of three main layers:
- The *surface soil* (Horizon A) is the upper layer, containing relatively large amounts of organic matter, which darkens the soil's colour. It is normally called the *topsoil*.
- The *subsoil* (Horizons B and C); Horizon B is a weathered mineral layer, often well structured, almost free of organic matter and usually lighter in colour than A. It is enriched with clay, silt, humus, iron and other nutrients leached from above. Horizon C is also a mineral layer, though less weathered than B and sometimes merging with the bedrock, which can be either shallow or deep, like people.
- The *bedrock* (Horizon R) is usually the parent material of the overlying soil profile.

The Water Table
At a certain level below the soil surface most soils are full of water. The top of this waterlogged soil is known as the water table, which rises and falls according to the amount of rainfall or irrigation, the time of year, the soil type and the drainage capacity of the soil.

If the water table rises too near to the soil surface for a period of time, the plant roots cannot absorb oxygen and the plants suffer as a result. Some species such as wheat have some tolerance of waterlogging. Rice has a hollow stem so that oxygen can reach waterlogged roots down the interior of the stem. See also **1Ad. "The Root System"**.

In general, arable and root crops need a fairly low water table to encourage their roots to develop fully, while grasses and cereals can tolerate a much higher water table.

Soil Improvement—*What can be done with poor, infertile soils?*
There is virtually no soil in the world that is not able to produce some form of healthy food crop if it is properly understood and cared for. Food growers should always try to grow the most suitable crops available, and the best available variety of that crop, while at the same time working towards maintaining or improving the fertility of the soil they cultivate.

Possible cures—some examples:
- acid soils can be cured with lime;
- alkaline soils can be cured with organic, basic fertilisers;
- saline soils can be cured with drainage and an adequate supply of rainwater and/or irrigation. Gypsum (Calcium Sulphate) may help the recovery of saline soils. In theory, several years of fallow can also cure salinity, though this is rarely practicable;
- waterlogged soils can be cured with subsoil cultivations or a good drainage system;
- soils which are deficient in the major elements can be cured with organic or inorganic fertilisers, and/or leguminous crops, and grazing animals;
- soils deficient in minor ("*trace*") elements can be cured either by adding those missing elements, either to the soil or crop, or by correcting other soil properties such as the pH (**1Cc**, pages 20–23);
- soils with trace element toxicities can normally be cured by correcting the imbalance in the soil of other, related nutrients.

Unfortunately most of these techniques tend to be rather expensive, and also the materials needed and the skills to use them are very often not available. However it is often possible to provide at least some of the resources needed to improve the soil's productivity, and the list above is included to demonstrate that all soils are dynamic and can be encouraged to grow food if the resources are available.

Food producers can make use of different kinds of plant and vegetable parts to improve their soil by converting them into organic matter, in compost heaps etc.

Organic Matter
This plays a vital role in the physical properties of soils. It consists of two components:
- undecomposed organic matter—fragments of plants and animal remains, and
- decomposed and partially decomposed organic matter, or *humus*—dark coloured material, which looks and feels like soil. It is derived from the more resilient undecomposed organic matter that is broken down, very slowly, by soil micro-organisms; it can persist in the soil for decades.

These two constituents have quite different physical properties. For example, undecomposed organic matter in the form of leaves, straw, manure etc. tends to keep the soil particles open, which is desirable in heavy clay soils but less desirable in light sandy soils. But with all types of soil the application of undecomposed organic matter does tend to reduce erosion and improve the capacity of the soil to retain nutrients and water.

The Benefits of Humus

Humus, which consists of organic matter that has partially or fully decomposed, improves the properties of all types of soil, in the following ways:

- protects soil from erosion by both water and wind;
- allows water to slowly and gently filter down through the soil, and reduces the run off and loss of water;
- improves light soils by increasing their water holding capacity, and other useful properties, while it improves heavy soils by giving them a "crumb" structure which is more easily cultivated (humus makes the soil become "lighter" ie more easily ploughed, cultivated);
- reduces water loss by evaporation from the soil;
- feeds earthworms and other beneficial insects;
- assists the soil to warm up more quickly in the spring, and to stay warm for longer in the winter;
- supplies essential nutrients to plants in a slow and gentle manner, and assists minerals in the soil to become more available to plants;
- reduces the speed of chemical changes when lime or inorganic fertilisers are added to the soil;
- releases organic acids which help to neutralise alkaline soils;
- "holds" nitrogen, in the form of ammonia and other compounds, so that the nitrogen is not denitrified by bacteria and so made unavailable to plants;
- reduces the damaging effects of fungal diseases and eelworms.

Maintenance of the Soil's Organic Matter Content

It is easy to see from the list above that the maintenance or improvement of the soil organic matter content is essential for satisfactory long term food production.

Unfortunately, in the presence of moisture, warmth and air, organic matter breaks down—it oxidises—and disappears in the form of carbon dioxide and water, leaving a small residue of minerals. This happens more rapidly in "hungry" soils—the light, sandy ones - than in heavier and colder soils.

Therefore the organic matter in the soil has to be constantly replenished by adding to it any form of organic material, such as manure, compost, green manure, leaf mould, peat, seaweed, sawdust, human excreta, crop residues and roots, weeds, grass etc. Indeed -

Anything which has lived before can live again!

The Compost Heap—three heaps good, one heap not so good

If there is enough available space, the easiest way to produce compost is to make three separate heaps, sited as close to each other as practical. The first heap is the one currently being filled, the second heap is full up and busy "making", and remains covered over with sacks, carpets etc. until the first heap is filled. The third heap is the one currently being used, or "mined", for its humus. As the third heap is used up, any undecomposed material in it is returned to the first heap for a second decomposing cycle.

As soon as the first heap is filled, it is covered over and left for a period; the third heap should then be emptied completely, and this then becomes the new "first" heap.

It is not necessary to turn over the material, nor to add artificial "aids", provided that:

- there is sufficient movement of air in at least some parts of the heap, especially at the base, which can be started off with twigs and small branches. The sides should be either slatted or open to the elements, to allow air to enter;
- the heap is kept moist;
- there are occasional layers of highly nitrogenous material, either manure or partly decomposed faeces, and, most important of all;
- the layers are continuously built up to form a flattish, horizontal top. The tendency is to put material into the centre of the heap. The centre should be reserved for mushy, edible material, while grasses and other green plants are placed around the edges so as to *maintain a flat top surface*.

Devils in the Compost Heap—Some Cautionary Tales

1. *Weeds*: the rhizomes of perennial grass weeds, and seeds of both weeds and crops, can re-grow after they have been through the heap and then returned to the soil unless the compost has been made very efficiently ie hot enough so as to destroy them.
2. *Human Excreta*: there is a health risk when using faeces, which should be either bio-digested or safely "stored" for several months before adding to the heap. A "soak" of dry soil, ash or sawdust should be sprinkled on top of every new addition No problem to mix pee and poo, if enough soak is added.
3. *Woody Material*: this may take a long time to break down, so it should be either cut up or smashed before being added to the heap, or put back onto the next heap.
4. *Goats & Rodents*: these often eat some plant material from the compost heap, which should then be either fenced off for goats, or de-ratted. Food scraps should never be composted; it is better to feed this to animals, birds etc.
5. *Water*: in hot dry regions it may be difficult to keep the compost heap moist enough to decompose properly. In this case the heap should be sited in a shady place, or a slatted roof can be fitted, and any spare water (or urine) should be put on the heap. In hot, dry weather it is also a good idea to cover the heap with old sacks, carpets and so on to reduce dehydration, at all three stages of making compost.

Soil Capping

Also known as *crusting*, soil capping is the formation of a hard "crust" or "cap" on a soil surface, usually after heavy rainfall and/or sunshine, or irrigation, which causes the soil surface particles to close up. This not only makes it difficult for seedlings to emerge, but also increases water run off and soil erosion; plant growth suffers as free gaseous exchange for the roots is reduced.

If a crust forms on soils already growing a crop, the crust should be broken up, with a hoe for example, to allow water to percolate down into the soil and to give the roots access to the air. The crop can be weeded at the same time.

Capping is more of a problem in clay soils, especially those with low organic matter content, and also occurs in soils with a high proportion of fine sand. It is caused by both physical and chemical processes, and can be reduced with applications of manure and other sources of organic matter. It can also sometimes be appropriate to cover the soil, with either a mulch (**1He**, pages 70–71) or a growing crop, during periods when either heavy rains or strong sunshine are likely to occur.

1Ca. Saline Soils

The UNEP has estimated that about 20% of the world's cultivated land and nearly 50% of all irrigated land is affected by salinity (Flowers and Yeo 1995).

This is not a recent problem; many historians maintain that the ancient Sumerian civilisation declined partly as a result of irrigation that caused salinisation—the toxic build-up of salts and other impurities.

The problem of saline soils is also discussed in **1N. "Irrigation"**, pages 93–99.

Saline soils have a high concentration of ions, both Sodium cations (Na^+) and Chlorine anions (Cl^-), and are generally unfavourable to the growth of most plants.

Alkaline soils above about 8.5 pH have only Na^+ cations and normally cannot be used at all for crop production. "pH" is discussed in **1Cc**, pages 20–23.

Some crops such as barley and cotton are quite tolerant of saline soils and can grow in soils with more than 5000 ppm TDS (Total Dissolved Salts/Solids). Other more sensitive crops such as beans and citrus trees suffer from salt stress in soils with only 960 ppm TDS. See list below, *Salt Tolerance of Plants* pages 16–17.

Salinity of soil is normally measured in terms of its Electrical Conductivity (EC, measured in deciSiemens/metre—dS/m) or Total Dissolved Salts (TDS—in ppm or mg/litre). EC is a measure of the conduction of electricity through water, or a water extract of soil. The EC value represents the amount of soluble salts in an extract, providing an indication of soil salinity. Saline soils are defined as those with an EC of greater than 1.5 dS/m for a 1:5 soil water extract and greater than 4 dS/m for a saturation extract. It can be interpreted in terms of the salinity tolerance of plants.

Conversion Rates: 1 dS/m = 640 mg/litre = approx. 640 ppm TDS.

These units are used to indicate the extent of the problem of osmosis, the ability of plants to take up water through their roots. The unit "ESP" (exchangeable sodium percentage) indicates the percentage of absorbed sodium ions to other cations that could be exchanged. Soils are categorised as being sodic with an ESP of 6–14% and strongly sodic with an ESP of greater than 15%.

Some soils are naturally saline; other soils become more saline, under one or more of the following conditions:
- when the irrigation water is saline, or when seawater inundates low lying areas;
- when the rate of evaporation is high (high temperature and/or wind speed);
- if the water table is high, as a result of poor drainage.

Some Notes on Saline Soils
- Most crops are more sensitive to soil salinity in hot, dry conditions than in cool, humid conditions.
- Choosing to grow salt tolerant species and varieties may be the simplest solution;
- Plants growing in poor, infertile soils may appear to be more salt tolerant than plants growing in fertile soils. In these cases it is the soil fertility and not the soil salinity which is the more important factor limiting plant growth.
- As more and more water is lost from the soil, by drainage, evaporation and transpiration, the soil moisture becomes more and more concentrated with salts. As a result, plants experience increased salt stress as well as water stress when the soil dries out.
- Crops are generally more sensitive when they are seedlings than when they are mature plants. This is partly explained by the fact that soils are normally more saline in the upper horizon, where the young seedling roots grow, than lower down in the soil.
- The level of soil salinity is constantly changing, due to changes in rainfall and/or irrigation, temperature and wind. Farmers who understand how and when these changes occur can sometimes produce crops on saline soils where others would fail.

What Can be Done about Saline Soils?
- Grow salt tolerant crops—see list below.
- Irrigate, applying extra water to *leach* (flush out) the salts.
- Install an underdrainage system to remove the saline drainage water away from the roots.
- Add hydrated calcium sulphate (Gypsum); this replaces the sodium in the soil, reduces the alkalinity and balances the salts in the soil.
- Add powdered sulphur; this makes sodium and chlorine more soluble, and allows other elements such as calcium and magnesium to replace them.

Salt Tolerance of Plants
Crops differ in the degree to which they are affected by soil salinity; some species such as barley can produce a reasonable yield in highly saline soils up to 18 dS/m, while others such as beans and carrots grow very poorly in soils with only 5 dS/m.

In general, plants with low drought tolerance also have low saline tolerance.

Tolerant

Barley *Hordeum vulgare*
Bermuda Grass *Cynodon dactylon*
Cotton *Gossypium hirsutum*
Date Palm *Phoenix dactylifera*
Durum Wheat *Triticum turgidum*
Guayule *Parthenium argentatum*

Jojoba *Simmondsia chinensis*
Leucaena *Leucaena leucocephala*
Saltbush *Atriplex* spp.
Silt Grass *Paspalum vaginatum*
Sugar beet etc *Beta vulgaris*
Triticale ***Triticosecale***

Moderately Tolerant

Chickpea *Cicer arietinum*
Fig *Ficus carica*
Mung Bean *Vigna radiata*
Oats *Avena sativa*
Olive *Olea europeaea*
Papaya *Carica papaya*
Pigeon Pea *Cajanus cajan*

Pineapple *Ananas comusus*
Sorghum *Sorghum bicolor*
Soybean* *Glycine max*
Taro *Colocasia* spp.
Tepary Bean *Phaseolus acutifolius*
Wheat* *Triticum aestivum*

Moderately Susceptible

Cabbage *B.oleracea* var. *capitata*
Casssava *Manihot esculenta*
Castor *Ricinus communis*
Foxtail Millet *Setaria italica*
Grape *Vitus* spp.
Groundnut *Arachis hypogaea*
Broad Bean *Vicia faba*
Irish Potato *Solanum tuberosum*
Lentil *Lens culinaris*

Linseed *Linum usitatissimum*
Lucerne *Medicago sativa*
Maize *Zea mays*
Pepper *Capsicum annuum*
Pumpkin *Cucurbita pepo*
Sunflower *Helianthus annuus*
Sweet Potato *Ipomoea batatas*
Tomato *L.esculentum*
Watermelon *Citrullus lanatus*

Susceptible

Almond *Prunus dulcis*
Apricot *Prunus armeniaca*
Apple *Malus sylvestris*
Avocado *Persea americana*
Carrot *Daucus carota*
Cassava *Manihot esculenta*
Cherry *Prunus* spp.
Currant *Ribes* spp.
Haricot Bean *Phaseolus vulgaris*
Lima Bean *Phaseolus lunatus*

Mango *Mangifera indica*
Okra *Abelmoschus esculentus*
Onion *Allium cepa*
Peach *Prunus persica*
Pear *Pyrus communis*
Peas *Pisum sativum*
Plum *Prunus domestica*
Rice* *Oryza sativa*
Sesame *Sesamum indicum*
Strawberry *Fragaria* spp.

* Some varieties of rice, soybean and wheat show some tolerance to saline soils.

Some Observations on Growing Food in Saline Soils

- **Varieties:** a few examples of significant differences of salt tolerance between different varieties of certain crops have been observed. This is not common, but it has been observed in: barley, Bermuda grass, berseem clover, birdsfoot trefoil, brome grass, creeping bentgrass, rice, soybean, taro and wheat.

- **Fruit Trees:** selection of the appropriate rootstock is important if fruit trees are to be planted in saline soils. Papaya may survive where other fruit trees perish in saline soil. Date palms can also tolerate high salt levels if the other growing conditions are favourable.

1Cb. Soil Analysis

Analysis of soils is the determination of the composition of soil by various chemical and physical methods. Analyses are normally done in the laboratory, though there are kits available which allow you to test for yourself if the soil is deficient in any of the major nutrients.

The pH value of soil and other materials can also be tested by the amateur—see below "Testing for Soil pH".

The CEC (Cation Exchange Capacity) of a soil measures its capacity to hold the major cations (calcium, magnesium, sodium and potassium—as well as hydrogen, aluminium and manganese in acid soil). It is a measure of the potential nutrient reserve in the soil and is therefore an indicator of inherent soil fertility.

Soil may be analysed both chemically and physically ("mechanically"). In both cases soil samples must be taken that are representative of the area, and sampled from a large number of sites using standardised procedures. One simple method to take soil samples is to imagine the field in the shape of the letter "W"; samples are taken at intervals all along this imaginary line.

- **Chemical Analysis**

The purpose of this is to determine the chemical composition of soil, including trace elements, and to determine its state of fertility. This information helps food producers to use the correct type and amount of fertiliser.

For the experienced observer a simpler, quicker and cheaper option to chemical analysis is to check the status of the soil's nutrients by observing how plants grow in it. Some of the symptoms of nutrient deficiency and toxicity are described in **1Cd. "Trace Elements"**, pages 23–29.

- **Mechanical Analysis**

The purpose of this is to determine the texture and physical properties of soil, which is mainly of interest for the classification of soils and their land-use.

Results of a mechanical analysis enable the soil texture or soil type to be classified, by means of the diagram shown on the next page, known as **The Soil Triangle**:

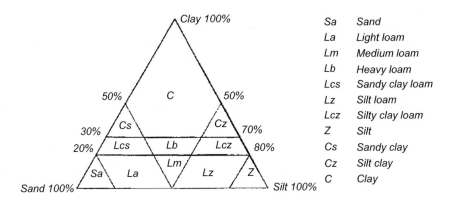

Figure 6. The Soil Triangle

Describing Soil Types/Soil Textures

It is not normally necessary to go to the trouble and expense of a mechanical soil analysis. Usually it is quite adequate to be able to describe a soil by taking some of the moist soil in your hand and feeling it between your thumb and the palm of your hand:

Sensation	Description	Particle Size (mm)
Small stones or gravel	Gravel	2.0+
Gritty feeling	Coarse sand	0.2–2.0
Also a gritty feeling between the teeth	Fine sand	0.02–0.20
Gritty feeling, and also makes your hands dirty	Loam	0.002–0.020
Smooth, silky feeling but the soil cannot be polished when damp between your fingers	Silt	0.002–0.020
Smooth, silky feeling and you can polish it between your fingers	Clay	<0.002

The colour of soil can also provide some information; dark brown or nearly black soils are normally rich in humus, while light brown or red soils often contain mineral oxides such as iron. More formal classification can be made using a Munsell Colour Chart.

The soil structure can be described by looking at the size of the soil particles:

Loose—when the particles are mostly separate, such as in sandy soils;

Friable ("Crumb")—when there is a mixture of separate particles and smaller lumps of soil, typical of loam soils;

Solid—when there are large blocks of soil.

Soil Types
Sand, silt and clay, as well as variable amounts of organic matter, are mixed together in soils in many combinations, giving soils their characteristic texture:

- **Sandy soils**—these are well drained and well aerated (oxygen is readily available to plant roots), but have low water holding capacity (retention). The sand particles may be either mineral fragments that slowly become available to plants by weathering, or worthless quartz, or a mixture of the two. Very sandy soils are generally poor soils, often low in nutrients and moisture.
- **Light soils**—these have a high proportion of sands. The name indicates the ease with which they can be cultivated, not their weight.
- **Heavy soils**—these have a high proportion of clay and silt. The name indicates that they are difficult to cultivate, and not their weight. Very heavy soils are poorly aerated and drained, and often produce plants with retarded/stunted root systems.

1Cc. Acid Soils/pH Value

Soils may become acidified over time as a result of a number of factors:
the bedrock, loss of calcium (lime) due to uptake by plants, the addition of nitrogen to the soil by either fertiliser or legumes (where N is converted to nitrates) and/or leaching.

An indication of the acidity or alkalinity of soil and other substances is given by its pH value, the measure of the concentration of hydrogen cations (H^+) and hydroxyl anions (OH^-).

Neutral substances have a more or less equal proportion of hydrogen ions (positive charge) and hydroxyl ions (negative charge), alkaline substances have a higher proportion of hydroxyl ions and acid substances have a higher proportion of hydrogen ions than hydroxyl ions.

pH is expressed on a scale of 0 to 14; pH7 is neutral, less than 7 is acidic and more than 7 is alkaline. In practice, soils with a pH of between about 6.6 and 7.3 are often described as being "neutral".

pH values are reciprocal logarithmic expressions to the power of 10, so that a soil with a pH of 5 is ten times as acid as a soil with a pH of 6, and so on.

Although all soils are inherently either acidic, alkaline or neutral, their pH can be altered a little. If the soil is too acid lime can be added; if it is too alkaline, organic fertilisers, sulphate of ammonia or leguminous green manure crops can be added.

Alkaline soils contain bases such as lime, soda or potash in large amounts.

Signs of Soil Acidity
Some indications can be observed by eye if a field, or parts of a field, are too acidic:
1. Bare patches, or poor growth, of acid sensitive crops such as barley, wheat and sugar-beet (see list of crops below).
2. Poor growth of clovers, and an accumulation of a *mat* of undecomposed organic matter in grassland.
3. A lack of earthworms.

4. The presence of Clubroot disease in Brassica crops.
5. The presence of acid tolerant weeds such as gorse, bracken, foxgloves, sorrel (sourdock), spurrey (sand weed) and Agrostis (Bent) species of grass.

The range of **Soil pH Values** is given below:

PH value	Description and typical location	Typical plants
3.5–4.0	Extremely acid. Rarely found, sometimes in forested areas	None
4.0–4.5	Also extremely acid. In humid forest areas, or where the soil is wet and peaty	None
4.5–5.0	Acid. Mainly in cold humid areas	Oats, lupins (and Irish potatoes above about pH 4.6)
5.0–5.5	Fairly acid. Typical of unlimed soil in wet climates	Millets, soybeans, potatoes, tomatoes, rye, cotton, tobacco
5.5–6.0	Slightly acid. Moderate climates with high rainfall	Wheat, grasses, beans, maize, cabbage, barley.
6.0–6.5	Neutral. Moderate climates with medium rainfall	Most crops
6.5–7.5	Neutral. Hot, dry climates	Most crops but not Irish potatoes or tomatoes
7.5–8.0	Alkaline. Semi-desert areas.	Lucerne (alfalfa), sainfoin (*O.viciaefolia*)
8.0 +	Extremely alkaline	None

Tolerance of Plants to Soil Acidity
Some food crops tolerate acid soils better than others, as indicated above. Further details are outlined below:
Sensitive
Barley, Cabbage, Carrot, Cauliflower, Chickpeas, Lucerne (Alfalfa), Mangolds, Mung Bean, Niger Seed, Red Clover, Sugar beet, Wheat.
Intermediate
Avocado, Cassava, Cotton, Cucurbits, Field Peas, Groundnuts (Peanuts)*, Haricot Beans, Kale, Lentil, Lima Beans**, Maize, Okra, Onion, Papaya, Sorghum, Soybean, Sunflower, Swedes, Tobacco, Tomato, Turnips, Wild White Clover.
Tolerant
Bananas, Buckwheat, Cowpeas, Guava, Irish Potatoes, Lupins, Mango, Oats, Rice, Rye, Sweet Potato, Watermelon, certain weeds***

*   Some varieties of Groundnuts have some tolerance of acidic soils.
**   Lima Beans tolerate acid soils better than most other beans and peas.
***   Refer to paragraph 5 of *"Signs of Soil Acidity"* on the previous page.

The list above offers only a broad guide, and there are often **wide differences between varieties**. For example, some varieties of sorghum are more tolerant of acid soils than some varieties of rice.

Lime

Agricultural liming material is defined as a material containing Calcium and Magnesium compounds that is capable of neutralising soil acidity.

Lime is in many ways the key element in the soil. Lime not only makes acid soils more alkaline, it also makes clay soils easier to cultivate, it reduces the loss of nitrogen from the soil, it releases phosphorus and potassium which become "locked up" in acid soils, and it renders harmless some trace elements such as manganese which may be present in the soil in excessive amounts.

Another attribute of lime, which may become increasingly relevant, is that it reduces the amount of Strontium-90 that is taken up by the soil.

Further details about lime are available online at: http://minerals.usgs.gov/minerals and http://www.aglime.org.uk

Application of Lime

It is often a good idea to apply lime to acid soils, especially in humid and high rainfall regions where large amounts of calcium are leached from the soil.

Even in normal conditions of arable cropping, or grazing by animals, the equivalent of about 120–500 kg/ha of lime is lost every year, from leaching and by being taken up by plants and animals.

In these situations many soils, especially sandy ones, slowly become more acidic unless lime is added to the soil. However if too much lime is applied, this can cause deficiencies in some other soil elements such as Copper.

Lime should not be applied together with manure, nor before Irish potatoes are grown.

Different Forms of Lime

Calcium carbonate ($CaCO_3$) in the form of ground limestone (dolomitic limestone is best), ground chalk, lump chalk, waste (factory) lime, slaked and hydrated lime (calcium hydroxide), quicklime (calcium oxide), marl, shells and by-products such as slag. Each of these sources has a different neutralising value. The lime is scattered onto the soil and then worked into the upper soil layers.

Below are some examples of the amount of lime (CaO, in MT/ha) needed to increase arable soil pH to about 6.5 (figures in brackets are for applications on to grassland)— based on soil depths of 200 mm (arable) and 150 mm (grassland); deeper soils require more lime.

If more than about 8 MT/ha is needed, this should be applied in two applications with the first ploughed in several weeks or months before the second.

pH	Sand & loamy sands	Clay loams and clays	Peaty soils (25% OM min.)
4.5	15 (7)	20 (7)	29 (7)
5.5	9 (5)	12 (6)	13 (4)
6.0	6 (2)	8 (3)	5 (0)

The figures above show that the quantities involved are large, and lime is not normally locally available in such large quantities.

The information is included here to show that although lime is one solution to the problem of soil acidity the remedy is neither cheap nor easy.

Testing for Soil pH

It is very easy for amateurs to discover the very approximate pH of a soil by using a Universal pH Indicator. Chemical indicators change colour in response to the pH of the solution. Most chemicals used as indicators respond only to a narrow pH range.

A universal indicator, however, is sensitive to a large pH range because it uses several indicator chemicals

A chemical is added to soil that is in suspension in water in a test-tube, and the change in colour is noted, as indicated below:

pH	Colour
Below 5	Red
5–6	Pink
7	Green
8–9	Blue
Above 9	Purple

Litmus paper or litmus indicator can also be used, but this is even less precise than a Universal pH Indicator.

Litmus turns to red in acidic soil water, and blue in alkaline or neutral soil water.

1Cd. Trace Elements

Also known as *micronutrients* (or *minor elements*) these are chemical elements that are required by plants and animals in very small quantities for the proper functioning of enzymes, hormones and vitamins.

The most important trace elements for plants are **manganese, molybdenum, zinc, boron, iron, copper and sodium. Magnesium, calcium and sulphur** are also known as trace elements, though because they are needed in relatively large amounts they are sometimes known as *secondary nutrients*.

Deficiency Diseases
Also known as *deficiency disorders*, these cause symptoms to be shown by plants when there is either an absence or a shortage of one or more of these elements in the soil.

Conversely, if these elements are present in excessive amounts some plants may suffer from *trace element toxicity*—for example, soybean leaves shrivel and fall off when there is excessive boron present in the soil.

There are two main types of deficiency diseases:

- **Direct**—when the reserves of the trace element in the soil are either too low or are not present at all, and
- **Induced**—when the trace element is present in the soil, but is either unavailable or less available for use by the plant due to another factor such as the soil pH or soil reserves of other nutrients—some examples are given below:

Soil	Deficiency symptoms shown
Acid (low pH)	Molybdenum (Mo)
Alkaline (high pH)	Manganese (Mn) and/or Iron (Fe)
With excess Phosphate (P)	Zinc (Zn)
With excess Potash (K) or Calcium (Ca)	Magnesium (Mg)
With excess Copper (Cu)	Iron (Fe)

Remedies
In most cases *direct deficiencies* can be corrected by soil applications, where the soluble salt of the missing element is sprayed or spread onto the soil surface and then worked in. Foliar treatments are also used, which is normally more effective when applied to young plants than mature ones.

Similarly, *induced deficiencies* can be corrected by foliar applications; a dilute solution of a salt, or a neutralised sulphate of the element concerned is sprayed onto the leaves of young plants.

Avoiding Trouble with Trace Elements
Soil which contains a healthy level of organic matter is unlikely to suffer from trace element deficiencies, and food growers who regularly give their soil either animal manure, compost or sewage sludge rarely have this problem.

If this is not practicable in the short term there is usually some remedy which can correct the deficiency or excess, as shown on page 28 under *"Correcting Trace Element Deficiencies"*. It is not a good idea to make a general practice of this, but in an emergency the deficiency can usually be remedied, as this list shows. In the long term however it is much better practice to prevent the problem rather than try to cure it, by maintaining an adequate level of organic matter in the soil.

The Effect of Soil pH on Trace Elements
In many cases the deficiency or toxicity of trace elements can be cured by altering the pH of the soil. For example Molybdenum can usually be made available to plants if lime is added to an acid soil, though in many tropical soils no Molybdenum at all is present. In this case Molybdenum must be added to the soil, for at least five years or even ten, in addition to lime.

Trace Elements in Context
Plants need three different types of nutrients in order to grow well:
1. **Major Nutrients**, also known as *macronutrients* or *primary nutrients*, these are: Nitrogen (N), Phosphorus (P) and Potassium (K).
2. **Secondary (Minor) Nutrients**, which are also essential for healthy plant growth but in much smaller quantities: Calcium (Ca), Magnesium (Mg) and Sulphur (S).
3. **Trace Elements**, which are also essential for healthy plant growth but only in very small quantities.

A Warning about Symptoms of Plant Nutrient Deficiencies or Toxicities
The symptoms of nutrient deficiency or toxicity that are displayed by plants are described in the following paragraph. Deficiencies may in fact be sub-clinical, where the crop does not manifest physical symptoms, but its performance is adversely affected.

To complicate matters further, these same symptoms may also be shown by plants when they are subjected to one or more of the following conditions, a situation known as *transient deficiency*: cold weather, drought, saline soil, mechanical damage, poorly applied fertiliser.

The Role of Plant Nutrients
☞ **READ THIS SHORT PARAGRAPH FIRST!**
An explanation of the following: each element, or "nutrient", is described in four ways: in terms of its function in plant growth ("Function"), the plant species which are most affected by the nutrient ("Indicator Plants"), the "Symptoms" shown by plants when they are grown in soils which are either deficient (D) or have excessive amounts (E) of that nutrient, and some "Remarks".
Nitrogen (N)
Function: to increase plant growth, the size of the leaves and yield.
Indicator plants: cereals, mustard, apples, citrus.
Symptoms: Leaves are chlorotic (yellow or pale green). Older leaves are yellow at the tips. Leaf margins green, but midribs yellow. Plants are stunted (D).
Remarks: Nitrogen may give dramatic and quick response by plants, but too much N causes too much vegetative growth with weak stems, sometimes causing lodging, and also increases the susceptibility of plants to disease, frost and drought.
Phosphorus (P)
Function: to develop root growth, to establish young plants and to help early ripening.
Indicator plants: legumes, root crops, maize, barley, lettuce and tomatoes.

Symptoms: although plants may be dark green, they are stunted. Older leaves may be purple in colour (D).

Remarks: Phosphorus ("Phosphate') is best applied a little and often, but there are no problems if excessive amounts are applied. It is most available to plants when the soil pH is 5.5–6.5. Often deficient in tropical soils.

Potassium (K)

Function: to improve plant quality, vigour, health, drought resistance, ability to store sugars and starches, and to form chlorophyll. Also to improve stomata function.

Indicator plants: legumes, root crops, maize, cotton and tomatoes.

Symptoms: lower leaf tips and edges are yellow to brown and appear burned on the outer edges. Internodes shorter than normal, and plants are stunted and often diseased.

Remarks: too much potassium makes plants slow to mature. Wood ash, compost and manure are good sources of Potassium. Not deficient in tropical soils as often as P.

Calcium (Ca)

Function: component of cell walls and membranes; to balance organic anions.

Symptoms: soft dead necrotic tissue at fast growing parts of plants, which wilt.

Magnesium (Mg)

Function: to allow chlorophyll formation and growth, to assist in nodulation of legumes, and in the utilisation of Phosphorus.

Indicator plants: most crops, including fruit trees and bushes, Irish potatoes, cauliflower and sugar beet.

Symptoms: older leaves are chlorotic and may have light grey stripes between the leaf veins. Leaves may have an unaffected, green edge, but these fall off prematurely (D).

Remarks: more common in light soils with high rainfall, and in soils with low organic matter content or with excessive Potassium or Calcium. Dolomitic limestone contains 5–10% Magnesium.

Sulphur (S)

Function: general plant metabolism, protein synthesis, formation of chlorophyll and nodule development.

Indicator plants: lucerne (alfalfa), clover and rape.

Symptoms: the leaves are chlorotic, the shoots are short and the stems are stiff, woody and thin (D).

Remarks: this is rarely a problem, and is more commonly seen in non-industrialised (ie less smoky) regions. Can be problematic with certain Brassicas, rapeseed and Canola.

Zinc (Zn)

Function: in the relationships of plant metals with enzymes, and in the utilisation of nitrogen and phosphorus.

Indicator plants: maize, cereals, potatoes, onions, avocado, apples, pears, citrus fruits and castor.

Symptoms: the young leaves are a light grey colour both sides of the midrib at the base of the leaf. It causes Mottle Leaf disease of citrus. Internodes are short. Branches die (D & E).

Remarks: Zinc becomes increasingly less available to plants as the soil pH and Phosphorus content rises.

Manganese (Mn)
Function: assists the transformation of nitrogen and the assimilation of other elements to form sugars and proteins, and in respiration.
Indicator plants: cereals, especially oats (causing Grey Speck disease), beans, chickpeas, peas (causing Marsh Spot disease), vegetables, radish, beet (causing Speckled Yellows) and fruit trees.
Symptoms: as above, and leaves are chlorotic; new leaves mottled (D). Stunted plants (D & E).
Remarks: symptoms are most marked in neutral or alkaline soils which are high in organic matter, but it is also harmful in acid soils (D & E). If in excess, Manganese scorches the root system, which develops poorly (E).

Molybdenum (Mo)
Function: in nodulation of legumes and in the reduction on nitrates.
Indicator plants: soybeans, lentils, brassica (especially broccoli and cauliflower, causing Whiptail), spinach, oats, citrus and pasture.
Symptoms: stunted plants. Leaves are chlorotic, droop at the tips & curl inwards (D).
Remarks: occurs in both acid and alkaline soils, but the cures are different (D & E). Less available in acid soils.

Boron (B)
Function: assists growth by contributing in the metabolism of Calcium. Also assists nodulation in legumes.
Indicator plants and Symptoms: vegetables (distorted leaves - D), soybeans (leaves shrivel and fall off - E), sugar beet (Heart Rot - D), swedes and turnips (Brown Rot - D), cauliflowers (become brown - D), celery (leaves are distorted, stems are cracked - D), root crops (hollow or rotten roots which turn grey and mushy in the middle. Leaves distorted - D) and citrus fruit (lumpy fruit - D).
Remarks: less available to plants in alkaline soils. Boron deficiency is common in sandy soils. Boron should be applied with care as there is only a small margin between deficiency and excess.

Iron (Fe)
Function: in the formation of chlorophyll and enzymes (eg respiratory) and in the metabolism of nucleic acid.
Indicator plants: sorghum, barley, cauliflower, citrus and peach.
Symptoms: leaves are chlorotic and often striped or mottled (D & E). With citrus plants the leaves remain green.
Remarks: iron is less available in alkaline soils and soils high in Copper. Symptoms are similar to toxicities of Copper, Zinc and other elements.

Copper (Cu)
Function: in photosynthesis, enzyme systems and Vitamin A formation.
Indicator plants: maize, cereals, vegetables, leaf crops, tobacco, avocado and fruit trees.
Symptoms: small or shrivelled grain in cereals. Leaves chlorotic and curled, and margins scorched. Multiple buds formed (D & E).
Remarks: not seen very often, but can occur after excessive lime applications. Soils take years to recover from copper toxicity.

Sodium (Na)
Function: to allow plants of the *Chenopodiaceae* family (beet, spinach etc.) to develop properly.
Indicator plants: beetroot, spinach, carrot and mangels.
Symptoms: plants seem to be in poor health (D & E).
Remarks: sodium is rarely a problem, and is not easily diagnosed.

Correcting Trace Element Deficiencies
Trace element deficiencies can often be corrected by either soil or, more commonly and efficiently, foliar (spray) applications.
    Great care must be taken with soil remedies as the quantities involved are very tiny and overdosage is usually toxic, so the application of an element on a field scale is in general only possible when the element is incorporated with a fertiliser mixture.
Magnesium
Soil: in acid soils, apply dolomitic limestone ($MgCO_3$), or Kieserite ($MgSO_4$), or special compound fertilisers, or Magnesian limestone.
Foliar: 28 g of Epsom Salts (Magnesium sulphate) dissolved in 4.5 l water per 0.8 m^2.
Sulphur
Soil: Sulphur granules, ammonium sulphate, sulphate of potash or superphosphate fertiliser.
Foliar: sulphur dust/wettable powder is used to control some mildews.
Zinc
Soil: 10–50 kg/ha of Zinc sulphate or Zinc oxide. Solid forms can be used in acid soils.
Foliar: 0.4% solution of Zinc sulphate (or oxide, carbonate etc.) + 3% lime, at 0.5–1.5 kg/1000 litres per hectare, or zinc chelate—also applied as a dust.
Manganese
Soil: 20–120 kg/ha solid form of Manganese sulphate, or spray 11.5 kg/ha dissolved in 450 litres water. For cereals, add 28 kg/ha when planting.
Foliar: 0.4% solution (28 g $MgSO_4$ in 3l water per 26 m^2, or 1 kg/1000 litres/ha).
Molybdenum
Soil: in acid soils, apply lime. In alkaline or neutral soils apply 100–500 g/ha of sodium or ammonium molybdate, or apply 56 kg/ha of molybdated gypsum (for legumes, the seed must be inoculated).
Foliar: 0.1–0.2% solution of Ammonium molybdate at 60 litres/ha.
Boron
Soil: 5–20 kg/ha of the solid form of borax or boric acid, or apply a special compound fertiliser or bonemeal.
Foliar: 0.2% borax (28g of borax or boric acid dissolved in 9 litres of water per 17 m^2).
Iron
Soil: on alkaline soils apply organic matter or acidifying agents to increase soil acidity, or chelates of Iron such as sequestrene, or ferrous sulphate at 10 kg/ha.
Foliar: 0.4–1.0% solution of Ferrous sulphate + 0.2% of lime, applied 2–3 times.
Copper
Soil: 10–50 kg/ha of Copper sulphate or copper chelates.

Foliar: 0.1% Copper sulphate solution (1 kg dissolved in 1000 litres of water/ha with 1 kg lime/ha), or 3 kg/ha Copper oxychloride.

Sodium

Although this element is very rarely deficient, agricultural salt (NaCl) is sometimes applied to soils, to increase yields of carrots, sugar/fodder beet and mangels. This is done two or three weeks before planting, but great care should be taken with salt as it can stunt or even kill plants; the practice is not very often recommended.

Chlorine, Silicon and **Aluminium** are also taken up by plants but do not appear to be essential for plant growth.

1D. FERTILISER

Anything which is added to the soil to increase the amount of plant nutrient (or plant "food") can be called a fertiliser, though the word is normally applied to inorganic (synthetic) chemicals which invariably have to be paid for in cash, as opposed to organic (non-synthetic) fertilisers—the manures, composts and so on which are normally produced by food producers themselves on their own holdings.

Chemical fertilisers provide the basic plant nutrients, mainly nitrogen (N), phosphorus (P) and potassium (K), either individually, as *straight* fertiliser, or more commonly as *compound* or *multi nutrient* fertiliser.

Organic fertilisers also supply these nutrients, but in much lower concentrations.

The actual amount of the nutrients contained in compound fertilisers is often expressed as:

- **Nitrogen**, as %N,
- **Phosphorus**, as P_2O_5 (phosphorus pentoxide, often called phosphate, sometimes incorrectly called phosphoric acid), and
- **Potassium**, as K_2O (often called potash).

The Main Types of Fertiliser

There are 3 main kinds of fertilisers, which are aggregated in various proportions by the suppliers to form the various compound fertilisers, some of which include other elements also, particularly Magnesium ($MgSO_4$).

1. Nitrogenous Fertilisers

Nitrogen is supplied either as nitrate (NO_3) or ammonium (NH_4), or in a combination. *Examples:* ammonium nitrate (NH_4NO_3)—the most widely used—nitrate of soda, sulphate of ammonia, urea, ammonium phosphate, potash nitrate etc.

They are sometimes applied in liquid form—ammonium nitrate/urea and other mixtures are used.

Anhydrous ammonia is sometimes applied as a pressurised liquid that gasifies when injected into the soil.

2. Phosphatic Fertilisers

These are either fast acting (water soluble) or slow acting (water insoluble).

Examples: ammonium phosphate, basic slag, (triple) superphosphate, mineral rock phosphates, bone meal, flour etc.

3. Potash Fertilisers
The main content of these are chlorides, also known as muriates.
Examples: muriate of potash, sulphate of potash, nitrate of potash (saltpetre), kainite etc.

The Purpose of Fertiliser
Fertiliser is used to replace elements in the soil that have been lost in a number of ways, including:
* leaching, where nutrients are carried down below the reach of plant roots;
* supporting the growth of crops and weeds;
* removed from the field within the plant material, either as the crop harvest or when grazed by animals—eg, a 4 MT/ha grain & 6 MT/ha straw crop of maize will remove about 110 kg of N, 40 kg of P, 130 kg of K and 15 kg of Calcium/ha.

The Effect of Soil Type
In general, soils with a high organic matter content require less fertiliser than light, sandy soils with low organic matter content. The main reason for this is normally that more nutrients are leached from soils with a low organic matter content.

Badly structured clay soils also need more fertiliser than loams or clay loams.

Excessive Use of Fertiliser
Sometimes too much fertiliser is applied to crops, perhaps in the hope of a bumper harvest, or as a result of bad advice or poorly adjusted application equipment. This is not only a waste of time, money and fertiliser but there can also be some adverse effects on the crop, some examples of which are set out in **1Cd. "Trace Elements"**, page 23.

Another example is when excessive nitrogenous fertilisers are applied to root crops such as manioc/cassava and sweet potato, when vegetative growth of the aerial parts—the leaves and stems—is increased, but at the expense of tuber growth which is reduced.

Organic Versus Inorganic Fertiliser
Inorganic, chemical fertilisers can and do very often produce lush green plants that give high yields. These fertilisers have increased productivity per hectare enormously over huge areas of the world, and they make a major and mainly useful contribution to agriculture.

However, excessive reliance on this form of plant nutrition can lead to some serious problems. The quality of crops which are grown in soils continuously fertilised with non-organic fertilisers tends to decline, water courses may become polluted by fertiliser runoff, the soil becomes gradually poorer in organic matter, and the plants which it supports begin to lose their capacity to tolerate attack from diseases and insects.

At this point farmers are liable to become more and more dependent on using certain fertilisers. They have to find the cash or credit to buy them—if they are available—and they have to master the sometimes difficult application techniques; often quite sophisticated machines are needed to do an efficient job, and these can be difficult to regulate and maintain.

Most subsistence farmers, and many others, find that fertilisers can be a mixed blessing. In the case of most very poor farmers it is normally more appropriate to promote the increased understanding and use of organic manures, composts, mulches and so on.

Approximate Value (as a percentage) of some Organic Fertilisers

Fertiliser source	N	P_2O_5	K_2O
Bone meal (raw)	4	22	
Blood meal	12		
Fish meal	5	5	
Greensand (glauconite)		1.2	0.1–3
Groundnut cake	7	1.5	1.5
Guano based	10	1.3	2.5
Coconut cake	1–3		
Cow manure (FYM)	0.4–0.6	0.1–0.3	0.4–0.6
Poultry & Pig manure	0.5–1.5	0.5–1.5	0.6–0.8
Slurry	0.5	0.1	0.4
Wood ash		1–2	5–8

These nutrients are not immediately available for plant growth, and only a proportion are utilised during the first growing season. Calcium, magnesium and sulphur are also often available present in significant amounts.

- **FYM**. On a dry matter basis FYM (farm yard manure) contains about 2–0.5–1.5, so 25 t/ha fresh manure would be needed to apply 160, 22 and 110 kg/ha of N, P and K respectively. The P and K in FYM is normally readily available, but the N becomes available over a period of a few years—only 25–30% of the N applied is available in the first season.
- **Slurry**. The composition of slurries varies widely, according to the dilution rate. Undiluted cattle waste contains about 10% dry matter, 0.5% N, 0.1% P, 0.4% K. Much of the N is present as ammonia which can evaporate after spreading, so slurries should be incorporated into the soil immediately after application, or applied just before rain.
- **Municipal composts**, derived from domestic sewage and other waste. The K content can be quite low, which can arise because during sewage treatment the liquid (which contains much of the K) is drained off.
- **Poultry manure** can substantially raise soil pH, and so should not be used if pH is already greater than about 6. It is often a good source of nutrients. On a dry weight basis, it may contain 2–5% N, 1–3% P, and 1–3% K. About half of the N in poultry manure can be available in the first season but much of the N may be lost if the manure is dried and then stored.

Fertiliser in the Dry Tropics

With irrigation, spectacular responses to fertilisers are often obtained, producing large increases in yield. But under rainfed conditions alone, response to fertiliser is highly variable and often uneconomic, for several reasons: there may be insufficient soil moisture in the root zone at the time when the fertiliser is present, the nutrients supplied may be unavailable, and there may be other damaging checks to plant growth caused by drought which impede the uptake of nutrients.

Apart from Phosphorus, which is often either absent or present only in very small amounts in "bagged" fertilisers, the use of fertilisers in the dry tropics is not recommended unless there is sufficient soil moisture. In other words **in dry areas it is often moisture and not nutrients which is the limiting factor to plant growth.**

Plant Food Ratios

A compound fertiliser which contains 22% N, 11% P_2O_5 and 11% K_2O has a plant nutrient ("food") ratio of 2:1:1.

Another fertiliser containing, for example, 13% N, 12% P_2O_5 and 20% K_2O has a ratio of approximately 1:1:1.5.

These ratios are normally printed on the outside of the fertiliser container.

Recommendations for applying fertiliser are usually expressed in terms of kilograms (kg) of plant food required by a particular crop per hectare (ha).

The number of kilograms of a particular nutrient in 100 kg of fertiliser is the same as the percentage of nutrient in the fertiliser, so a 50 kg bag of "straight" ammonium nitrate (34%N) contains 17 kg of nitrogen. A 50 kg bag of 12:12:18 "compound" fertiliser contains 6 kg of N, 6 kg of P_2O_5 and 9 kg of K_2O.

If the fertiliser also contains other elements in addition to N, P and K this fact should also be written on the bag, so 12:12:18 + 6S means that there are 3 kg of sulphur present per 50 kg bag in addition to the N, P and K. In UK agriculture there is some confusion regarding "kilograms" versus "units". Whilst the level of elemental N is quoted, the levels of P_2O_5 and K_2O are normally quoted—rather than elemental P and K. Sulphur is quoted both ways.

Metric/Imperial Conversion

See also **Section 3. "Conversion Tables and Statistics"** pages 313–316.

$$1 \text{ lb} = 0.454 \text{ kg} \quad 1 \text{ kg} = 2.205 \text{ lb}$$
$$1 \text{ acre} = 0.405 \text{ ha} \quad 1 \text{ ha} = 2.471 \text{ acres}$$
$$1 \text{ lb/acre} = 1.121 \text{ kg/ha} \quad 1 \text{ kg/ha} = 0.892 \text{ lb/acre}$$

Fertiliser Response

If the yield of a crop is increased due to the application of a fertiliser, that crop is said to *respond* to that fertiliser under those particular conditions. However if those conditions change, such as the rainfall pattern, the timing of the fertiliser application, the previous crop grown and so on, then that crop may not respond at all, or it may respond in a different way, to the same application of fertiliser.

An example is quoted below to illustrate how difficult it can be to predict the response of a crop to an application of fertiliser, and the need to understand something about the soil, the climate and the crop nutrient needs before recommending the use (including timing and quantity) of any fertiliser.

Example: If wheat is grown in rotation with a fallow, then nitrogen will accumulate in the soil during the fallow year. The yield of wheat grown in the same place the following year may therefore be limited by the amount of available phosphate instead; so the wheat will respond to a phosphatic fertiliser.

But if the wheat is grown continuously there is little or no accumulation of nitrogen, and so the wheat will respond to a nitrogenous fertiliser, until the available phosphate is used up. But the wheat will not then respond to phosphate by itself, because it is now nitrogen that is the limiting factor. A compound fertiliser containing both nitrogen and phosphate would then be appropriate.

Economic Return from Fertiliser
A value:cost ratio of 1 means that the value of the increase in yield equals the cost of the fertiliser used (including transport, application and other costs).

A value:cost ratio of 2 represents a profit of 100%, and so on.

Thus the term *economic response of a crop to a fertiliser* means that the value:cost ratio is greater than 1. In other words it would be advantageous to apply fertiliser in these conditions—see chart below.

Another commonly used term is the "fertiliser response ratio", the number of additional kilograms of additional crop produced per kilogram of additional plant nutrient applied. In countries with low to middle level yields and in areas which are not too drought prone, typical ratios are 8–12 for cereals, 4–8 for oilseed crops and 30–50 for roots and tubers. These high ratios do not however always translate into economic incentives for using fertiliser, especially when the fertiliser is expensive, product prices are low and marketing opportunities are low.

The chart below shows the results of a fertiliser trial with cabbages in Columbia:

Fertiliser N–P–K	Yield kg/ha	Yield increase		Value $/ha	Net return $/ha	Value : Cost ratio
		Kg/ha	%			
Control	53,500	—	—	685	—	—
45–0–0	75,267	21,767	41	946	261	16.1
45–45–0	75,367	21,867	41	936	251	9.5
45–45–45	94,433	40,933	77	1172	487	14.2

Nitrogen Flush
In many dry tropical soils the amount of nitrogen that is available to plants can fluctuate very much during the year. For example, at the start of the rainy season there is often a *flush* of available nitrogen, which can be taken up by wild plants and also any crops that have been planted early enough in the season.

This useful phenomenon is called the *Birch effect* or *Nitrogen flush* and is one good reason why it is often a wise idea to plant as early as possible in the dry, tropical climates, especially those with distinct rainy and dry seasons. If soils samples are taken at this time, misleadingly high levels of soil nitrogen may be recorded, leading to less than optimum applications of nitrogen fertiliser at planting time.

In addition, a small amount of nitrogen, about 2–4 kg/ha, can often be added to the soil by rainfall, as oxides of nitrogen created by lightning (**1Ae**, page 6).

Plant Food (Nutrient) Requirement of Food Crops

The amount of food required for each crop varies widely, and also depends on the soil type (including any inherent deficiencies in major or minor deficiencies), the yield target, the season and the previous cropping and fertiliser use.

A few examples are listed below, but only for illustrative purposes—professional advice should always be taken before the amount and type of fertiliser is used or recommended.

Crop	Kg/ha		
	N	P	K
Winter wheat	120	50	50
Barley (feeding)	100	50	50
Maize	120	45	45
Sorghum	85	30	30
Legumes	0	50	50
Irish Potatoes	220	250	300
Carrots	60	125	95

Split Dressing

Winter sown crops in temperate climates are not normally given much nitrogen at planting time because most of it would be leached before it could be used by the crop. Instead, the fertiliser is *split-dressed*—for example, phosphate and potash is applied at planting time, and nitrogen in the spring.

A split dressing is also often advisable for a number of crops in other climatic regions. With maize for example some fertiliser is often applied before or during planting, some a few weeks later, and the last application just before pollination.

Storing Fertiliser

Fertiliser should be stored as cool and dry as possible. If it is in plastic bags, these should be stacked neatly and flat to minimise rips in the bags. Take care to keep water well away; some fertilisers are also a moderate fire risk.

For the benefit of illiterate people involved, the bags should be marked with a clear design to indicate their potential danger.

Fertiliser is sold by the manufacturers in various forms: "Big Bags" of half or one tonne (MT), in bulk, in liquid or gas form, or in 50 kg plastic bags. At the village market level it is often sold in very small quantities, sometimes for prices well above its actual economic value (see above).

Applying Fertiliser

There are four main ways to ensure that most of the fertiliser is used by the plant:

1. Broadcasting	The fertiliser is spread, or sprayed if in liquid form, on to the soil surface, the *seedbed*, before planting and then worked into the soil with a hoe, rake or harrows. This is a quick method, but much of the fertiliser is lost to weeds or to leaching before the crop can utilise it.
2. Placement	The fertiliser is put in the soil together with the seed, either manually or by machine (a combine seed drill). The fertiliser should never be in contact with the seed, but should be placed either on one or both sides of the seed and/or about 2.5 cm below it. This is a much more precise method than broadcasting, but slower. It is very often the best method to use.
3. Top dressing	The fertiliser, normally N, either liquid or solid, is applied directly on to the growing crop. Winter sown crops and grassland are very often top-dressed with nitrogen in the spring. Small-scale farmers and gardeners often place the fertiliser by hand around individual plants and trees, which is very efficient but also very time consuming.
4. Foliar application	The fertiliser can be applied together with irrigation water, a technique suitable for dry fertilisers, ammonia, liquid compound fertilisers and gypsum. But foliar sprays, as supplied by overhead sprinkler systems, are only suitable for small quantities of fertilisers because leaves are damaged by high concentrations of salts. Normally trace elements rather than major elements are applied this way; if large amounts of fertiliser are to be applied, surface (flood) irrigation is normally more suitable.

Residual Value of Fertilisers

Fertiliser, and lime, which is applied to a crop is usually not completely used up by that crop. A *residual value* of fertiliser or lime is then available for the following crop.

Potash is mostly used up after two crops, while phosphate can remain for three or four. Nitrogen is not usually carried over in this way, but is rapidly either used by plants or leached out of reach of plant roots. This is just one of the many factors to consider when planning a crop rotation, discussed further in **1Ha. "Rotation"**, page 65.

For example, legumes and cereals are often planted after potatoes so that they benefit from the large amount of fertiliser that is often given to potatoes.

The value of land sold or rented may increase if it has a high content of residual fertiliser.

1E. THE PLANT

1Ea. Annual/Biennial/Perennial

An **annual** plant is one that completes its life cycle in one growing season; the plant develops, flowers, fruits and dies, usually within one year. Examples include beans, wheat, maize and rice.

A **biennial** plant is one that normally develops during its first year then flowers, fruits and dies only in its second year. Examples include onions, carrots and beetroot.

Sometimes biennials behave like annuals, often as a result of some form of stress, and produce seed in their first year. The plants are then said to have *bolted,* and the plants themselves become known as *bolters.* These plants normally cannot be eaten.

A **perennial** plant is one that continues to grow from year to year. There are two kinds of perennial plants:

- Herbaceous perennials, such as Perennial rye-grass and couch grass, in which the stems and leaves die down to ground level or close to it in the autumn or at the end of the rainy season. In spring, or with the onset of rains, new shoots develop from underground storage organs.
- Woody perennials, such as fruit trees and bushes, in which each year's new growth arises from the woody stems, so that height and breadth are increased every year.

1Eb. Botanical Classification/Taxonomy

The classification, or taxonomy, of plants is based on the similarity between various parts of their anatomy, particularly the flowering parts.

All plants are divided into **monocotyledons** ("monocots"), which have a single cotyledon in the seed, and **dicotyledons** ("dicots"), which have two. These two plant types have different growth habits; monocots grow from the base of their leaves, pushing their leaves up and out from this base, while dicots grow outwards from the edges of the leaves.

All known plants are classified by botanists under the following headings: **Divisions, Classes, Orders, Families, Genera** and **Species**, described below. Sometimes these are further subdivided into sub-orders, sub-families etc.

Botanists are constantly changing or modifying the classification of individual plants as they find new evidence either for or against relationships between different plants. This often leads to the same species being classified in more than one way, leading to two or more *synonyms* (syn.). The synonyms of some of the most important food crops are included in **Section 2** *"Description and Characteristics of the Main Food Crops"* and also in **Section 3A** *"Naming and Classification of Food Crops"*, pages 288–309.

Disagreement amongst botanists can even extend to the naming of plant families, the *Cruciferae* family for example sometimes being classified as the *Brassicaceae* family.

A basic understanding of how and why plants are classified is very useful; if there is an international understanding among all people who work with plants and seed this helps to reduce confusion. Plant classification in the common language of Latin allows an international agreement about the general characteristics of plants, and how they can be described. Comprehensive details on plant taxonomy are available online at the USDA's website GRIN—the Germplasm Resources Information Network.

Recent research involving gene transfer (modification) and identification has lead to a greater understanding of the similarities and differences between species, so the never ending renaming of species and sub-species will inevitably continue.

Division
All flowering plants belong to one of two Divisions:

- **Angiosperms**—plants that have their ovules and seeds enclosed within the pistil, and which develop into a seed-bearing fruit. Almost all agricultural and horticultural plants, including all of those described in this document, are Angiosperms.
- **Gymnosperms**—plants which have "naked seeds". Their seeds develop on the surface or at the tip of an appendage, and are not enclosed in it. Examples: pine trees, juniper and the ferns.

Class
Plants are also classified into one of the two plant Classes:

- **Monocotyledons**—often called "monocots", their seed has only one cotyledon- see above. The veins on the leaves are normally parallel, or nearly so. Virtually all of the true cereal crops—rice, wheat, barley, maize etc.—and sugarcane and grasses—are monocotyledons.
- **Dicotyledons**—often called "dicots", each seed contains two cotyledons ie virtually all of the other food crops and non-food species. The veins on the leaves normally have a type of a network pattern.

Order
An Order consists of several related families. For example the Families *Papaveraceae* (Poppies), *Fumariaceae, Cruciferae* (Brassicas) and *Resedaceae* are all members of the Order *Rhoedales.*

Family
Similar genera are grouped into families. Any group of plants that can normally cross-pollinate and reproduce with each other are classified into the same family. This divides them from plants of other families with which they do not normally reproduce.

Most of the important field crops are in two families, the ***Leguminoseae*** (legumes) and the ***Poaceae (Gramineae)*** (grasses and cereals). Other plant families of interest to food producers are: *Solanaceae* (eg tomatoes, Irish potato, eggplant), *Crucifereae* (eg cabbage, broccoli and other brassicas), and *Chenopodiaceae* (eg sugar beet and spinach).

Genus
A genus consists of a number of similar species. The plural form of Genus is "Genera"; similar Genera are grouped into Families. For example, some of the important Genera of

the legume (*Leguminoseae*) family are: *Trifolium* (clovers), *Glycine* (soybean), *Phaseolus* (field beans), *Pisum* (peas) and *Vigna* (cowpeas).

Species

Two plants are considered to be the same species when they can reproduce (ie their gametes can fuse) and produce fertile offspring. The normal abbreviation for species is "sp." (plural "spp."). Plants that are very similar botanically can be further subdivided into sub-species, or "ssp.".

Plant names are normally classified using the binomial system—the first, or "generic", name describes the Genus of that plant (eg *Glycine*), while the second, or "specific", name describes its species (eg *max*) ie *Glycine max*, the soybean.

Variety

This word has two different meanings in agriculture and horticulture:
1. Equivalent to *cultivar* (abbreviated to cv. and cvs.), and
2. A botanical distinction between two very similar species.

The difference between the two meanings and the way they are used is described below.

1Ec. Cultivar/Variety

1) In the agricultural and horticultural sense, a variety is a group of plants of the same species that are:
- **distinct**—in the sense that plants of one variety are botanically identical,
- **uniform**—in the sense that the variations in plant structure which do occur between plants of one variety are capable of definition, and
- **stable**—in that the variety will remain largely unchanged when it is grown on for a number of years.

In this way the word "variety" is interchangeable with "cultivar", the former being favoured by the farming community and the seed trade, the latter more common with botanists. The word cultivar is used internationally to describe varieties and strains of cultivated plants.

Example: *Homo sapiens* describes all human species, with Mongoloid people being a Chinese cultivar or variety and Caucasian people as a European cultivar or variety.

2) Botanists have their own meaning for the word "variety", normally abbreviated to "var.", which can be described as a botanical variety as opposed to an agricultural variety described above.

Example: the crop species *Brassica oleraceae* includes several distinct botanical varieties ("var."):

B. oleraceae var. *botyritis*—broccoli, cauliflower,
B. oleraceae var. *italica*—sprouting broccoli (Calabrese), and
B. oleraceae var. *germifera*—Brussels sprouts.

Each of these botanical varieties normally has several agricultural varieties that are grown by food growers and farmers. To avoid confusion between the two meanings of the word it can be useful to distinguish them as follows:

– *Cultivar (cv.)*—for botanical distinctions between closely related species,
– *Variety (var.)*—a group of agricultural or horticultural plants which are essentially identical.

The word "cultivar" should not be confused with the word *cultigen*, the name given to a plant species which is found only in cultivation and which did not originate from a wild type.

1Ed. Photosynthesis / Respiration

Photosynthesis is the process by which green plants synthesise carbohydrates from water and carbon dioxide. The energy from sunlight is utilised to achieve this, and it takes place in the *chlorophyll* (green pigment) of the plant. The process is sometimes known as *carbon assimilation* (ie a gas, carbon dioxide, is converted to a solid, carbon).

Respiration is the process of the oxidation of carbon and hydrogen and the release of energy, carbon dioxide and water—ie the reverse process to photosynthesis, and is continuous in living plant cells. Respiration provides the energy for plant growth during the hours of darkness when photosynthesis virtually comes to a halt. As a result, the dry weight of a growing plant decreases slightly each night, but this loss is more than replaced by photosynthesis the next day.

The heavily shaded lower leaves often consume more carbohydrates by respiration than they can manufacture by photosynthesis. In these cases removing the lower leaves (of tomatoes, for example) may help the plant, as well as facilitating weeding, irrigation and harvest.

1Ee. Transpiration

Transpiration is the process by which water is extracted from soil by root hairs, moves upwards through the plant and is given off through the leaves and stems as water vapour which enters the atmosphere.

Transpiration may occur either through the plant's *cuticle*, the moist membrane on the outer layer of the plant, or through the *stomata*, very small holes, found most commonly on the underside of leaves.

New evidence suggests that transpiration may be a leaf cooling mechanism to prevent the denaturation of leaf protein.

1Ef. Wilting

A plant appears to *wilt* when its cells lose turgor (rigidity) as a result of water loss; its leaves and young stems become weak and drooping. Wilting normally occurs when water loss by transpiration exceeds water intake through the roots.

Young plants are more susceptible to wilting than older plants because their root systems are less well developed, especially soon after they have been transplanted.

Wilting is not always caused by water loss; it can also occur as a result of certain wilt diseases, caused by both fungi and bacteria. Plants grown in *saline soils* can also often become wilted even though the soil may have enough moisture, discussed in **1Ca**.

1Eg. Shade Plants/Sun Plants

All plants need light in order to grow, but the amount of light they need varies according to the species and the growth stage of the plant. For example groundnuts (peanuts) like to have plenty of light during germination and early growth, but during flowering and maturation they can grow well in some shade. In practice this means that groundnuts can be intercropped with taller plants such as wheat provided that the wheat does not shade the groundnuts early in their life cycle.

Other plants such as pigeon peas and most of the cereals require full lighting during their entire life cycle if they are to grow well—these are the so-called *sun plants*. Their leaves display a higher light saturation point and maximum rate of photosynthesis.

By contrast *shade plants* such as the peppers (*Capsicum* spp.) will grow well in lightly shaded places such as under trees. In general, their leaves are thinner, have more surface area and contain more chlorophyll than the leaves of sun plants.

The effect of Temperature
Sunlight is only one factor which influences plant growth. Temperature is another. Although some crops such as tomatoes and onions prefer full sunlight—the "sun plants" described above—they also suffer from excessive temperatures. So in very hot regions tomatoes will actually grow better in lightly shaded places protected from the full heat of the sun, even though they are normally considered as sun loving plants.

Legumes also prefer to grow in full sunlight, but in very hot sunny places they will also grow well if they are lightly shaded.

Sometimes even different varieties of the same crop species have different requirements for light and temperature.

This is a good example of the subtlety of understanding required by people who grow food. They cannot blindly follow rigid rules and expect consistently good crops, but should be aware of the different needs of different plants for light, temperature, water, air, soil etc. and then attempt to regulate these factors as far as practical.

1Eh. Vegetative Reproduction

Also known as *vegetative propagation* (*multiplication*) or *asexual reproduction,* this is the form of reproduction in plants that does not involve a flower, seed or any sexual process but involves removal of part of the plant. This part is then placed in soil, or water, where it develops into a new plant, identical to its parent.

This form of reproduction perpetuates plants of uniform appearance, quality etc. for many generations (although occasionally a mutation occurs to change this, such as the sudden unexpected appearance of a red tuber in a white variety of potatoes).

There are both pros and cons of vegetative propagation as opposed to propagation by seed, described below:

Advantages of Vegetative Reproduction

Plants are established quickly. Also, since all the progeny are identical to each other food growers know exactly what type of plant will grow as the plants will have all the same characteristics as the parent, which the farmer may have selected in the first place.

With propagation by seed there is always the danger of cross-pollination, especially with certain species (eg rye and most brassica species), so that seed can sometimes produce some quite unpredictable plants. Such plants are usually less useful than their real parents, so in the case of a few crops such as rye, and to a lesser degree most other crops, "new" seed should be bought every year from a professional seed producer.

Disadvantages of Vegetative Reproduction

The plant parts which are used for vegetative reproduction—the stems, roots etc—are bulky, moist and do not store well, and they cannot be transported as easily as seed.

Some diseases, and also some insects, which are present in the parent plant may be passed on to its progeny. This is a problem with Irish potatoes, for example, which transfer at least two virus diseases, Leaf Roll and Mosaic, from the parent plant to the tubers.

Examples of Vegetative Reproduction

- **Root**—rarely used by food growers, although rhubarb, sweet potato, cassava and carrots can be propagated by their roots;
- **Rhizome**—an underground (usually) stem that often sends out roots & shoots from its nodes. Very shallow, or even above ground. Also known as creeping rootstalks, or rootstocks. *Examples*: asparagus, couch grass, meadow grass, some irises;
- **Tuber**—the enlarged tip of a rhizome, typically high in starch, used for food storage. Described in **1Ei.** *Examples*: Irish potato, Jerusalem artichoke;
- **Stolon**—also called *runners*, they are horizontal, above ground stems, unlike rhizomes, which are normally below ground. *Examples*: Trefoil and the clovers;
- **Stem cutting**—propagation by cuttings is commonly known as "striking". Root formation may be assisted by applying auxins, such as IBA (Indole-3-butyric acid). *Examples*: sweet potato, cassava and sugarcane;
- **Bulb**—a vertical underground stem. *Examples*: onion, shallots, garlic, leek, lily;
- **Corm (Bulbo-tuber)**—a short, swollen, fleshy underground stem—*Cormels* are new, small corms arising vegetatively from a parent corm. Corms are normally flattened on the top and bottom. *Examples*: taro, banana and crocus;
- **Graft**—a woody stem taken from one perennial plant and fixed to another one. Most fruit crops are grafted at some stage of their lives.
- **Adventitious buds**—develop into above ground stems and leaves, forming on roots near to the ground surface and on damaged stems. Adventitious roots form on stems where the stems touch the ground surface.

Most of the crops mentioned above can also reproduce sexually by producing flowers and seeds and thus have two ways in which to reproduce themselves. Onions for example can be either grown vegetatively from bulbs (*sets*) or sexually from seed.

Clone is the name given to a group of plants (or animals) that are all genetically identical as a result of having been propagated vegetatively (asexually).

1Ei. Determinate/Indeterminate

These terms describe the two ways in which annual plants produce flowers and are mainly used when describing legumes.

Different varieties of the same species may differ in this respect. With soybeans, cowpeas and tomatoes for example there are both determinate and indeterminate varieties.

Determinate varieties of plants increase in growth very little, if at all, after they have started to flower. The terminal or central flower is the first to open. The main advantage, especially for mechanised harvesting, is that almost all of the grain matures at about the same time.

Indeterminate varieties continue to grow and increase in height after flowering has started. The terminal flower is the last to open. They tend to have fewer branches than determinate types, and are normally grown in areas with short day length ie where flowering begins early on in the life cycle of the plant. They can be very useful in subsistence agriculture because they produce a fairly regular food supply, harvested by hand, over a long period of time. But with mechanical harvesting they can cause problems because on any given day only a proportion of the seed will be mature and ready for harvest. The rest of the grain will be either immature, and too moist, or it will be lost because it has *shattered* ie previously matured and fallen to the ground.

1Ej. Day length/Photoperiodism

Many plants, including most of the legumes, are *day length sensitive*. This means that their reproductive stage, ie flowering, is triggered by their response to day length, the number of hours sunshine in a 24 hour cycle.

Photoperiodism in plants can be defined as the effect on flowering of the relative lengths of day and night. Plants respond to day length in one of three ways:

- **Short-day plants** stop flowering as soon as the day length has passed a critical value:

Examples: rice, most soybean varieties, maize, sorghum and millet. They generally require only 8 or 9 hours of sunlight (ie a dark period of at least 12 hours) in order to flower;

- **Long-day plants** start to flower soon after the day length has passed a critical value:

Examples: carrot, pea, lentils, oats and most varieties of wheat. The day length on the Equator is normally not long enough for these species to flower;

- **Day-neutral ("ever-blooming") plants** are unaffected by day length.

Examples: tomato, cucumber, cotton, sunflowers and buckwheat (some varieties are long-day).

Some plant species have different varieties with different photoperiodic responses. For example some cowpea varieties are short-day, some are long-day and others are day-neutral.

The key to photoperiodic response is in fact the length of darkness during a 24-hour period. If the light (daytime) period is interrupted with a period of darkness, this will not affect flowering. But if short-day plants are exposed to even short periods of light during the night, flowering is prevented even though the period of day length is correct.

The number of 24-hour cycles of suitable photoperiods to which plants need to be exposed to trigger flowering is different for each species. One exposure is enough for some species while other species need three or four.

Moving Seed from One Country to Another

This photoperiodic response of plants is more than just an interesting botanical feature. Anybody who is involved with seed programmes should have some understanding of the practical relevance of photoperiodism if seed is moved from one country (or latitude) to another.

Photoperiod sensitive crops such as maize, sorghum and soybeans will very often not grow successfully if they are introduced to a region that is very different in latitude from the region to which they are adapted.

Example: if a variety of soybean which is adapted to grow in latitudes far from the Equator is introduced and grown in a latitude much closer to the equator, it will start to flower much earlier than normal. This means that its vegetative growth stage will be much shorter than normal, and it will probably produce a lower yield as a result.

1Ek. Growth Period

Also known as the *length of growing period* (LGP), the *number of days to maturity* and the *maturity time*, the *growth period* of a crop is the term used in this book to describe the number of days from planting to harvesting. This figure, usually expressed in days, varies with growing conditions, temperature and so on.

The growth period is a useful piece of information to know when looking at any particular crop, for reasons given below:

- If a farmer who is trying to grow food in *marginal conditions* chooses to grow a variety of a crop which has a short growth period (it is *early, quick growing, precocious* etc.) then they will very often be more likely to produce at least some food than if they choose to grow a *late, unimproved* or *local* variety with a longer growth period.
- For subsistence farmers, food today is much more interesting than food tomorrow. The earlier that a farmer can harvest, and feed himself and his family—and then plant a second crop—the more he is attracted to that crop.

- Varieties with a short growth period can often be planted at a time of the year which avoids periods which are difficult for plant growth, such as in the dry season or when insect or disease attack is likely to happen.

So we can see that farmers who decide to grow an early variety, and who probably accept a lower yield as a result, often make a good decision, especially if they can then immediately plant a second crop which would not have been possible if they had planted a later but higher yielding variety.

Plants have a number of elaborate ways of adapting to their environment, one of which is by changing their growth period. In general, plants which are growing in unfavourable conditions (too hot or dry, or in infertile soils for example) will grow faster, ie will have shorter growth periods, than plants growing in more favourable conditions. The plant is stimulated to reproduce itself, by producing at least a few seeds, as a result of the unfavourable conditions—if the plant delayed maturity any longer it might die before it could produce any seed at all. So the plant shortens its growth period and thus becomes more likely to succeed in reproducing itself.

If the plant is grown in more favourable conditions its growth period will probably be longer. These conditions may be changed either by man or by a change in climatic or field conditions. For example, if a variety of a crop is taken from one altitude to a higher altitude its growth period may be reduced. Or if heavy applications of fertiliser or irrigation water are supplied, the growth period may be extended because the plants are then provided with more favourable conditions.

Plant breeders spend a lot of time and energy developing varieties that mature early ie which have a short growth period. The challenge facing plant breeders is how to incorporate this precocious behaviour with acceptable yields.

Late maturing varieties, with longer growth periods, have their own advantages, such as larger and more dependable yields, especially when neither the shortage of land nor the immediate supply of food is a serious concern for the farmer.

1El. Tuber

A tuber is an enlarged, swollen tip of an underground stem (or *rhizome*). Tubers have buds, or *nodes*, in the axils of rudimentary leaves or scales. They typically have a high starch content. *Examples*: Irish potato and Jerusalem artichoke (*Helianthus tuberosus*).

Tubers contain stored food, mainly starch, and are used by food growers in the vegetative propagation of some crops. They are often confused with "roots"—the Irish potato, which actually produces tubers, is normally called a *root crop*, though technically they produce tubers and not roots.

The true root crops are those where the root itself, and part of the hypocotyl, is enlarged and which do not form leaf buds on the root. *Examples*: carrots, turnips, cassava and sweet potato. Tubers are also described in **1Eh. "Vegetative Reproduction"**, page 40.

1Em. Rogue Plants

A rogue plant is any plant that is a visibly different type from other plants of the same crop growing together in the same field. They are not always a problem, but they are of particular relevance to producers of seed crops, who try to produce a uniform and standard product. This type of rogue, in a seed crop, is also known as an *off-type*, a term which can apply to either the plant or seed. In this sense, to *"rogue a field"* means to remove these off-type plants, normally by hand, before they flower and so leave a more uniform quality of seed to be harvested.

For most farmers the word *rogue* means "weed", such as wild oat rogues in a cereal crop, which are also normally pulled up by hand since chemical control is unreliable or very costly.

1En. Volunteer Plants

A volunteer is a plant of one species growing within a crop, or uncultivated land, of another species. Volunteers usually come from the previous year's crop, as a result of seed or tubers being left behind in the field by mistake. Potato plants for example are often found growing in the following year's cereal or legume crop. Volunteers can also be brought into the field with compost or manure, or by birds, animals or human feet.

Volunteer plants can be viewed from two points of view: either as a welcome extra source of food in a mixed or hand harvested cropping system, or as a weed and a nuisance in mechanised farming systems.

The benefits of crop rotations which are planned to control pests or diseases may be reduced if volunteer plants appear in a crop because these pests and diseases may be carried over in the volunteers and so survive and continue from one season to the next.

1F. THE SEED

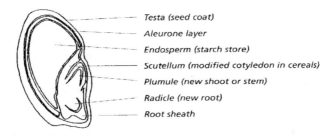

Testa (seed coat)
Aleurone layer
Endosperm (starch store)
Scutellum (modified cotyledon in cereals)
Plumule (new shoot or stem)
Radicle (new root)
Root sheath

Figure 7. Cross-section through a typical cereal seed (barley)

The word *seed* is normally used for the part of the plant that is used for sowing, while the word *grain* refers to the same part of the plant, but when it is used for consumption by humans or animals.

One physical feature of a seed which can be useful for identification purposes, of legume varieties for example, is the part known as the *hilum*, the point of attachment to the funiculus, which leaves a scar, normally oval in shape, on the seed where it has broken away.

1Fa. Germination

Germination is the process whereby a seed is transformed into an independent, established seedling. It is a complex and intensively researched topic, but in essence a seed will only germinate when it is has sufficient *air, moisture* and *warmth.*

1. Seed Dormancy
Sometimes a healthy, living seed is unable to germinate even though all three of the conditions are favourable—the seed is then said to be *dormant.*

Seed dormancy can be a very beneficial survival strategy of plants in the wild state because it helps to ensure that the seed only germinates when the conditions are favourable for plant growth. But in cultivated crops dormancy can sometimes be a problem for food producers. For example some species of *Beta vulgaris* and some of the legumes such as clovers sometimes will not germinate for weeks or even months, even when all the conditions appear to be favourable.

Sometimes if seed is planted which has only recently been harvested it germinates very slowly or not at all. Sorghum seed for example is dormant for about a month after harvest, and most *indica* type rices need a one to three-month "rest period" after harvest in order to reach maximum germination and vigour (*japonica* type rices normally do not have a dormancy period).

Fortunately this characteristic has been more or less eradicated from the commonly grown food crops due to selection pressure from plant breeders. The dormancy period for these crops is normally passed either just before or just after the seed falls from the seed head.

Mechanisms of Seed Dormancy
Dormancy is imposed on seeds by a number of mechanisms, which fall into two categories:
* ***Embryo (Internal) Dormancy***—when the seed embryo holds the mechanism, and
* ***Hard Seed (Coat Imposed or External) Dormancy***—when the seed *testa* or outer membrane is hard and does not readily *imbibe* (take in) water.

In some species germination is either stimulated or inhibited by the seed being **exposed to light,** either in continuous light or even short bursts. Examples of food crops that will not germinate in the light include chives, garlic and amaranths.

Other species such as maize, the smaller cereals and many legumes such as beans and clovers will germinate in either light or darkness. The variety *Grand Rapids* lettuce is inhibited by "far red" light (wavelength about 730 nm) but is stimulated by red light (wavelength about 660 nm).

Breaking Seed Dormancy
To avoid the problem of dormancy seed may have to be processed in some way to *break dormancy*, such as by using machines to scratch the testa or by leaving the seed moist and at temperatures between 1°C and 10°C for several weeks.

As soon as the food producer or seed company understands the dormancy mechanism of a certain species then it is relatively simple to mimic the appropriate conditions of temperature, water and light so as to deceive the seed into breaking dormancy.

2. Germination and Temperature
Seed will only germinate between certain temperature ranges. For example, if lettuce seed is planted in soil at 2°C or at 30°C it will not germinate, or only poorly. This is not true dormancy but a characteristic of plants to improve their chance of survival by not starting to grow in hostile conditions.

In hot conditions: to overcome this problem, which is at its most critical 3 or 4 hours after sowing, plant the seeds in the cooler, evening temperature. Vice-versa in cold places.

A dramatic example of the effect of temperature on germination is shown below.

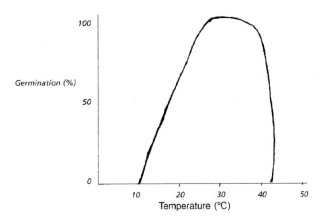

Figure 8. The effect of temperature on germination of tomato seed

3. Styles of Germination
As shown in the drawings below, seed develops into seedlings in one of two ways:

- *Hypogeal* germination, where the cotyledon(s) remain below the soil surface, within the testa and only the apex and first leaves are raised upwards.
  Examples: horse (broad) bean, peas, maize and most monocotyledons.
- *Epigeal* germination, where the cotyledon(s) and shoot apex are lifted above the soil surface; they turn green and become the first leaves of the new plant.
  Examples: castor beans, haricot beans and most dicotyledons.

With onions, a sharply bent, tubular cotyledon breaks the soil surface and slowly straightens. The cotyledon base encloses the shoot apex, and the first leaf emerges through a small opening at the base of the cotyledon.

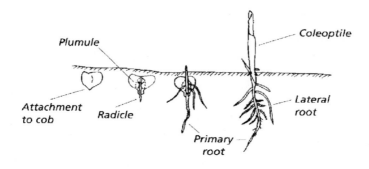

Figure 9. Hypogeal germination in Maize

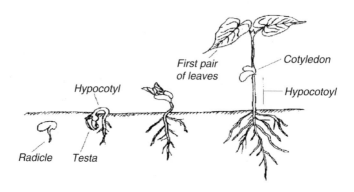

Figure 10. Epigeal germination in Haricot Bean

Some plants such as the runner bean (*Phaseolus multiflorus*) and peas (*Pisum sativum*) are normally *hypogeous* ie they display hypogeal germination, although their cotyledons arise out of the ground and display epigeal germination.

Germination Tests
Before seed is distributed or planted it is essential to have it tested for germination if this is possible (time and facilities available, cost etc.). Testing is normally done by the seed supplier and/or governmental seed authority, in which case a written and dated certificate should be obtained as a matter of priority.

Independent, unbiased germination tests are particularly important if, for any reason:
- the seed supplier tested the seed some long time previously,
- the seed supplier's test is not considered to be reliable,
- the seed has been stored in hot and/or humid conditions,
- the seed has been treated (*dressed*) after the test, or
- the seed has been transported a long distance or has been handled roughly.

Poor Germination Test Results
With larger quantities of seed, if the test result is poor, say less than about 80% for most crops, the seed should not be written off as there are some steps to take, such as:
- refer to the seed purchase contract, and take appropriate measures with the relevant party—seed supplier, lawyer etc. Recourse to the law may be necessary (see **Section 3B. "Seed Purchase Procedures"**, pages 310–312).
- if appropriate, plant the seed as soon as possible—before or during the planting season—unless the seed has a very low germination rate, in which case it may be usable as food (unless it has been treated, of course).

Seed with a low percentage of germination with low vigour usually produces weak seedlings and plants.

Testing Seed for Germination
At least 400 seeds should be tested, divided into lots of 100. To make sure that a representative sample is tested, small amounts of seed should be sampled from several seed bags from all over the seed store, then mixed together and the test sample removed.

In practice, and if at all possible, it is a good idea to have the seed tested by professionals. If possible, find a local agricultural college, university, independent seed supplier or government seed laboratory to test the seed for germination.

Informal germination testing can easily be done **in a warm place** at home or in the office. Place the counted seed onto flat plates with absorbent, unbleached, white tissue. Place more paper on top of the seed—use plenty of tissue so that the seed does not dry out too quickly. Thoroughly moisten the paper with clean water, then cover the plate to reduce evaporation. Continue to check the dampness of the seed/tissue, especially for the first few days when the seed is inbibing (absorbing) large amounts of water. Obviously, the larger the seed the more water is absorbed; twice daily inspections/top ups are usually required for the first two or three days. Keep the seed warm at all times.

When the seed is sent for testing, it can be useful—and avoid a lot of confusion—if an **Analysis Label** is tied to the seed bag.

```
┌──────────────────────────────────────────────────────┐
│                                                      ● │
│  Analysis Label                                        │
│  Name:                                                 │
│  Address:                                              │
│  Crop:                                                 │
│  Variety:                                              │
│  Ref No.:                                              │
│  Total weight:                                         │
│  Germination (Min.):    %. Inert Matter (Max.):     %  │
│  Hard Seed (Max.):      %. Other crop seed (Max.):  %  │
│  Purity (Min.):         %. Weed seed (Max.)         %  │
│  Dates of test: start.......... finish..........       │
│  Origin:                                               │
│  For distribution in:                                  │
│  Seed Treatment/Dressing:                              │
│  Weight of this sample: ......kg                       │
│  Comments:                                             │
└──────────────────────────────────────────────────────┘
```

Acceptable Germination Rates

This percentage varies according to the law of the country in which the seed is sold. A few examples are given below of the "acceptable", ie legal minimum, germination rates according to UK law: Wheat and barley 90%, oats 85%, peas and rye 80%, maize 75% and carrots 60%. The rate is different for each category of seed eg 98% for "Basic" or Grade 1 wheat seed, 90% for "emergency"/for free distribution wheat seed.

4. Seed Vigour and Viability

Seed which has approximately 70% or higher germination rate is said to be *viable*. The speed with which the seed germinates determines its *vigour*.

The chart below shows this difference between the vigour and the viability of a seed lot:

Seed Lot		$10°C$	$25°C$
A. *Low Vigour*	% germination	82.5	85.3
	T_{50}*	12.1	2.8
B. *High Vigour*	% germination	81.2	84.5
	T_{50}*	5.2	2.3

* T_{50} is the time taken, in days, for 50% of the seed to germinate.

Both lots of seed are viable, but Lot A has lower vigour than Lot B because it takes longer to germinate. This characteristic of low vigour often only becomes apparent in poor conditions for germination - in this case, low temperature.

☞ **A Word of Caution about Germination Tests**

The example above shows that even when a seed lot has a good germination rate when tested in ideal conditions in a laboratory or at home, it will very often germinate either less or more slowly when it is planted out in the field. With sorghum, for example, a laboratory germination test of 90% may give only about 50% germination in field conditions.

5. Seed Longevity (Storage Life)

In ideal storage conditions such as those maintained in a plant gene bank (about -20°C and 5% relative humidity), seed of most species will remain viable and vigorous for many years. But in real life seed is never stored this way, and it is also often transported long distances and handled roughly, which can be very damaging to fragile seed such as soybean or groundnuts.

In general, seed loses its germination more rapidly when stored in hot and humid places than in cool dry ones. Seed with a high moisture content also loses viability more rapidly than dry seed.

As usual in agriculture there are no hard and fast rules, and for this reason the following points should be noted:

- test seed before it is distributed, if at all possible;
- try to store seed for as short a time as possible, especially in hot or humid conditions;
- transport seed to more favourable storage conditions if it has to be stored for a long time;
- handle seed very gently—it is a living organism. Avoid throwing or dropping sacks etc.

Approximate Seed Longevity:

1–2 years:	maize, okra, parsley, parsnips, soybean
3 years:	asparagus, beans, leeks
4 years:	beet, chard, pepper, pumpkin
5 years:	broccoli, cauliflower, celery, cucumber, eggplant, lettuce, muskmelon, radishes, spinach, squash, turnip, watermelon
5+ years:	cucumber, finger millet, teff

6. Recalcitrant Seed

As noted above, seed remains viable for longer when its moisture content and storage temperature is reduced. However some species have what is known as *recalcitrant seed* which does not obey these laws and which is killed when their moisture content is reduced below a relatively high value. All such species are perennial.

Examples: cocoa, palm-oil, coconut, avocado, mango, mangosteen, durian, rambutan, langsat, jackfruit, oak, chestnut and horse chestnut.

1Fb. Seed Rate

This term is used to indicate the amount of seed that is needed to sow (or to "plant") an area of land. The seed rate is normally expressed in kilograms per hectare (kg/ha). *Example*: a farmer has a 1 hectare field to sow and decides to grow only one crop. If he decided to grow barley he would need about 50 kg of seed, but if he chose to grow pearl millet instead he would need only about 5 or 10 kg.

The seed rate is determined by the desired target plant population. This is defined as the appropriate plant population to achieve optimum crop performance. The seed rate is calculated by first assessing the thousand grain weight (TGW), then considering the likely percentage survival of the seeds to achieve that target population. Factors to consider in this estimation include: sowing date, soil type and temperature, and the farming system itself—for example, most organic systems require higher seed rates than intensive regimes due to higher losses from insect pests and diseases.

Seed rate is similar to most aspects of food production—there are very few firm rules to follow, and the seed rate for any one crop may vary enormously. With field peas, for example, an early variety, with small seed that is planted in good time on infertile soil in a dry region may need a seed rate of about 50 kg/ha. But a later maturing variety of peas, with large seed and which is planted late in fertile soil in a region with good rainfall may need a seed rate of 250 kg/ha.

Farmers very often increase the seed rate when they are sowing fields that are known to be heavily infested with weeds in the hope that the weeds will be more crowded out and shaded.

Effect of Germination Rate on Seed Rate
If the germination rate (discussed in the previous paragraph on germination) of the seed to be planted is less than 100% it is obviously a good idea to plant correspondingly more seed.

Some examples of seed rates (in kg/ha) of 3 crops with germination rates of between 60% and 90%+ are shown below:

Crop	90%+	80–89%	70–79%	60–69%
Sorghum	4–5	5–6	6–8	8–12
Soybean	45–60	50–70	60–80	70–90
Oats	80–90	90–100	100–120	120–150

1Fc. Hybrids

Hybrids are the offspring of parents, either plant or animal, of different species or varieties (inbred lines). The second and subsequent generations which result from growing hybrid seed are not regarded as being hybrids.

Hybrids can be either sterile or fertile; they are more likely to be sterile if the two parents are very different genetically, such as with the mule, the result of a hybrid cross between a horse and a donkey.

Hybrid Vigour
This is the genetic phenomenon where the characteristics shown by the offspring are more marked than the average of the two parents of a species. This is a strong argument in favour of using hybrid seed, which can often produce plants that are hardier and grow more strongly than either parent.

Hybrid Seed
This is produced by specialist seed producers, mainly for sale to technically advanced farmers. Although it can produce much higher yields and the plants may be stronger and more uniform, *hybrid seed, especially of cereals, should be regarded with suspicion by subsistence farmers*, for the following reasons:
- it is more expensive than non-hybrid seed;
- it should only be *once-grown*. If the seed produced from hybrid seed is grown the following season (*twice-grown*) the resulting crop is likely to be very mixed in appearance, growth period etc. and will generally perform less well than a comparable non-hybrid crop. This means that the farmer has to find the money to buy seed every year instead of growing it for himself, free of charge;
- it requires a high level of crop management for several years before the full economic potential of hybrid seed is reached, and
- many hybrid varieties have a lower content of edible protein than open pollinated varieties.

In the UK the current performance of hybrid wheats and oilseed rape has not been significantly more cost effective than non-hybrid equivalents—the seed is still rather overpriced.

Availability of Hybrid Seed
Despite the problems noted above, the potential for the increase in global food production through the use of hybrid seed is enormous. Hybrid seed is a powerful force in modern agriculture and it is likely to become increasingly so in the future and plant breeders are working hard to improve the agronomic characters of hybrids, to complement their higher yields.

The most commonly grown hybrid seed is maize, and some wheat and rice hybrids are also available. Other commonly used hybrids include sunflower, sugar beet and a wide range of vegetables including tomato, cucumber and many of the Brassica family. Legume hybrids such as pigeon pea and chickpea are also available.

1Fd. Composite Varieties

A composite variety of maize or any other crop is not strictly speaking a variety (or cultivar) but is a population of plants containing genes from many previous parental generations.

By selection of plants over many years composites reach a state of equilibrium with the environment in which they are grown. As a result they are often a fairly mixed collection of plant types, each type having proven to be adapted to the growing conditions in a particular locality or climatic region.

Method of Composite Seed Production

Pairs of individual plants which normally self-pollinate, such as maize, are artificially *crossed* (or made to pollinate) by a plant breeder. The plants that result from this cross, which now contain genes from two parents, are then also artificially crossed together. The same process is repeated the following year, and sometimes also the year after that.

This results in plants that contain genes from eight or more of the original parents. These plants are then grown in the area for which they are intended, and allowed to self-pollinate. This population of plants is then grown on for a number of years in a certain type of environment. Those plants that are best adapted to that environment will be selected, initially by the seed producer/plant breeder and later on by food growers themselves.

These "survivor plants" gradually evolve to dominate the population. The result is a series of *landraces*—a population of naturally selected plants—that are adapted to grow in certain conditions. These plants become increasingly well adapted every year by a process of progressive selection, and a certain proportion are always likely to survive difficult conditions.

The fact that composite seed can be grown on by farmers from year to year is also obviously a major advantage over hybrid seed.

Influence of the Food Grower on Composites

A composite population of plants, often known as a *composite variety*, has a continuously changing genetic makeup, and natural selection is the principal force which acts to produce genetic change.

Food growers are the secondary force, who every year select from their composite plant population those plants which are best adapted to the conditions on their farm. If they have a good eye for selecting plants, their composite crop should become increasingly well adapted to the local environment.

1Fe. Inoculation/Nitrogen Fixation

Leguminous plants (the "legumes" or "pulses") normally grow better if they have *root nodules* which are present and functioning ie pink in colour. The nodules utilise *Rhizobia* bacteria to "fix" atmospheric nitrogen, which becomes available to plants.

A healthy, well-nodulated crop growing in a soil with low nitrogen content can fix 200–300 kg/ha of nitrogen, which represents a large—and, in effect, free—source of fertiliser. In some cases even the protein content of the grain may be increased.

When a new plant species, or a new variety of an existing species, is introduced to an area it is important that the appropriate *Rhizobia* bacteria are present in the soil to allow nitrogen fixation to occur in the root nodules.

Some of the nitrogen that is fixed in the nodules is used by the host plant, and the rest is released into the soil when the nodules are old and disintegrate, ready for the next crop. This ability to enrich soils by fixing nitrogen in them is of great value to food producers and deserves to be better understood and fostered by all concerned.

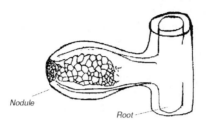

Figure 11. Field pea root nodule

Rhizobia
Cereal crops need a lot of nitrogen to grow well so a mixed cropping system of cereals and legumes is often a good idea. The benefits of this type of cropping have been known for thousands of years, and have been exploited in South America with maize/bean farming systems and in Ethiopia with cowpea/sorghum and cowpea/millet systems since earliest times.

Many different kinds of bacteria are capable of fixing nitrogen from the air and making it available to plants, both legumes and non legumes, but it is the *Rhizobia* kind which are the most relevant for food production.

There are many different strains, or species, of Rhizobia present in leguminous crops. The Rhizobia enter the plant roots, which are stimulated to produce lumps of various shapes and sizes ranging from 1 to 10 mm in diameter where the bacteria form a symbiotic relationship with legumes.

In some soils, Rhizobia bacteria are naturally present; in others they have to be artificially introduced, or *inoculated*, into the soil by the gardener or farmer. Rhizobia have two main features that are relevant to food producers:

- They are **specific to certain Genera or species of legumes**. In other words, one strain of Rhizobium that is effective for soybeans may not also be effective for lima beans, for example.
- They can **only survive for a short time** when they are not in either seed or soil.

Inoculum
Inoculum (or "inoculant") is a live, biological product—a bacteria—which should always be stored in a cool, dry and shaded place—about 4°C is optimum.

Because soils are often specific to certain groups of legumes, they frequently have to be inoculated to get the best crops. The Rhizobium culture is contained in certain inert substances such as peat and is inoculated into the soil directly or, more commonly, via the seed itself.

Seed should be planted immediately after inoculation, and protected from direct sunlight.

Although you can buy seed that has already been inoculated, this has a very short life and it is normally farmers and gardeners themselves who apply one of the four types of inoculum:

- **Powdered**—the most common form, in which finely ground peat is normally used as the inert substance;
- **Granulated**—micro granules are produced from clay flakes and powdered inoculum. It is applied into the seed furrow and is used when the seed is treated with a fungicide or insecticide, which can be harmful to Rhizobia. Seed dressings of Thiram are generally safe to use with inoculum;
- **Liquid**—either applied into the seed furrow or added to the seed before planting;
- **Pellet support**—placed together with the seed, either by machine or by hand.

Failures with Inoculation
Inoculation can fail to stimulate active nodules for several reasons:
- incorrect strain of Rhizobium for either the crop or the conditions;
- old inoculum, or stored or applied at high temperatures;
- seed kept too long between inoculation and planting;
- if treated seed is also subsequently inoculated, unless precautions are taken;
- soil too dry;
- soil too acid (especially when below about pH 5.5);
- excessive nitrogen in the soil;
- deficiency of phosphorus, boron or molybdenum.

Deciding Whether to Inoculate
There are no firm rules to follow when inoculation is advisable, but if a leguminous crop has been grown often and successfully in a field then it is usually not necessary.

However if a new leguminous crop is grown for the first time in that field, especially one which is specific in its Rhizobium requirement, then inoculation is very often worthwhile and cost effective.

The crop species which are most specific in their requirements are: chickpeas, lima beans, soybeans, lupin, lucerne and subterranean and Kenya White clovers. Pigeon peas are less specific, while cowpeas are fairly nonspecific and can normally use the strains of Rhizobia which are commonly found naturally in tropical soils.

The whole process of inoculating seed can be quite daunting, expensive and logistically challenging so if possible conduct some field trials in the first growing season which comes along. In this way, if you can wait for a few months (one growing season) to find out the answer to the question "Is inoculation a cost-effective exercise?".

It may be a good idea to plant out a number of trial size areas, half with inoculated seed and half with uninoculated seed. During the growing season make observations of the relative nodulation, plant growth/health and finally yields of the different trial areas, and then decide if inoculation is a worthwhile operation for future years.

How to Obtain Inoculum

Ideally the seed suppliers themselves are able to provide the appropriate strains of Rhizobia inoculum. However it does not travel well, and it is destroyed at temperatures of about 35°C, so if possible inoculum should be obtained locally. Local agricultural research centres are sometimes able to advise on sources etc.

If it becomes necessary to make a special order request for inoculum, the timing is of course of the essence as the inoculum should be new and fresh at the time of planting. Several weeks are usually needed to prepare a significant volume of inoculum.

The variety names of the crop to be inoculated should be known, and provided to the people producing the inoculum.

Even in optimum storage conditions inoculum should not be used after six months at the longest. In cool but not refrigerated conditions it should not be kept for more than about 3 months.

Seed Inoculation Procedures

If the inoculum is in its normal powder form, one relatively easy and effective method is described below. The normal application rate is 7–10g of inoculum per kilogram of seed—the longer the delay before the seed is planted the more inoculum should be used:
1. Mix 1 part of inoculum with 2.5 parts of water. **2.** Pour this over the seed.
3. Mix together so that every seed is covered. **4.** Wait until the seeds do not stick together, then mix again. **5.** Sow as soon as possible, within three or four days if possible, though the inoculum may still be viable after seven days if the conditions are favourable.

1Ff. Seed Treatment (Dressing)

Also known as *seed dressing* it is the chemical treatment of seed, especially cereals, with fungicides and sometimes insecticides or other products, described below. The principal purpose of treating seed is to reduce the spread of soil-borne and seed-borne pests and diseases, and also to promote good seedling emergence and to minimise yield loss. Seed is almost always treated by the seed supplier, though food growers and farmers can quite easily treat their own seed with the appropriate equipment, which can be made at home using a drum welded asymetrically onto a frame.

A coloured dye, normally red or blue, is usually added so that the seed can be seen clearly by everyone to have been treated. Treated seed must never be used for human or animal food, but should be safely stored in a dry place for the next planting season.

Inoculation of seed that has been previously treated requires that the inoculum should be sprayed in suspension into the seed furrow just before planting, then covered by soil together with the seed.

The traditional dressing for pest protection was gamma-HCH (formerly called gamma-BHC), which is on the restricted use list. In the late 1990s Bayer released a less toxic insecticidal seed dressing "Imadacloprid", under the trade name "Secur", mainly for use with cereals. This has a mainly systemic action, against aphids (and other virus vectors) in winter wheat and oats, but it also suppresses wireworm and leatherjacket activity, deters slugs from hollowing out the grain, and also deters birds. Bayer expects to replace Secur with the new "Deter" range, based on a new insecticide clothianidin. A further range of seed dressings, "Redigo", "Redigo Twin" and "Raxil Pro" was launched by Bayer in 2005.

In addition to fungicides and insecticides, seed may also be treated with nematocides to control nematodes, or with bird and animal repellants, or with growth stimulants (such as Ergostim for maize) or with chemicals to protect the seed from herbicides.

Further details: see the *UK Pesticide Guide* (The "Green Book") published by the British Crop Protection Council/CABI, and the PNW Weed Management Handbook (available online).

Advantages of Seed Dressing:
- *logistics*: the plants are protected at the appropriate time, when they are young and weak;
- *environment*: pesticides etc are applied to the seed only, not blanket sprayed on the soil or into the air;
- *economics*: less seed per unit area is needed, and the extra yield/quality often more than pays for the cost of the dressing.

Composition of Seed Dressing:
The chemicals normally used include:
- active Agent—the fungicide or insecticide;
- glutinous Agent—to make sure the chemical sticks to the entire seed surface;
- diluting Agent—to improve the distribution of the chemical so that the seed is covered uniformly;
- warning Colour—a coloured dye.

Sometimes the seed is *coated* before treatment with a glutinous agent such as textrine linseed oil. This protects the seed to some extent from attack by wireworms and birds.

Seed Dressing for Disease Prevention
Seed treatment is mainly used to reduce the incidence of diseases which are found on the surface of seed, for example Bunt (*Tiletia cares*) of wheat. Some other diseases can also be controlled to some extent, such as:
- diseases inside the seed, eg Loose Smut (*Ustilago nuda*) of wheat and barley;
- diseases which attack the seed from the soil, eg the seedling Blights;
- diseases transmitted by wind, eg Powdery Mildew (*Erysiphe graminis*).

Fungicides
There are both systemic and non-systemic forms of fungicidal seed dressings, used according to the fungal disease to be controlled.

The older type of organomercurial seed dressings are slowly being replaced, and fortunately are now banned in many countries due to their toxicity to birds, mammals and so on. Among the evils of this type of poison is their long persistence in a potentially lethal form.

More modern non-mercury dressings tend to be more specific in their action, so the trend nowadays is to use a cocktail of various dressings. Unfortunately the non-mercury dressings tend to be more expensive.

Insecticides

Insecticidal seed dressings are more recent than the fungicidal dressings and only became widely used following the development of the organochlorines in the 1940s. The cyclodienes (Aldrin, Dieldrin, Heptachlor, Chlordane and Endrin) as well as DDT have been withdrawn from use in most countries for environmental and health reasons.

1Fg. Vernalisation

Vernalisation is the subjection of plants at the seedling stage to a period of cold conditions, which triggers mechanisms in the plant which lead to flowering.

Some varieties of some plant species such as winter wheat and barley will fail to flower if they are planted in the warm weather in the spring. These varieties need to be vernalised by a cold winter period before they will produce flowers, and consequently seed. Varieties that display this characteristic are known as *winter varieties* and are planted in temperate regions in the autumn. Normally, temperatures of 0–8°C for a period of one or more weeks are necessary to stimulate flowering.

By contrast, *spring varieties* require no cold period, or vernalisation. They are planted in the spring, and flower in the summer of the same year. Some varieties of cereals (including triticale) are said to be *facultative* or *intermediate*; these varieties can be planted either in autumn/early winter if temperatures are not too low, or in spring if the growing season is long enough.

Vernalisation can be achieved artificially, by keeping the seed moist and between 1°C and 2°C for up to eight weeks, or even longer for some species.

1G. CROPS

1Ga. Cropping Calendar

If it is possible to construct an accurate cropping calendar of the area where you are studying the food production systems then you are a long way towards understanding the agricultural crop situation in that area.

A cropping calendar, or Crop Year Table, is simply an easily understandable display of the crops grown in any particular location. It shows very clearly which crops are grown, and during which months they are planted, cultivated and harvested.

Other information such as average rainfall, temperature, etc. can also be included on the calendar.

It is also possible to indicate on a cropping calendar which crops are the most important, from any particular point of view (area cultivated, economic value, nutritional value, harvest yield etc.). This can be done by not only listing the crops in order of preference (most important at the top, least important at the bottom) but also by use of different sized blocks—see *Example*; the blocks can even be given different colours—blue for harvest yield, red for economic yield etc.

Cropping Calendar (Crop Year Table)

Exact Location: West bank of Limpopo River, between Ashenagook and Timbucthree.
Importance of Crops: In terms of area cultivated, from top to bottom (Sweet potato most important, Sorghum least important):

	J	F	M	A	M	J	J	A	S	O	N	D	Area / value (1)	(2)
Sweet potato	P * H	P * H	P * H	P * H	P * H	P * H	P * H	P * H	P * H	P * H	P * H	P * H		
Barley			P	P	*	*	*	*	*	H	H			
Field pea				P	*	*	*	*	*	*	H	H		
Linseed				P	P	P	P	P	P	H	H			
Maize	H		P	P	*	*	*	*	*	H	H	H		
Sorghum	H	H	P	P	P	*	*	*	*	*	*	*		

KEY
P   = planting time
* * *   = growth period (rainy season(s) can also be included)
H   = harvest time
(1)   = area cultivated (ha)
(2)   = economic value (eg $/ha)

In this example, column 1 on the right ("Area") could indicate that, for example, almost 50% of the land area is used for cultivating sweet potatoes, 30% for barley etc. Column 2 ("Value") might show, for example, that barley produces a higher income/value than sweet potatoes.

The total amount of area shaded in the right-hand columns should of course equal 100%.

1Gb. Break Crops

Sometimes called *change crops,* for reasons explained below.

In many parts of the world food growers have a marked preference for one crop more than others; this may be a cereal, often maize or sorghum, or a root crop such as cassava or sweet potato, or even sometimes an oilseed crop or a vegetable.

If one of these crops is grown continually on the same land for a number of years, this farming practice is known as *continuous cropping* or *monoculture.* The danger with this practice is that certain kinds of weeds, diseases and insect pests tend to build up, and certain soil nutrients also become depleted. As a result, crop yields tend to diminish after a few years.

Food producers can respond to these falling yields in different ways. They may accept the situation fatalistically, or they could launch a chemical attack on the weeds, diseases or insects, or they could introduce a break crop—in other words another type of crop. There are a number of **advantages** to break crops:

- diseases which are carried over from one season to another, either in the soil or in crop residues, are less able or unable to survive and consequently damage subsequent crops;
- insect pests are similarly often, but not always, reduced in numbers;
- the soil is depleted of a different proportion of soil nutrients, some of which may have become exhausted under a monoculture—legumes would contribute nitrogen, for example. A break crop may even sometimes increase the soil's organic matter content;
- there is no expense, trouble nor possible danger of applying chemicals.

The best type of break crops are those which are equally useful as the crop which is normally grown. The break crop should provide the same amount or more of food or income. It may also have useful by-products (eg stalks/stems for building or fodder) which are not produced by the normally grown crop, and it should not have an excessively long growth period ie it should "fit in" with the other crops on the farm.

There are three main types of break crop:

- **Seed Crops**—peas, beans and oilseed crops that are planted every few years between a cereal monoculture can be very beneficial. Even oats or maize can be used as break crops to reduce disease build-up in other cereal crops.
- **Root and Forage Crops**—these also usually help to slow the build-up of weeds, diseases and pests when they are grown after a cereal monoculture.
- **Grassland**—this can be a very valuable longer term (2–10 years) break crop, especially if it is well used, for grazing, hay or silage, and if it can be easily reconverted back to an arable crop when necessary. Efficient ploughing and even herbicides may be necessary to successfully change grassland back to arable cropland.

Despite the advantages of break crops there is often resistance and reluctance by farmers to grow them. The reasons may be based in tradition and cultural beliefs - fear of using "new" techniques or crops, possibly not previously attempted in the area - or the reasons may be more practical, such as the lack or expense of the appropriate seed or machinery. Another unsolved mystery.

1Gc. Catch Crops

Catch crops are normally fast growing (early maturing) crops or varieties which are planted late in the season, either too late for a normally grown crop or to replace a crop which has failed.

Catch crops are sometimes sown densely so as to suppress weed growth, in which case they are also known as *Smother Crops* (**1Gh**, page 64). Sometimes they are ploughed in to the soil when still young and green, in which case they are also known as *Green Manure Crops* (**1Hd**, page 69).

In temperate climates forage crops are planted as a catch crop in the autumn, between cereal crops for example, so as to provide grazing early in the spring.

Examples of species which can be used in temperate regions are: buckwheat, swedes, turnips, rape, Italian ryegrass (sometimes mixed with winter rye) and mustard. In warmer, drier regions millets, teff, grass pea (vetch) or other fast growing legumes can be used.

1Gd. Climbing Crops

Any crop which can grow upwards on vertical or near vertical supports is known as a *climbing crop*. Examples include lima beans, runner (pole) beans, gourds, cucumbers and yams.

Often these crops which could be grown as climbing crops are allowed to grow horizontally, which is generally less useful, because climbing crops have a number of advantages:

- They occupy a smaller area of land surface because they utilise vertical space.
- They support their edible pods and leaves above the reach of predators such as slugs, insects, rodents and other animals.
- They keep the edible plant parts off the soil surface so they dry out more easily and so are less liable to become diseased or splashed by mud, or smothered by weeds.
- They allow light and air to surround the plant, which helps its growth and allows the edible parts to mature and colour more evenly.

Leucaena leucocephala

Sometimes plants themselves are used as *support crops* to provide support to other, climbing crops. Maize and sorghum, for example, commonly support various types of climbing plants such as beans.

Another efficient system is to use the Horse Tamarind, White Popinac, Leadtree or Ipil-ipil *Leucaena leucocephala* (described in **Section 2G**, page 275) as a support crop.

The climbing crops benefit from the nitrogen supplied by the trees, from the support the trees provide and are also protected from wind and direct sun. In addition the land area is used to produce a crop of food at the same time as the trees are growing. Various types of climbing beans are suitable for this.

1Ge. Companion Crops

Also known as *nurse crops*, these are two or more crops that are grown together in the same field. At least one of these crops, and often more than one, benefits from the presence of the other. Examples include onions, garlic and marigolds planted together with vegetable crops to protect the vegetables from some harmful insects.

Carrots for example can benefit if onions or other *Allium* species are grown close to them because the Carrot Fly is repelled or confused by the smell of the onions. Similarly, many insect pests of tomatoes and cabbages are kept away if marigolds are grown close to them.

Trap crops are similar in that they are planted specifically to attract certain harmful insects; they are then ploughed under or destroyed as soon as this has happened. Trap crops can also be used to stimulate *Striga* weed seed to germinate, the trap crops themselves not being harmed by this weed (**1L**, pages 88–90).

Mixed Cropping (**1Hb**, page 66)—two or more crops which are grown together in various systems of intercropping or multiple cropping almost always behave as "companions" towards each other.

For example when maize and beans are grown together the maize benefits from the nitrogen produced by the beans while the beans often (but not always) benefit from the shelter which the maize provides, such as from hot dry winds, and vertical support.

1Gf. Cover Crops

When a crop provides protection to either a **second crop** or to **the soil** it is known as a *cover crop*. For example, wheat or barley "covers" the more delicate grasses or clover which are often planted under these cereals (the cereals are said to be *undersown* with grass/clover), or grass planted under fruit trees "covers" and protects the soil from leaching and erosion.

In the tropics Kudzu Vine (*Pueraria lobata* syn. *P. thunbergiana*, *P. montana*), a hairy, leguminous perennial climbing plant, is often used as a cover crop, which is used later as forage for animals. Once established, the vines can grow 20m in one season, about 30 cm per day! It only grows well in warm areas & is ploughed in after 2–3 years.

In most climates, but especially in wet, dry or hot conditions, it is usually a good idea to cover the bare soil to protect it from erosion and loss of nutrients. Either a mulch or a cover crop can be used, though cover crops have the advantage over mulches of "holding" or "locking up" soluble plant nutrients in the soil, which reduces the leaching of these nutrients.

Cover crops are often ploughed under a few weeks before planting the main crop, in which case they serve the dual purpose of cover crop and green manure (page 69).

1Gg. Pioneer Crops

Pioneer crops are normally associated with grassland or with recently cultivated "new" land. When old grassland has become exhausted or needs renewing for some reason, it is ploughed up and a pioneer crop is then often planted. A typical pioneer crop in temperate zones is a mixture of Italian ryegrass, rape and turnips.

Pioneer crops are normally fertilised and/or grazed heavily by animals so as to accumulate dung and urine in the soil, and then ploughed under. This process may be repeated if the old grassland or "new" land is in a very low state of fertility.

1Gh. Smother Crops

A crop which is planted densely (ie at a very high plant population) and which grows vigorously to occupy most of the growing space is known as a *smother crop*. The purpose of smother crops is to retard or smother the growth of weeds; if legumes are used then they will also increase the soil nitrogen content as well as crowding out the weeds.

Smother crops (as well as trap crops and cover crops) are often ploughed under after they have served their function, in which case they also serve the purpose of providing green manure.

Examples: alfalfa, foxtail (Italian) millet, buckwheat, rye, sorghum and Sudan grass.

1Gi. Shade Crops

Some plants prefer shady places to grow in, other plants like to be in the full sun. The difference between the two is discussed in **1Eg. "Shade Plants / Sun Plants"**, page 40.

There are many and varied ways in which taller, sun loving plants can be planted so as to provide shade and protection for smaller shade loving ones. In addition to providing shade the sun plants also usually provide some form of additional protection, from wind, animals and so on. Permanent shade trees for example - which are often planted to provide shade for coffee or cocoa - also provide a range of advantages: their branches even out the temperature fluctuations between day and night and produce a more favourable microclimate, they extend the life of the shade plants below them, and they reduce dieback and biennial bearing habit.

Shade crops can also reduce evaporation and transpiration, and damage from hail and weeds; the leaf mulch is taken more efficiently into the soil, and gives longer protective cover for the soil; their root systems may assist drainage and aeration; and they may provide firewood, timber, honey and other useful products. Not bad.

On the other hand great care should be taken in planning or managing shade trees since they can also harm crops by removing excessive water and nutrients. They need careful management so that there is a well-matched balance between the species chosen for each purpose.

Trees are not the only species grown to provide shade; annual plants are also frequently used as shade crops. Examples in temperate climates, where shade crops play a lesser role, include broad beans protecting lettuce, and sunflowers protecting green peppers.

In hot climates a wide range of plant species are used, including pigeon peas, bananas, *Crotalaria* spp., *Tephrosia* spp.(may also deter rodents, due to the rotenone it contains), papaya, cassava and tree cassavas, tannias (*Xanthosoma* spp.), dasheens and eddoes—the so-called "cocoyams" (*Colocasia* spp.) and *Albizia* spp.

1H. FARMING SYSTEMS

1Ha. Rotation

Rotation is the name given to a cropping system in which different kinds of crops are grown in more or less a fixed sequence on the same land. Rotations are of the greatest benefit when crops are grown as pure stands and are rarely used when crops are intercropped. One basic form of rotation is known as slash-and-burn (or *swidden*), used in many simple agricultural systems.

The Benefits of Rotation

- reduces the accumulation of disease and insects, such as Club Root in brassicas, Brown Streak virus in cassava, Take-all in cereals, Eelworm (nematodes) etc.;
- reduces the accumulation of weeds;
- maintains or improves soil fertility;
- distributes labour requirements (for cultivations, planting, harvesting etc.) more evenly through the year, reducing labour, financial (paying for everything at once) and storage bottlenecks;
- avoids creating a sub-surface *pan* by not cultivating the soil to the same depth every year;
- makes use of a greater depth of the topsoil because different crops occupy and use different levels of the soil profile;
- ensures that fertilisers are used to the best advantage.

Sometimes a part of the farm or garden is left *fallow* as part of the crop rotation. This is called *bare fallow* when the soil is left unplanted, or *grass fallow* if grass is planted. Provided that weed growth is controlled, a fallow period helps to naturally reconstitute the soil's fertility.

Rotations should be flexible so that new crops can be introduced and other small changes made according to changes in needs, markets etc.

In many parts of the world cereals are grown almost continuously, with *break crops* of, say, legumes or potatoes occasionally grown to rest the land. In these situations, fertiliser has replaced crop rotations as the principal means of maintaining soil fertility. This type of farming system depends heavily on chemicals, including fertilisers, in order to maintain yields. The long term negative impact of such activity on a global scale is clearly unsustainable and other means, which include sensible crop rotations, should be adopted whenever practical.

Principles of Crop Rotation

Planning a crop rotation, and adapting it when necessary, requires paying attention to a number of factors, including the following:

- legumes like alkaline soils or plenty of lime in more acid conditions, while Irish potatoes prefer more acidic soils;
- Irish potatoes like a lot of manure, while other root crops such as carrots do not because manure can often cause their roots to become *forked* ie they split;
- the Brassica family also like lime, but only after it has been in the soil for about a year; some crops should not be grown on the same land two years in succession, mainly to avoid the build-up of diseases. Ideally some years should elapse between, for example (with number of years in brackets)—peas (3–5), sunflowers (4) and linseed (5–6);
- if lime is put on the land for a crop such as soybeans, then the next year's crop can be one which likes some but not a lot of lime, such as maize or tobacco. The third year can be a crop which likes only a little lime, such as cereals or Irish potatoes;
- in low rainfall areas more drought resistant crops such as millet may be the only ones to survive following a crop that takes a lot of water from the soil, such as lucerne (alfalfa);
- if grassland is ploughed in, even if it has only been established for a year or so, this will increase the soil organic matter and will also improve the tilth of soil which may have become compacted by heavy rain, machinery, trampling by animals etc.;
- the type of soil—some can successfully grow crops of sorghum, barley and so on for many years, while other soils would rapidly lose fertility or become infested with diseases and/or pests, and so need frequent break crops of legumes, root crops etc.;
- the deep root penetration of some crops such as lucerne and clover brings up nutrients from deep in the soil and also helps to increase soil aeration which can be highly beneficial to the following crop;
- market prices often have a greater influence on the choice of crop which is grown than the proper use of land;
- cultural and traditional beliefs, the need for food security and the availability of seed or planting material are some of the other factors which play a part in the use, or abuse, of crop rotations as an invaluable tool in the food producer's armoury.

1Hb. Mixed Cropping

Mixed, or *multiple*, cropping is the name given to farming systems where more than one crop is grown on the same land in a year. The most common types of mixed cropping systems are various combinations of *intercropping*, but there is another important type known as *double (multiple) cropping* in which either:

- successive but different (fast growing) crops are grown in the same field in the same season, or

- one crop is planted in small areas over a period of time so that only a little matures at a time. In this way a continuous supply of that crop is provided over a long period.

Intercropping

Intercropping is the farming practice where two or more crops are grown together simultaneously.

There are four main types of intercropping:
Row intercropping—where the crops are grown in distinct rows;
Mixed intercropping—where the crops are grown mixed together, without rows.
Multistorey intercropping—where the crops which are grown are very different in height, such as peppers growing under bananas.
Relay intercropping—where the crops are grown in sequence, one or more of them being planted before, during or after the harvest of the others. An example of a three-crop relay system is shown below.

An example of relay intercropping, with 3 crops

October November December January February March April May
```
*************************************************************************
 _____   ❧ ❧ ❧ ❧ ❧❧ ❧ ❧ ❧ ❧ ❧❧ ❧
*************************************************************************
 _____   ❧ ❧ ❧ ❧ ❧ ❧ ❧ ❧ ❧ ❧ ❧ ❧
*************************************************************************
```

KEY

* = local sorghum (200-210 days)—planted early-mid October, harvested mid-late May
_ = haricot beans (80 days)—planted late October, harvested mid-January
❧ = sesame (100 days)—planted late January, harvested mid-late May

This kind of information can be clearly displayed on a **Cropping Calendar** (Crop Year Table), described previously in **1Ga**, page 59.

Advantages of Mixed Cropping

The advantages of mixed cropping outlined below are particularly relevant in poor soil in drier regions:
- The different leaf arrangements and heights of the different crops utilise the sunlight more efficiently.
- The different root systems and their depth of penetration utilise the available water and nutrients more efficiently.
- If legumes are grown, the other crops benefit from the fixed nitrogen.
- Crops suffer less from pests and diseases, which cannot spread so easily when their host plants are separated by other species.

- The total yield per unit area can be higher than with monocultures, and soil fertility is maintained for a longer period.
- The soil is often protected more efficiently against erosion.
- Labour requirements are spread more evenly over the year.
- The food supply is also extended over a longer period, and is more varied.

Disadvantages of Mixed Cropping

Compared with monocultures:

- Mixed cropping is generally not appropriate for mechanised harvesting.
- The utilisation by plants of fertilisers and other agrochemicals is less efficient.
- The different plants can become tangled and grown together which may make harvesting slow and tedious.
- Immature crops (eg the sorghum in the example above) may be damaged during harvest of the mature crops (the haricot beans in the example above). In this example, the sesame and sorghum are harvested at the same time.

Maize is often grown together with cassava (manioc). Research has shown that in humid areas the yield is often the same whether maize is grown with cassava or not. In other words the few tonnes per hectare of cassava are a virtually free bonus.

In a study of one arid area conducted by ICRISAT (International Crops Research Institute for the Semi Arid Tropics) near Hyderabad it was calculated that the probability of total or nearly total crop failure of pigeon peas grown as a monoculture is 1 year in 5, and when sorghum is grown as a monoculture, 1 year in 8. When both crops are grown, in separate fields, the chance of both crops failing falls to 1 year in 13. But when the two crops are intercropped the failure rate falls to only 1 year in 36.

1Hc. Alley Cropping

This farming technique has been developed at the IITA (International Institute for Tropical Agriculture) in Nigeria since 1976 and has subsequently spread to most parts of the world.

The basic system consists of various field crops growing between rows of trees, particularly leguminous species such as *Leucaena leucocephala* and *Sesbania drummondii* (Rattlebox, Poison Bean, Coffee Bean, Siene Bean, Rattle-bush etc).

Foliage from the trees improves the soil organic matter content, and the nitrogen fixed in their roots—as well as other nutrients present in their foliage—increases soil fertility. In addition, soil erosion by both wind and rain is reduced.

Alley cropping systems adapt the principles of ancient systems of bush-fallow, but there are two major differences:

- the trees are planted and replanted at regular intervals, and not just allowed to slowly re-grow
- the trees are improved strains of fast growing legumes, and not just the local wild species of trees, shrubs, etc.

Alley Cropping—Method

Rows of trees are planted 4–8 metres apart. During the main growing season, which usually coincides with the main rainy season, they are pruned back, and often not allowed to reach their normal full height. The prunings of branches, leaves and twigs can be used as either fodder, mulch, or ploughed in.

When the trees reach the size required they are cut down, providing fuelwood, poles and stakes.

The foliage makes a fairly good fertiliser. For example, 15–20 MT/ha per year of leaves and twigs can be produced in warm, humid regions. This can be either fed to animals, used as a surface mulch, or ploughed or dug back into the soil—in which case providing up to about 160 kg nitrogen, 15 kg phosphate and 150 kg potash per hectare.

In more arid regions, more drought resistant or locally adapted tree species can be used, planted at wider spacings.

Alley cropping can provide a useful alternative to a fallow period, allowing farmers to continuously cultivate the same area of land. There is also the additional advantage of reducing or eliminating the need for chemical fertilisers.

1Hd. Green Manure

A green manure crop is grown to be ploughed under into the soil when the plants are young and green ie succulent. Sometimes the green manure crop is cut down or pulled up and then left on the soil surface as a mulch; after some time the plants rot and are taken down into the soil by termites, earthworms etc. There are both positive and negative aspects of green manure, described below.

Positive Aspects of Green Manure

- The soil is enriched with nitrogen and other plant nutrients;
- The organic matter of the soil is increased;
- Nutrients are brought up from the subsoil by green manure crops with deep root systems such as alfalfa (lucerne);
- Leaching of plant nutrients during periods between regular crops is reduced;
- Soil erosion is reduced;
- Sometimes crops such as sweet potato and cotton are less damaged by fungal diseases of the roots when grown after a green manure crop.

Negative Aspects of Green Manure

- Volunteer green plants can be a nuisance in the following crop, so must be effectively removed by either a stale-seed bed and/or cultivations;
- Land is occupied which is then not able to grow a cash crop or a food crop;
- Only a little nitrogen will be added to the soil if the plants are old and fibrous, although the soil's organic matter content will still increase;
- The soil's organic matter content will only increase a little if the plants are young, though the soil's nitrogen content will be increased;
- Good ploughing techniques, which need a high investment in machinery, fuel and/or labour, are needed to thoroughly incorporate the plants into the soil.

Cover crops and smother crops are commonly used as green manure after they have served their purpose.

Legumes are the most commonly used green manure crops as they provide the biggest increase in soil nitrogen.

Examples—Crops used for Green Manure:

Alfalfa, cassia species, clovers, cowpeas, lupin, mustard, rye, siratro (*Macroptilium atropurpureum*), sunflower, sunn hemp (*Crotalaria* spp.) and velvet beans.

- two or more of these crops, or different varieties of the same crop, are often planted together. This spreads the risk of one failing and also increases the range of benefits.
- sometimes these crops are planted together with a food crop, such as siratro with millet. Siratro can fix Nitrogen, which it passes on to the companion millet.

1He. Mulch

The mulch is placed on the soil surface where it forms a kind of cover or protection. Mulching of a soil surface can improve the fertility of the soil, reduce weed growth and water loss and increase crop yields in almost all farming systems, although the effect is often more dramatic in dry regions than in humid ones.

The mulch itself can be almost any material, biodegradable or otherwise: crop residues (stems, leaves, harvested pods etc.), grass cuttings, branches or young trees, pulled up or hoed weeds, leaves, sawdust and even jute, or plastic sheeting. The prunings of crops such as bananas, oil palm, tea and coffee can also be used as mulch. Old carpets work well, on a small scale, if they are made of natural material.

Green manure crops can also act as a mulch. They can be either pulled up, hoed or cut down and then left on the soil surface to act as the mulch. They can be ploughed in after some time, and so complete their dual-purpose role of green manure and mulch.

Advantages of Mulch

- The soil is protected—from drying out, and consequent loss by wind erosion, from heavy rain—and loss by water erosion—and—from high temperatures;
- As the mulch rots down into the soil it provides nutrients for subsequent crops;
- More rain water is able to filter into the soil due to reduced evaporation and runoff;
- Weed growth is reduced; this can compensate for the fact that mulching needs more labour than burning;
- The soil retains more moisture because evaporation is reduced;
- It is claimed that nitrogen fixation can be increased by free-living soil bacteria, and also that the effect of fertilisers can be improved.

Disadvantages of Mulch

- The spread of diseases and/or insect pests can be increased if infected or infested plants are used as mulch. In theory at least this disadvantage can be eliminated if

clever crop rotation is practised to break the life cycle of pests and diseases. Slugs love mulches;
- Mulch material can interfere with subsequent land preparation, such as by obstructing ploughs with long woody stems, if the mulch has not become well rotten and broken down. The mulch may be so dense that it has to be removed from the field before land preparation can begin. As a result labour input can be higher than if the crop residues had been removed after harvest and/or burned;
- The material which is suitable for mulch often has other valuable uses, such as for fuel, construction or animal fodder - particularly valuable in dry regions;
- Large and bulky volumes are involved. 20 MT/ha or more may be needed to make a reasonable job of mulching;
- In dry areas the breakdown of mulch into the soil can be very slow.

Mulch as Free Fertiliser

Obtaining the appropriate chemical fertiliser is very often a major problem facing food producers. Even if it is available close enough to be transported to their farm, they often do not have enough cash or credit. But there is always the possibility of using mulch as a source of plant nutrients—ie fertiliser.

On the negative side, the demands on labour may be very high, and there may only be a relatively small gain of nutrients.

Crop residues such as stems and leaves contain 40–90% of the nutrients which crops remove from the soil. These residues can contain the equivalent of 50 kg/ha or more of the major plant nutrients, which can be made available to crops which grow later on in that field if the crop residues are incorporated into the soil and become broken down.

Living Mulch

As well as using dead or inorganic material the mulch can consist of living, growing plants, when they also act as a *cover crop*, described on page 63.
Examples: perennial peanuts (*Arachis prostrata*) under maize, and *Desmodium* species (the tick clovers or beggars weeds) growing under rice can dramatically improve crop growth and may also lengthen the growth period of crops, for the following reasons:
- soil water is conserved;
- soluble nutrients are "locked up" and not lost by leaching;
- weeds are smothered;
- soil temperature fluctuations are reduced.

Mulch—a Summary

The benefits of mulching are greatest in hot, dry regions, and unfortunately it is these very regions which have the least amount of available mulching material—and where there are the most pressing needs for mulch material, stems in particular, to be used as building material, fuel, animal fodder and so on.

In more humid regions where mulching material is more plentiful, increases in yield may not be so impressive, but the advantages of mulching the land—such as reducing water erosion and stifling weed growth—almost always outweigh the consequences of leaving the land bare and exposed to the elements.

1Hf. Silage

Silage is a type of animal food made from green crops that are cut in the field and preserved by fermentation for later use. A crop that has been treated this way is said to have been *ensiled* (or *ensilaged*).

Ensiling a sward of mixed grass species allows a farmer to optimise the energy production, because the grass can be cut at the time of maximum sugar content and digestibility.

The digestibility of grass is measured in units known as the "D-value" of the grass, which gives an indication of its digestibility to ruminants. Good quality grazing grass should have a D-value of about 65D or above.

A large number of different types and varieties of grasses and clovers growing together is recommended, so that seeding (heading) dates are spread over a longer period. This ensures that some proportion of the grass crop to be ensiled is at the correct stage of growth over an extended period.

The types of crops used to make silage are basically the same as for making hay (**1Hg**, below); grass/clover mixtures are the most common, and maize is also commonly used. The plant material should be clean (without soil or faeces), and wilted (**1Ef**, page 39–40) to reduce the water content.

Grass is transformed into silage by two main **types of bacterial fermentation**:
• Carbohydrates in the plant material are converted into organic acids;
• Proteins are converted into aminoacids, which act as preservatives.

Silage that has been properly made is a yellow/brown colour and contains mainly lactic acid, and some acetic acid, and should have a ph of 4 or less. The best silage is often made from crops which have been cut after a period of dry, sunny weather, when the sugar content of the grass is at its highest. The cut grass is normally left for some hours to *wilt* ie lose moisture.

Sometimes additives such as molasses are added to increase the proportion of sugars and formic acid, the latter being useful in increasing the acidity of the silage which suppresses the growth of harmful *Clostridia* bacteria. In recent times, many farmers routinely inoculate their silage (either clamps or big bales) with formulations of freeze-dried *Lactobacilli* to stabilise the fermentation process.

1Hg. Hay

Food for animals, often referred to as *"fodder"*, often has to be preserved during or just after the rainy season, for use in the dry season, or in cold, hard winters, when animal grazing is limited.

Hay making is a less complicated way to conserve fodder than silage making, and good hay can be produced with just a sickle, rake and a few days of warm dry weather. Grass/clover mixtures are the most common form of hay, but it can also be made from:
• **Monocotyledons** - cereals and grasses such as wheat, oats, Sudan grass, pearl millet and other millets, and ryegrass.
• **Annual legumes** - vetches, clovers and field peas.

- **Perennial legumes** - lucerne (alfalfa), sainfoin (*Onobrychis viciaefolia*)
- **Cereal/legume mixtures** - vetches and oats. This usually gives poor quality hay due to the unequal rate of drying of the two species.

Sainfoin plants grow upright and make good hay, but only one cut per year is taken (at the half to full bloom stage). Higher moisture content than lucerne, but cures better than the clovers. Normally grown alone, it competes poorly with aggressive grasses.

Cutting Hay

Plants that have been cut before flowering are normally more nutritious and digestible than more mature plants. However the yield of dry matter increases as the plant matures, so the decision on the best time for cutting a hay crop is always a compromise between yield and quality.

For most hay crops the most appropriate time is at flowering. Lucerne in particular should be cut early in the season, as soon as the first flowers appear, as at that time the protein content falls rapidly and the cellulose content increases rapidly.

Curing Hay

The quality of the hay produced is largely determined by the curing process, the objective of which is to rapidly reduce the plant moisture content from about 60–75% to about 25–30%.

After cutting, the plant material is left in the *swath* or *swathe* in the field until it has wilted due to exposure to wind and sun. It is then turned over and concentrated into narrower bands of plant material known as *windrows*. This reduces the area of cut plant material (forage) which is exposed to sun and rain but still allows air to circulate which further reduces the plants' moisture content. In mechanised farming the windrow also assists hay baling operations.

Losses During Haymaking

Even under ideal conditions about 10–15% of plant nutrients are lost, and the protein content also often falls, by 30% or so. These losses, and other losses described below, are caused by a number of factors:

- *Biological factors*—sugars and starches are reduced as the plants slowly die after they have been cut.
- *Light and heat*—the digestibility and palatability of hay is reduced by prolonged exposure to sun, especially when the hay is dry.
- *Rain*—any hay which is in contact with the soil may rot; also soluble nutrients are leached, and the hay becomes discoloured and unpalatable.
- *Mechanical*—if the plants are cut too high up the stem, potential hay is left behind in the field (known as *stubble*). Secondly, because the leaves dry out more quickly than the stems many of the leaves are lost when the plants are being turned and dried.
- *Storage*—if hay is stored at a moisture content above about 30%, heat can build up which causes losses due to fermentation, heating and moulds. In severe cases the

hay may even spontaneously combust. Mouldy hay, ie hay that is infected with fungi, should not be fed to livestock.

1Hh. Land Area Measurement

Most subsistence farmers, and many others, do not know the exact number of acres or hectares that they cultivate—nor do they need to know. But even if they do know they are quite understandably generally reluctant to tell anyone, for obvious reasons. Farmers may show you some of their land, but keep quiet about other areas they farm on the other side of the hill.

   If for any reason it is absolutely necessary to measure a farmer's land, great care should be taken to explain your reasons. The amount of land that a farmer and his family cultivate is a sensitive issue that should be approached carefully, with tact and diplomacy.

Method for Approximate Land Area Measurement
1. Divide the area to be measured into convenient numbers and sizes of 90° (right angled) triangles. This can be done either by eye or by using a tape measure (or measured rope or string) and 3 stakes—labelled A, B and C in the diagram below:

2. One person stands at stake A and directs a second person to place two more stakes, one at D and one at E, in line with AB and AC, at the boundaries of the field.

3. Measure, or pace out by foot, the lengths in metres of AD and AE (Also DE, as a check or if greater accuracy is needed).

4. Calculate the areas of each triangle (½ × base × height).

5. Add to this figure the small areas outside each triangle which have not been measured. The proportion can be judged by eye—for example "Area of the triangles measured = 3.26 ha, plus approximately 15% more for unmeasured areas = total area approximately 3.75 ha".

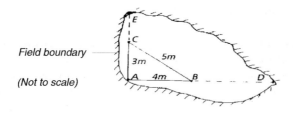

Figure 12. Land area measurement—approximate method

More Accurate Measurement of Land Area
1. Use the same technique to mark and measure the area into triangles, as shown below:

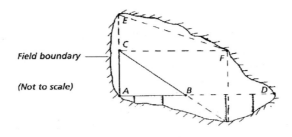

Figure 13. Land area measurement—more accurate method

2. Using either a tape measure or a surveyor's metre chain measure all the lengths (shown as ‡ on the diagram above) at right angles from the line AD to points on the field boundary where this changes direction. Note down the following readings (for example):

D	92	0
	81	12
	64	18
	26	5
	8	3
A	0	0

The first column of figures are the metre marks on the chain or tapemeasure (line AD), starting at zero at point A. The second column of numbers are the distances from the line AD to the field boundary.
Repeat this operation for AE, EF etc.

3. Using the readings in 2. above (for DA), draw a scale field map on graph paper, and calculate the areas outside the triangles by counting the squares on the graph paper. Repeat this calculation for AE, EF etc.

1I. THE HARVEST

1Ia. Maturity

Crops that are grown for their grain, the "seed", such as most of the cereals and oilseeds and many of the legumes, are normally harvested when their seed is mature. A mature seed is one that has developed sufficiently so that it would germinate if it were in the right conditions of temperature, moisture and air.

Many other crops, including most fruits and vegetables, are harvested before they mature. Many of the leguminous crops are also harvested before maturity, when the fresh green pods or grains are taken and eaten as a vegetable.

The number of days from planting the seed of a crop until the plant is mature and ready for harvest is often called the *growth period* of that crop.

Some crops such as oats, rye, sesame and vetch (grass pea) *Lathyrus sativus* should be harvested before the plant is fully mature—at a point when the seed is mature, but still has a relatively high moisture content. The reason is that with these crops the seed-heads or pods *dehisce* or *shatter* readily ie they open and so allow their seed to fall to the ground and become lost. The seed then has to be dried before it can be stored.

1Ib. Yield

In agriculture and horticulture the *yield* of a crop is the term used to describe the amount of food produced per unit area. If 1000 kg of sorghum grain is harvested from one hectare of land, the yield is 1000 kg/ha.

Yield is normally expressed in kilograms per hectare (kg/ha) or tonne per hectare (MT/ha or tonne/ha).

"Average Yield"

The yields of food crops depend on a whole range of factors and so it can be very misleading to quote the "average yield" of any particular crop. Even the same variety of a crop can produce widely differing yields if they are grown under different conditions of soil type, climate etc. The Ministry of Agriculture or its equivalent can often provide information about the average yields of the crops which are grown in their area, but the information should be viewed with caution as it is often inaccurate, for reasons explained above.

Mixed (or *multiple*) cropping systems, very commonly found in poorer parts of the world, can also confuse estimates of yields per unit area. It should be made clear if any average yield figures apply to the crop when grown alone (in *pure stand*) or when it is mixed with others.

The average yields of most of the important food crops are quoted in **Section 2, "Description and Characteristics of the Main Food Crops"**.

Reliability versus High Yield

In subsistence farming situations food producers are often more interested in the reliability of a crop than in producing high yields. They prefer to produce at least some food every year than to produce a high yield one year and very little or nothing the next year.

In dry regions the sorghums and millets are often more useful in the long term than maize. Although maize may sometimes yield more than these types of cereal, especially in good years, it may often yield very little or it may even fail completely. Thus, even though the average yields of sorghum and millets are generally lower than maize, it is often better to plant them because they are more dependable than maize in many marginal conditions.

Yields of Human and Animal Food

Many crops are dual-purpose; they produce both grain for human consumption and also leaves and stems for animal food (*fodder*). Stems (or *stalks*) are also often used for fuel and for construction material.

For this reason it is useful to have an understanding of all the uses for the various crops. For example, very tall varieties of sorghum with strong stems may have low yields of grain but are more useful than higher yielding short varieties in areas where stems are prized for building or for fuel.

1Ic. Haulm

Haulm (sometimes spelled halm) is pronounced "horm" and is the name given to the stems and leaves of crops, especially after harvesting. Haulm also often includes the *hull* or *husk,* the outer coverings of the grains, especially of cereals, and also the empty pods of crops such as peas and beans. Haulm is most commonly used as animal fodder, either ensilaged, made into hay, or grazed by animals in the field after harvest.

The haulm of Irish potatoes is often *burned off* with chemicals in modern farming systems, to allow the tubers to harden off in the ground, to facilitate mechanical harvesting, and, most importantly, to limit the spread of Blight disease. Pulses, rape and linseed are also sometimes burned off to help harvesting, especially if they are very weedy.

The Spread of Pests, Diseases and Weed Seeds in Haulm

Haulm can often be a source of diseases or insect pests that have been allowed to live from one season to the next on diseased or infested plant stems, leaves etc. Partial control of these pests and diseases can be obtained therefore by removing the haulm from the field immediately after harvest and either burning it or feeding it to animals.

Examples include Stem Borer of maize and sorghum, and the Wilt and Rust diseases of lentils. Weed seed is also often spread in haulm, such as wild oats and blackgrass in temperate cereal growing areas.

1J. INSECTS

Crop Losses due to Insects
Insects are frequently the most serious problem for food and livestock producers in tropical agriculture. The percentage of crop yield lost on a global scale as a result of damage by insects has been estimated as follows: Europe 5%, North and Central America 9%, South America 10%, Africa 13% and Asia 21%. These figures represent enormous quantities of food, and underline the importance of having some understanding of insects and how to control them.

Life Cycles of Insects
Most insects have four stages in their life cycle: egg, larva, pupa and adult. The larval stage (the *caterpillar* or *maggot*) is the principal feeding stage of most insects and is often the most damaging, for plants and animals. Adults are often merely the insect's means of reproduction and/or dispersal.

Other insects such as aphids and slugs have a three-stage life cycle. The eggs hatch out into nymphs, which are similar to the adult but smaller. The nymphs feed and develop, and often shed several skins as they grow larger. When they have become fully fed the nymphs have passed into the adult stage, and the life cycle continues.

Eelworms
This is the term used for several worms of the class Nematoda (phylum Aschelminthes). They are tiny, from 400 micrometers to 5 mm in length, and are found throughout the world. Some species occur in both animals and plants.

Eelworms are notorious soil pests, and they have their own form of life cycle. The eggs hatch out in the soil and the young female larvae then feed on roots of plants. After fertilisation by the male she produces eggs but keeps them in her body until she becomes a bag full of eggs, known as a *cyst*. The eggs are released into the soil when the cysts break open. Control is normally by crop rotation to break the life cycle.

Many eelworms are *specific pests* because the damage they cause is specific to certain crops—for example, the potato cyst eelworm (*Heterodera rostochiensis*) attacks only plants of the *Solanaceae* family. Eelworms increase in the soil particularly when a susceptible crop is grown too often in the same field, so they can be controlled with crop rotation ie not growing that crop in that field for many years.

Eelworms are not all bad, however. Beneficial nematodes are increasingly being used on a small scale to control a range of insect pests. Details of this form of biocontrol are widely available online.

Aphids
Aphids are also known as green fly or black fly. They do some damage to crops by feeding on them but usually cause more damage because they transmit virus diseases from plant to plant. The best way to control aphids is with systemic insecticides, although sometimes more harm than good is done due to the phenomenon of *resurgence*—see page 80. Safer control methods are described below:

Controlling Aphids

Compared with other insects, aphids are relatively easy to control. With light infestations on small areas the plants can be sprayed vigorously with a soap and water solution, or even water alone. This knocks the aphids to the ground; normally they do not climb back onto the plants, and the soap acts as a deterrent to further attack.

With more severe infestations of aphids, mild non-persistent insecticides such as derris (a climbing leguminous plant *Derris elliptica* from SE Asia that contains rotenone in its roots) and pyrethrum (*Chrysanthemum cineralifolium*) give good and safe control.

Aphids very often congregate together at the tips of plants such as broad (horse) beans. If the plants are big and strong enough the infested tips can be removed by hand, then destroyed.

Preventing Aphids

On a small, garden scale a repellent can be made by boiling in water the leaves and stems of plants such as rhubarb, anise, chrysanthemums, coriander, asters and marigolds, or mixtures of these plants. A strong solution of this sprayed onto plants acts as a fairly reliable deterrent. Soap and water sprays can also work.

Insecticides

Most modern insecticides, or pesticides, are synthetic organic compounds such as chlorinated hydrocarbons, organo-phosphorus compounds and carbamates. They are available in either granular, liquid or powder (dust) form.

Systemic Insecticides

These work by being absorbed by plants, remaining within the plant tissues. They are used to control sap-feeding insects such as aphids and capsids. Many of them also kill by contact and give protection for a week or so in this way, until the active chemical is absorbed within the plant.

Contact Insecticides

These rely on coming into physical contact with the insects.

The third group of insecticides, which includes gamma-BHC and Malathion, are essentially contact insecticides but also have some systemic action.

Online information is available from the Pesticide Action Network website.

☞ Insecticides are Deadly!

Even under rigidly controlled safety conditions insecticides are highly toxic chemicals which can cause suffering and death to fish, insects—beneficial as well as harmful—livestock, wild animals and people. If it is *absolutely necessary* to use insecticides then the greatest care must be taken to observe all safety precautions, which are normally clearly described on the chemical container. In Sri Lanka, and elsewhere, one of the commonest causes of farmer death is from pesticide poisoning.

In Europe, the mandatory Pesticide Safety Precautions Scheme (PSPS) and registration scheme ostensibly minimises hazards.

Some Alternative Pest Control Methods

Integrated Pest Management (IPM) considers both crops and pests as part of an ecological system and combines natural factors that limit pest outbreaks while using insecticides as a last resort. In most situations where insects are a problem, some form of IPM is now considered to be the most appropriate approach to their control.

Rotenone, a natural product extracted from the roots, seeds and leaves of certain tropical legumes such as Derris, is commonly used as a safe form of insect control. Another example of biocontrol is the Trichogramma—the larvae of which eat insect pest eggs. Further information on biological control of pests is available on the Wikipedia website and elsewhere.

When it has been decided that insects are causing economic damage, or they are liable to do so, there are a number of solutions which can be adopted on a small scale which avoid the need to use expensive and potentially lethal insecticides:

Biological	Ladybirds, chalcid flies and fungi can be encouraged to combat aphids and others.
Nicotine	Either in the form of home-grown tobacco or cigarette ends nicotine can be an effective remedy for insects such as aphids, scale insects, weevils, caterpillars and leaf miners. The nicotine mixture is made by soaking tobacco leaves and/or butts of cigarettes in water for some days. Unfortunately this is a powerful insecticide which may also kill harmless or useful insects (and do no favours to the gardener).
Trap Crops	Crops such as onions can be useful in deterring insects from damaging other crops. This method of insect control is discussed in **1Ge. "Companion Crops"**, page 63. Alternate hosts for the insects, and also "green bridge volunteers" should be eliminated.
Neem Tree	This large tropical tree, *Azadirachta indica*, also known as Nim, or Margosa, can be used as a safe, natural and free insecticide, especially when mosquitoes and malaria are prevalent. Sometimes the dried leaves are burned inside storage containers and buildings to deter insect infestation. Details are available online from "One Tree's Arsenal Against Pests".
Pyrethrum	This is a botanical insecticide, mainly produced in the flowers of a perennial herb, *Chrysanthemum cinerariaefolium*, which contain pyrethrins, used as a safe and effective insecticide.
Soap & Water	On a small scale this can be used to deter insects such as aphids, and also to give some protection to the plants that have been sprayed, until the next rain.
Timing	A minimum interval of time between ploughing and planting reduces the risk of pest survival (eg. six weeks for Frit fly).

Insect Resurgence

When persistent organophosphate insecticides are used against a certain insect pest very often the natural enemies, the predators, of that insect are also killed. Subsequently the insect pest may recover faster than the predator, and so cause more damage than if the insecticide had not been used in the first place. This is known as insect resurgence.

Secondary Pest Outbreak
Sometimes the use of an insecticide enables a normally harmless insect to become a pest. An example is the White Wax Scale insect which lives on citrus trees and which is normally kept under control by other natural insect predators. However if the citrus trees are sprayed with insecticides such as parathion to control the Red Scale insect then the predators of the White Wax Scale insect are also killed so that its population increases to damaging levels.

Another example is from West Africa, where the Cocoa Shield Bug developed into a serious pest in cocoa plantations where there had been intensive use of chlorinated hydrocarbon insecticides to control the Mirid Pest *(Sahlbergella singularis)*. The change in status of the Shield Bug, from harmless to harmful, is due to the elimination of its parasites and predators by the indiscriminate use of insecticides.

Resistance to Insecticides
Resistance can build up if one insecticide is used very frequently, and is another good reason for minimising the use of insecticides. As resistance to a particular chemical develops, the insect population returns to the numbers where damage occurs more and more rapidly. The chemical then has to be applied more often and/or the dose rate increased, or a different chemical has to be used.

Insecticide Treadmill
The three examples quoted above indicate the less obvious but possible dangers of using insecticides. If these chemicals are used indiscriminately and without any understanding of the ecosystem they can create more problems than they solve. They can leave food growers on an insecticide treadmill that is very difficult to get off.

1K. DISEASES

In global terms plant diseases are estimated to cause almost 10% loss in crop yields. In fact most plants of cultivated crops develop some form of minor disease symptoms at some stage during their life but these symptoms are very often below the level of economic interest, and so pass almost without notice.

Diseases can usually be kept relatively under control by using appropriate, disease resistant varieties and good husbandry techniques such as rotations, maintenance of soil fertility and destruction of diseased plants. So it is unusual for a crop to be entirely wiped out by disease, though this does occasionally happen.

Some crops cannot be grown economically in certain regions because one or more diseases are prevalent there. For example, rubber cannot be grown profitably in South America because of the fungal disease Leaf Blight, even though rubber as a species originated in South America.

This situation is similar for many other important tropical crops, which are now mainly grown in regions far from where they originated. Soya is now mainly grown in the Americas, though it originated in northeast Asia. Groundnuts are now found mainly in India and China even though they originated in South America. Pressure from disease and insects is the main reason for this global redistribution of crops.

Causes of Plant Diseases
Fungi, **viruses** and **bacteria** can cause diseases in plants - three harmful organisms that live as parasites on plants. Fungal diseases are very often the biggest problem for plants.

When and Why do Diseases Develop?
- Weak plants are always more liable to be damaged by diseases (and insects), so unfavourable growing conditions, such as poorly drained or infertile land, drought, overcrowded plants, heavy weed growth and so on will lead to a higher incidence of disease.
- If susceptible crops are grown. Some crops, and some varieties of these crops, are *resistant* to diseases (they are not affected) or are *tolerant* (they are not seriously affected). Others are *susceptible* (they are likely to be affected). Unfortunately this resistance or tolerance is often quickly overcome by the disease organisms, and a new variety which is released on to the market by plant breeders as "resistant" or "tolerant" when it is launched may become "susceptible" in just a few years.
- If the climate is favourable for the spread of disease; in general diseases spread more quickly in hot and humid weather than in cold and dry weather. Some fungi however, such as Powdery Mildew and other wind-borne diseases spread more in dry weather.
- If there is an imbalance of plant nutrients in the soil, for example if there is too much or too little nitrogen or potassium.
- If the same crop, or crops that are related, are grown continuously on the same land, ie without rotation, or if the rotation is badly planned.

Disease Control
Farmers and gardeners have several ways of controlling and preventing diseases; in most cases, preventing diseases is easier and cheaper than curing them, by using:
- Resistant or tolerant crops or varieties.
- Crop rotations—soil-borne diseases and diseases carried over in plant residues can often be controlled if their life cycle is broken by an appropriate crop rotation.
- Timing—crops can often be planted at certain times of the year so as to avoid harmful temperatures and/or rainfall at critical stages of plant growth.
- Crop husbandry—strong and healthy plants can escape diseases altogether, while weak plants in the same growing conditions may be badly infected. All infected plants and residues should be removed and destroyed. Heavy infestation of weeds also can increase the spread and damage done by diseases.
- Clean seed—disease-free plants should always be chosen by farmers, either from their own crops or when seed is bought in from outside, or "off the farm".

- Fungicides—as a last resort these can be used, as a preventative measure as well as a cure.
- Seed dressing—this use of chemicals, both fungicide and insecticide, can be a very valuable and appropriate method of protecting the seedling and plant. However great care should be taken to ensure that treated ("*dressed*") seed is not eaten by people or animals.

Treated or Untreated Seed—to Treat or not to Treat?
There is always the possibility that a proportion of the seed that has been treated with chemicals will not be planted but will be eaten, by people or by birds or animals.

Seed that is distributed in refugee or rehabilitation programmes is often sold locally, and may then be eaten if the danger of the seed treatment has not been understood. So it is important to carefully consider whether seed should be treated or untreated before it is distributed. The advantages and disadvantages are summarised in the table below:

Treated Seed	Untreated Seed
Advantages	
<ul><li>Improved yields due to better disease and pest control, and therefore healthier plants.</li><li>Recipients of the seed are encouraged to plant the seed and not to eat it.</li><li>Pesticide is only on the seed, not sprayed on the whole field</li><li>Less seed/area is needed.</li><li>Better root structure, leading to better canopy development.</li></ul>	<ul><li>Can be safely eaten - not necessarily an advantage.</li><li>Can be used as a substitute for a food distribution if unsuitable as seed (low germination, inappropriate crop or variety, delivered too late etc.)</li><li>Lower purchase price.</li></ul>
Disadvantages	
<ul><li>Health hazard if seed is eaten.</li><li>Health hazard to all involved during transport and distribution due to inhaling chemical dust.</li><li>Trucks may need special cleaning before being used again to transport food etc.</li><li>Possible delays in delivery time.</li><li>Higher purchase price.</li><li>Increases selection pressure on resistant varieties.</li></ul>	<ul><li>Loss of yield due to (controllable) diseases and pests.</li><li>The seed is more likely to be eaten and not planted.</li></ul>

The topic of seed dressing/treatment is also discussed in **1Ff**, page 57.

Sometimes during seed distribution/sale projects some of the seed is eaten by the people who are receiving it, which may indicate:

1. Either the seed is not appropriate—the crop species supplied is either unknown or unwanted, or a good local supply of seed already exists—or the seed was provided at the wrong time of year, and/or
2. Food, or more food, should be provided at the same time as the seed, often referred to as "seed protection".

If there is enough time it can be a good idea to make a few enquiries about the wisdom of having the seed treated before it is planted:

- Identify the pests and diseases that cause problems with the currently grown crops, and find out about their economic significance to farmers in "normal" years.
- Find out if these most harmful pests and diseases can be controlled with seed dressings, or if there are more appropriate control methods.
- Make enquiries, normally from the seed suppliers, about the seed dressings that they have available, cost of application versus untreated seed, and the actual health hazards arising if people eat some of the seed that has been treated.

Types of Plant Diseases
Diseases of plants can be caused by fungi, viruses or bacteria:

A. Fungi
A fungus is a living plant which does not have chlorophyll; it cannot therefore manufacture its own food from sunlight and so has to feed on the carbohydrates, protein and other food produced by either plants, animals or decaying (rotting) plant or animal material.

Most fungi are either *saprophytic* or *parasitic*: saprophytes feed on dead plant or animal tissue, while parasites depend on a living plant or animal.

However, other types of fungi such as *Botrytis* which causes Chocolate Spot of beans and other diseases adopt both lifestyles. They live saprophytically, on dead plant tissue, for part of their life cycle, and then become parasitic for the rest of their life.

Transmission of Fungal Diseases

Fungi spread and multiply by means of *spores*, which are the fungal equivalent of seeds. Their spores are dispersed in one of the following ways:

1. *Soil-borne*   The spores fall off infected plants and are washed into the soil, or they can enter the soil together with infected plants. They can remain in the soil, viable, for many years. When a susceptible crop is grown the fungus enters the plants, produces more spores, and so repeats the life cycle. Control is by crop rotations, resistant varieties, seed dressing and the removal and destruction of infected plants. *Examples*: Take-all and Eyespot of cereals, Common Scab of Irish potatoes and Clubroot of brassicas.

2. *Wind-borne*   The spores are carried in the air, sometimes for thousands of kilometres,until they land on susceptible plants, where they develop. Control is mainly by growing resistant varieties, fungicides and the destruction of infected plants - seed dressings can control only a few, such as Ethirimol and Triadimenol for the control of Powdery Mildew. *Examples*: Mildew and Rust of cereals, Potato Blight of Irish potatoes and tomatoes, and Chocolate Spot of beans.

3. *Seed-borne*   The spores are carried either on the seed coat (*exodermal* diseases), such as Bunt and Covered Smut of wheat, or inside the seed (*endodermal* diseases), such as Loose Smut of wheat. When the seed germinates and grows the fungus may either kill the seedling or it may develop together with the plant, and so cause the disease symptoms and produce more spores. These fungi can normally be controlled with seed dressings.

4. *Insect-borne*   Insects carry the spores from plant to plant, either within their body or on the outside of it. This is a bigger problem with virus diseases than with fungal diseases.

5. *Trash-borne*   True Eyespot (Septoria) can be carried over in stems etc of winter wheat, and Potato Blight can also be carried over on potato haulm.

Fungicides

Fungicides are chemicals that will kill fungi and prevent the spread of disease, **but they should be applied in good time ie as early as possible.**

Fungicides are used to control fungi that are spread either by the wind, in seed, soil or trash, and by rain-splashes.

There are several different groups of fungicides, classified according to their chemical structure and mode of action. They can be either *CONTACT*—acting at the first point of contact with the plant, or *SYSTEMIC*—transported away from the point of entry, to act elsewhere in the plant.

Successful fungicide programmes prevent and/or eradicate disease as appropriate. Visible disease symptoms often develop some time after infection has occurred—this is called the *Latent Period*, the period between infection and *expression* ie the time from when a spore lands on a leaf and the time a lesion is produced with further spores. The Latent Period varies enormously; in wheat, Leaf Blotch (*Septoria triticii*) has about four weeks latent period. If a fungicide is applied that has only preventative but no curative (or *kick-back*) action, it will prevent further infection but not eradicate the pathogen already present. So if wheat has been recently infected with Leaf Blotch, but the symptoms have not yet appeared, a **strobiluron** fungicide would not be appropriate, and a **triazole** would be the type to use.

Strobilurons are a recently developed fungicide group. They are powerful synthetic mimics of naturally occurring chemicals produced by the subterranean hyphae of some fungi to prevent the development of other "rival" fungal pathogens competing for their substrate. The first strobilurons to be developed had no kick back activity, but the new generation does have kick back.

Triazole **fungicides** are amongst the most commonly used in Europe, and are both curative and protectant, with a systemic mode of action.

Morpholine fungicides are another group which are contact fungicides only, and have no kick back.

Sometimes fungicides are applied to a crop as a routine, preventative measure even before any symptoms are seen, when certain diseases are known to usually occur every year. This is normally done with high value crops such as tomatoes and other vegetables, and also with Irish potatoes to control Potato Blight.

A famous old fungicide is known as ***Bordeaux Mixture***. This is one of the oldest and most effective of the fungicides, giving good protection against Potato Blight and other diseases. It can be bought, or even made at home, as follows: **Recipe**—Dissolve 225 g of copper sulphate in about 23 litres of water. Make a creamy mixture of 150g of quicklime mixed with a little water. Pour the cream through a fine sieve onto the copper sulphate solution. Test the solution with a clean knife—if the blade comes out coated with copper, add more cream to make the copper dissolve totally. This mixture should be used up within no more than two days.

B. Viruses

A virus is a submicroscopic living organism, a single nucleic acid surrounded by a protein coat. They do not produce any kind of spore but are spread from plant to plant (or animal to animal) within their cells. They are cell parasites and are usually specific to both host cell and host species.

Transmission of Viral Diseases
*   Aphids (*greenfly* and/or *blackfly*) are the most common and important vectors of plant viruses. Most insect vectors have sucking mouthparts, used for feeding on plant sap. A vector, or "carrier", is an organism that carries a disease producing microorganism from one host to another.
*   Eelworms (nematodes) have also been shown to spread some soil-borne virus diseases that damage Irish potatoes, strawberries and other crops, including barley

and wheat with BaYMV (Barley Yellow Mosaic Virus), which is becoming a major threat in Europe. Control of eelworms normally is crop rotation

- Virus diseases (eg Lettuce Mosaic) are occasionally transmitted in or on the seed.

Some Examples of Diseases caused by Viruses

- Irish potatoes are damaged by three viral diseases: Leaf Roll, caused by one specific virus, Potato Mosaic, caused by various different viruses, and Spraing and Mop Top, caused by soil-borne viruses. Control is mainly by using virus-free "seed" ie tubers, which is grown in areas where virus diseases do not occur. If symptoms do occur, the plants and tubers should be removed and destroyed as soon as possible.
- Mosaic and Virus Yellows are often found on sugar beet and mangels but rarely cause significant damage. Control is mainly by controlling aphids, the only vector of these diseases.
- Cabbage Black Ringspot also attacks cauliflowers, turnips and Brussels sprouts. It is spread by aphids, so control of aphids is the best control for this disease.
- Cauliflower Mosaic. All volunteer plants and plants left in the field after harvest of the brassica family should be removed and destroyed to prevent this and other diseases.
- Turnip Yellow Mosaic is spread by one of the flea beetles (*Chrysomelidae*), which is unusual as this is a biting insect.
- Groundnut Rosette Disease—described in Section **2B. "Legumes"**, page 165.
- Mosaic diseases of soybeans and several other crops.

C. Bacteria

Bacteria are essential to any ecosystem, because they are responsible for:
- the decomposition of organic matter built up by plants and animals;
- the production of essential nitrogenous plant nutrients;
- the reduction of cellulose-containing tissues to humus.

Unfortunately bacteria can also be very destructive (ie pathogenic), and cause diseases in man such as typhoid, cholera, TB and diphtheria. In plants, bacteria cause diseases when they become parasitic on those plants. Bacteria can multiply with fantastic speed, by simple division of individuals into two halves.

Examples of Bacterial Diseases

Bacterial wilt of potatoes and tomatoes, Fire-blight of pears and apples, Blights and Leaf Pustule of soybean, Blackleg of potatoes, Soft Rot of turnips and Crown Gall of sugar beet.

Bacterial diseases are best controlled by the destruction of infected plants.

Bacteria and Legumes

The accumulation, or *fixing* (ie gas to solid), of nitrogen by the roots of legumes is performed by a group of highly beneficial bacteria, and is discussed on page 54, **1Fe**.

1L. WEEDS

A weed can be loosely defined as any plant that is growing in a place where it is interfering with the growth, harvesting or marketing of other plants. Cowpeas for example can be a serious weed in soybean and maize crops that are mechanically harvested, since the long cowpea stems can block the harvester; cowpea seed is also not easily removed from soybean seed.

Weeds can be the limiting factor that determine the area of land that a farmer can cultivate; in these cases, correctly applied herbicides may enable more land to be cultivated.

Losses due to Weeds
The percentage of crop yield lost due to the growth of weeds has been estimated to be about 16% in Africa, 11% in Asia, 8% in North, South & Central America and 7% in Europe

Excessive weed growth damages crops by competing with them for moisture, nutrients and light, and may also have other negative effects such as increasing the damage caused by pests and diseases. However a moderate weed infestation often has little or no effect on crop yields.

Some weeds, such as Couch Grass ("Twitch", "Quick Grass" or "Devil's Grass" *Agropyron repens*), produce exudates that retard the growth of other plants.

Weed seed that is present in harvested grain may spoil it for consumption or sale, or make the grain unusable as seed.

Weed Control
This is an important factor in crop management, especially when the crop is young. There are many different techniques available, including crop rotation, mulching and mixed cropping, as well as mechanical and chemical methods. There is a great deal of recent interest in the use of natural enemies of weeds (and insects) in a branch of agriculture referred to as *biological control* or *biocontrol*, outside the scope of this handbook. In general, strong and healthy crops will tend to outgrow and dominate weeds while weaker crops will succumb to them.

Hand weeding
Pulling up weeds by hand, or using various types of hoes, is the normal method for small scale and subsistence crop producers to deal with weeds. Herbicides and the equipment to apply them are expensive, and often not available. Hand tools enormously increase the efficiency of weeding by hand, as discussed on page 92, **1M. "Tools"**.

Circle weeding
Fruit trees and bushes generally benefit if weeds are removed from under them, especially when the plants are young and/or the growing conditions are not favourable. In these cases, recently planted trees and bushes should have weeds removed in a circle of about one metre around them. In poor land, or with very old or weak trees, weeds should also be removed in a circle about the same size as the crown of the tree (the spread of the branches). Take care not to damage the tree or bush roots by hoeing too

deeply, nor to produce a saucer-like depression around each tree, unless this is a deliberate form of micro-catchment to catch rainwater.

When to weed

In general, the earlier the better. Small, young and tender weeds are easier to uproot or cut than older, tougher weeds. Also the root system of the crops that are being weeded are less disturbed.

Small weeds are also less likely to re-grow after they have been uprooted and left on the soil surface, and they have less time in which to damage the crop by removing nutrients etc.

Obviously it is a good idea to uproot weeds before they have developed, and released, their seed.

Striga—"Witchweed"

More information is available online in the IDRC Archive publication "Controlling the Noxious Weed Striga" and elsewhere.

Striga is a parasitic plant weed that can devastate crops and cause 80% or more loss of yield in sorghum, maize, rice, millet, sugarcane and other crops. It is found in many parts of Africa, Asia, India, North America and Australia. The weed penetrates the roots of flowering plants, mainly of the cereal family, and digests the root system of its host. In some parts of Ethiopia it is known as *Harama Hazabe* ("Weed from Hell").

There are about 60 species of Striga, of which *Striga lutea* is the most important. Striga can badly affect the following: sorghum, maize, rice, wheat, oats, rye, pearl millet, finger millet, Italian millet, teosinte (*Euchlaena mexicana*) and sugarcane.

One species, *S. gesneroides*, is sometimes found in Africa and Asia and unlike other Striga species can parasitise tobacco, several legumes and some species of Ipomoea.

Striga is an annual, each plant producing up to 500,000 very tiny seeds (0.0045 g) which can remain viable in the soil for up to 20 years. Normally the seed will only germinate when a susceptible crop is grown in the same soil.

Controlling Striga

Alas this is virtually impossible, and the spread of this weed can normally only be limited to some extent, by using a combination of the following methods:

* Crop rotation of 15 years or more without growing a susceptible crop. In the rotation *trap crops* are grown such as cowpeas, soybean, field peas, groundnuts, sesame, sunflower or sunn-hemp (good in poor, sandy soil); these trap crops stimulate the Striga seed to germinate but are not themselves seriously damaged. The Striga plants must then be destroyed before they set seed. Sometimes the trap crop itself is ploughed in together with the Striga, before it sets seed.
* The increased use of fertilisers may have some effect, but unfortunately this also tends to increase the vigour of Striga.
* Wider plant spacing of some crops may help them to withstand Striga infestation.
* Careful hand weeding before Striga sets seed can be done in small areas; weeding should be repeated several times every season.
* Some herbicides such as phenoxy and benzoic acids, 2,4-D, ametryne and atrazine can give good control in some crops if the soil and weather conditions are favourable.

- Some varieties of sorghum such as *Gubiye, Framida, Dobbs* and *Radar*, and a few sugarcane varieties have some resistance to Striga. Purdue University has been active in developing and distributing a number of Striga resistant varieties. Maize resistant varieties are being developed at IITA, Nigeria.

Striga in Seed

Rapid check for presence of Striga seed in a sample:

Put a random sample in a plastic bag, stand the bag upright and agitate/shake the bag as often as possible over a 24 hour period. While still upright, remove the upper 75% of seed. Examine the lower 25% for a brownish dust/powder. If present (if possible, inspect under a hand lens or microscope), reject the lot for seed. If absent, the whole lot can be used for seed, or, to be more careful, use only the upper part for seed and the lower part for food.

More precise check for presence of Striga seed in a sample:

Sow the sample in a field known to be free of Striga, or in sterilised soil. In about 45 days the Striga will show itself.

Herbicides

Detailed technical information is available in the *PNW Weed Management Handbook* (available online).

Herbicides, or weedkillers, are chemicals that kill or damage plants by their phytotoxic properties. Their method of action is either *selective*—killing or stunting only weeds and leaving the crop undamaged (eg MCPA and 2,4-D) or *total (non-selective)*—killing or stunting all vegetation (eg Paraquat and Diquat /Reglone).

Some observations on Herbicides

- In general, seedlings and young plants are more susceptible to herbicides than mature plants.
- Some "total" herbicides become "selective" when they are diluted sufficiently.
- Even the so-called selective herbicides can have some negative effects on the growth of the crop, and so are normally used at low concentrations.
- Both selective and total herbicides can be applied either as foliar sprays (the *contact* or *translocated* sprays), or directly onto the soil where they have a residual effect.
- Some herbicides, normally the selective ones, are available in granular form.
- 2,4-D and MCPA are taken up (*translocated*) by plants very strongly, and very small quantities can damage certain sensitive crops such as coconuts and palms.

Types of Herbicides

**1. *Residual*—mainly for the control of perennial weeds. They are applied to the soil and then worked into it where they persist and so kill weeds as they grow or germinate.

**2. *Contact*—scorch, stunt or kill nearly all parts of the plant that they touch. They often do not kill perennial weeds, but just slow down their growth. They are normally sprayed in combination with other soil-surface (residual) herbicides; these kill off the weed seedlings that germinate some days after application of the herbicides. This dual effect can eradicate weeds for long periods.

3. *Hormone*—a form of translocated or growth regulator herbicide. They are absorbed into the plant tissues, and move about within the plant. They are most effective on weeds that are growing rapidly. Normal plant growth is upset, and the plants become twisted or distorted but not always killed. The underground plant parts are also often damaged. This check in the growth rate of the weeds allows the crop plants to become dominant and thus overcome the weeds.

Some Commonly Used Herbicides

MCPA and 2,4-D—very selective, used against broadleaf weeds in cereals, grass and sugarcane. The application rate is about 1 litre/hectare of the commercial solution. Can be mixed with contact herbicides such as sodium chlorate for circle weeding or on weeds between rows of tree crops. Some crops such as the palms and coconuts are very sensitive to them.

2,4,5-T—for woody plants; it kills most trees. Principal constituent of the defoliant Agent Orange. Banned in most forward thinking countries due to its persistence and toxicity.

Amiben (Chloramben)—used in legume and vegetable crops, and also in paddy rice just after transplanting the seedlings.

Dalapon—kills grasses by direct contact, at 4–12 kg/ha. It should not be sprayed directly onto crops. Similar to TCA, both of which persist for several months.

Simazine, Atrazine and Ametryne—selective, used in maize, pineapples and in circle weeding as pre-emergence sprays at 1–4 kg/ha. Persist for 6 months or longer.

Monuron (CMU) and **Diuron (DMU)**—mainly used for young weed seedlings and germinating seeds. Used either after cultivation or with contact herbicides. Persist for up to 6 months.

Paraquat (Gramoxone)—contact, pre-emergence. Becomes inactivated when in contact with clay particles in the soil. Used widely in minimal (zero) cultivation systems, when the land is not ploughed before planting. Toxic to animals and kills all plant tissues, especially grasses. Rate of application is about 0.25–2 litres/ha of the commercial solution.

Diquat (Reglone)—similar to Paraquat but more effective against broadleaved weeds. Used to remove potato haulm. Both Diquat and Paraquat persist (ie remain active) for only a few days.

Glyphosate (Roundup)—slow-acting, translocated. Used for clearing grass and broadleaved weeds before planting, and for some tree crops. Becomes rapidly inactivated in the soil.

TOK (CMU)—Selective. Used in rice in a granular form (7% a.i.—active ingredient). Applied after transplanting, but before weeds have germinated, at about 20 kg/ha. It is claimed that TOK is relatively harmless to fish.

Sodium Chlorate and **Sulphuric Acid**—total, with a wide effect on all plants. Used at 10-15 kg/ha in dry conditions. Often mixed with selective herbicides such as 2,4-D and MCPA. It is inflammable when it is dry.

☞ **Safety Precautions**

Herbicides are toxic chemicals that are potentially dangerous to all life forms. Particular care should be taken to prevent herbicides from reaching rivers or other water sources.

- Herbicides must be applied at the correct growth stage of both the weed and the crop, and at exactly the recommended dosage. It is imperative to read the label on the container very carefully and also to follow all the instructions very carefully.
- Storage of herbicides, and all other chemicals, must be in a place protected from both the elements and thieves.
- Some herbicides are persistent, both in the soil and in the produce of the crop. For example, a herbicide which is used on a cereal crop may persist in the cereal straw, which can be dangerous if it is fed to animals, or if it is used as a mulch on a sensitive crop.
- In windy weather nearby crops can be damaged by spray drift, which can also affect people and animals.

1M. TOOLS

In vast areas around the globe the land and crops are still mainly cultivated with hand tools. These tools have been slowly evolved, developed and improved over generations of use so as to be the most suitable for the local conditions of soils and crops, and they are often made only from locally available raw materials.

Ideally hand tools should be easy to make, and easy to maintain and to repair, and they should be made from locally available materials. Local blacksmiths are often responsible for these functions, though the shortage of raw materials, especially metal, can often be a problem. In some countries blacksmiths have a low social status, though this situation is slowly changing as local people increasingly appreciate the value of these valuable artisans in their community.

HDRA (the Henry Doubleday Research Association based in Coventry, UK) conducts research on locally made tools and equipment using local materials and skills.

Rate of Work: The human power unit is limited to about 0.1 HP (0.075kW), so farmers who have to rely on hand tools are often limited by them as to the amount of land they can cultivate and hence the amount of food they can grow.

Distribution or Sale of Hand Tools

Many farmers tend to be rather conservative and traditional people who are reluctant to adopt new ideas. Their choice of tools is no exception. In agricultural development or rehabilitation programmes it is essential that any tools which are provided are the same, or very similar, as the tools which are normally used by the recipients. It is well worth spending some period of time in conversation, and observation, to discover exactly which tools are the most appropriate.

For example, if the blades of the hoes that are provided are too heavy, or too light, or if the handles are too long or too short, then many farmers will not use them.

If it is not possible to buy the correct tools in sufficient numbers it is often a good idea to have them made locally. This way you will be more likely to get locally acceptable tools, and you will also introduce work and money into, or close to, the project area. The local blacksmiths or people who normally make tools should be involved with this at all stages—they can teach young apprentices their skills, and they will know which materials are available and which types of tools, and handles, are acceptable and in demand in the area.

"Tool Handles—to have or not to have, that is the question".

It is not always easy to decide to provide tools with or without handles. There are no rules to follow, but the chart below summarizes some of the issues to help decide the "with or without tool handle dilemma"—an apparently minor detail, but one which can make or break the success of a tool distribution programme.

With Handles	Without Handles
Advantages	
• Better quality handles, normally than locally made ones. • Tools can be used immediately.	• Lower costs of purchase, storage and transport. • Farmers can make exactly the right handles to suit them. • Delivery time from the supplier may be reduced.
Disadvantages	
• Higher costs of purchase, transport and storage. • May be unsuitable quality (type of wood, length of handle etc.) • Delivery time may be longer, especially if different suppliers supply the tool and handle.	• Increased deforestation in the project area and surroundings from tree felling for supply of handles. • Some families will be unable to make good quality handles • Time delay before the tools can be used.

1N. IRRIGATION

Irrigation is the practice of applying water to plants. In arid regions irrigation can, and often does, supply all or almost all of the water needs of plants. In temperate regions irrigation is usually supplementary to the natural rainfall, sometimes known as *top-up irrigation*.

Water is becoming an increasingly valuable commodity. Many irrigation schemes involve numerous, varied and contentious issues surrounding **water rights**. The "ownership" of water is usually a contentious topic, which should be clearly and definitively agreed as far as possible before any physical work starts. The issues are often complex; sharing of water, both between different uses (domestic and/or irrigation and/or livestock) and between different users with different claims, must be clearly resolved as a first priority.

Water use committees may be established to decide who gets how much water and on which days. Agencies that facilitate the organisation of such committees may make a valuable contribution to food producers, sometimes more valuable than technical or financial support.

Methods of Irrigation
There are three main types of irrigation systems: surface, overhead (sprinkler) and drip.

1. *Surface Irrigation*
Again, there are three main types of surface irrigation;
- *Flood irrigation*—water covers the soil surface in a continuous sheet. In practice, it is almost impossible to maintain uniform water coverage, so some parts of the field always receive too much or too little water.
- *Furrow irrigation*—water runs down furrows between rows of plants. Very labour intensive, and also salts often accumulate in the crop beds between furrows.
- *Corrugation irrigation*—water is applied in small furrows running down the slope. Some overflowing of furrows normally occurs. Also labour intensive.

Surface irrigation normally requires the least amount of manmade materials, and the water source may be cheap or even free, such as a river or water storage pond. There are many ingenious ways of flooding fields, some of which are high above the water level.

These systems require the water to be lifted first, either with "cheap" human powered devices such as bamboo pumps in Bangladesh and elsewhere, or with diesel/petrol driven pumps.

Currently about 80% of the irrigated area around the world involves some form of surface irrigation.

2. *Overhead or Sprinkler Irrigation*
Normally powered by powerful machinery, but with sufficient ingenuity and head of water gravity alone can sometimes provide enough power for simple systems. Water is delivered through a main line from the source of supply to the lateral lines, then discharged above the crop through sprinkler heads on riser pipes attached to the laterals.

Each sprinkler head supplies water to a circular area, the size of which depends on the nozzle size, wind and the water pressure. To ensure uniform coverage, the patterns are overlapped from 35% to 70%, depending on the type of sprinkler and the wind.

Advantages—compared with the other two irrigation systems, water is distributed more evenly and in a more controlled way, so less water is needed for a given land area; land preparation is reduced, and land on quite steep slopes can be irrigated; fertiliser can be applied with the water; labour input is lower; crop damage from heat or frost can be reduced; and the accumulation of salts in the soil is reduced.

Disadvantages—very high cost, to buy, to install and to run; a filtration system is also normally necessary; over-watering can easily happen, leading to water runoff and erosion; and the normal problems associated with operating and maintaining sophisticated equipment.

3. Drip (Emitter or Localised) Irrigation

A highly efficient system if the water supply is clean or can be cleaned (filtered). Small plastic nozzles are installed at intervals, to coincide with plant spacing. Depending on the conditions, from 1.2 to 4 litres per hour are delivered directly above the plant root zone. The Leak hose is another form of localised irrigation, where a porous rubber hose leaks water along its entire length. This product can withstand high pressure and ice, and requires low pump capacity, but is usually not economical to run because it is haphazard and has a high flow rate.

With drip irrigation the amount of soil which becomes wet is a function of the soil texture; sandy soils become wet in a long, carrot-shaped section underground, about 0.8–1.5 m in diameter, while clay soils become wet in a shallower, onion-shaped section, about 2.8–4.3 m in diameter.

The chart below shows the approximate ranges of wetted area per spot location, for three soil types:

Soil type	Area wetted (m^2)	Diameter wetted (m)
Sandy (Coarse soil)	0.45–1.84	0.76–1.53
Loam (Medium soil)	1.84–2.16	1.53–2.75
Clay (Fine soil)	2.16–3.36	2.75–4.27

The soil is wetted by a process known as *capillarity,* the movement of water in the soil due to the attraction of soil for water. The rate of movement depends on the size of the pore spaces in the soil, and the soil moisture content. The smaller the pore space, the further (and slower) is the movement of water.

Other Irrigation Methods

For high value crops, or when the supply of water is very limited, special techniques can be used:

- **Drip Hose/Leak hose**—see above.
- **Bubblers**—as with sprinklers, underground pipes serve a rigid pipe extending above ground. The water comes out as it would from a tap, flowing in a continuous stream from special brass or plastic nozzles that provide water to individual trees or bunded areas.
- **Air-wetting Methods**—used in greenhouses to provide a humid atmosphere. Used for the propagation of plant cuttings, to raise young seedlings and sometimes with soft fruit and high value vegetables.

- *Foggers & Misters*—relatively high water pressure is needed, using very fine nozzles located on a pipe network near to the plants.
- *Humidifiers*—fans draw air across a greenhouse, passing through a wet membrane, so that the whole greenhouse becomes humid.
- **Subsoil Irrigation**—such as the Durwick or Cell Systems. An artificial field is made with an impermeable membrane of PVC sheeting, butyl rubber or similar material placed about one metre below the soil surface. Water—which should have a low salt content—floods the area and is then drained away to be pumped again and reused.

Information Required for the Design of an Irrigation System
In order to design an efficient irrigation system, as much as possible of the following data should be collected:

- **Climate and Geography**—daily maximum and minimum mean temperatures, rainfall, relative humidity and wind speed. The longitude, latitude and altitude may also be useful information.
- **Soil Texture**—this gives information on infiltration rates and water holding capacity. The soil texture, or type, also influences to some extent the size of nozzles used; smaller nozzles should be used on fine sands and silts to reduce the problem of soil capping (crusting).
- **Soil Chemistry**—a soil analysis to determine the following data can be made to help decide the correct irrigation method and scheduling: pH, EC (Electrical Conductivity), adjusted SAR (the relationship between sodium, calcium and magnesium), ESP (Exchangeable Sodium Percentage), chlorides and sulphates.
- **Water Quality**—pH, EC (or TDS—total dissolved salts/solids), adjusted SAR, sodium, calcium, magnesium, chlorine, nitrogen, phosphorus, potassium, carbonates, bicarbonates, sulphates and organic matter. If one or more of these elements is present in excessive amounts, especially chlorides or sodium, the balance of all nutrients in the soil may be upset and so make other essential nutrients unavailable to plants. The Boron content in water can be a problem if it exceeds 1ppm, but it is very expensive to analyse.

Overhead irrigation with water containing more than about 125ppm of sodium or chloride may cause the leaves to burn and fall off the plants.

Calculating the Quantity of Water Required
As a *very approximate guide*, the following daily amounts are required for field crops—figures are the maximum ("peak") daily requirements, for the driest season:

Arid desert	Semi-arid/ mediterranean	Temperate
16 mm/m^2/day*	8-10 mm/m^2/day	3-4 mm/m^2/day

* 16 mm/m^2/day = 16 litres/m^2/day = 160,000 litres/ha/day

Example of the calculation of monthly water requirements.

Calculation of soil moisture deficits from average evapotranspiration and rainfall data for a typical sandy loam soil in a temperate region in the northern hemisphere.

All figures are in mm.

Month	Rainfall (P)	E*	P–E	Storage	Excess	Deficit
March	88	16	+72	50	+72	0
April	47	45	+2	50	+2	0
May	49	80	-31	19	0	0
June	40	105	-65	0	0	46
July	53	100	-47	0	0	93
August	51	88	-37	0	0	130
September	50	40	+10	0	0	120
October	57	16	+41	0	0	79
November	88	—	+88	9	0	0
December	100	—	+100	50	59	0

E* = Evapotranspiration

Controlling the Temperature of Crops

The practice of cooling of crops and protecting them from frost is often associated with irrigation since the same overhead systems can be used to alter the temperature of crops:

1. Cooling of Crops—in very hot weather it has been proved that some crops benefit from being cooled with water, while benefiting from irrigation at the same time. Cooling of crops normally uses less water than protecting crops from frost.

The critical temperature at which many plants benefit from cooling appears to be 32°C, and ideally water should be applied until the temperature falls to this. In practice however, some crops in some climates should only be cooled for a few hours per day, and allowed to remain at high temperatures—above 32°C—for the rest of the day. In fact the sugar content of some crops may fall if the plants are maintained at 32°C.

2. Frost Protection—to avoid plant damage when frosts occur water can be applied by sprinklers to crops if they are at a vulnerable stage of their life cycle, such as flowering.

The flowers are protected in a remarkable way, governed by basic laws of physics. When water changes into ice, heat is released and the local air temperature also rises as a result. At the same time, ground heat is held to some extent by the fog or mist that is created.

When the temperature then rises to above zero, heat is absorbed and the ice melts. The heat is regained from the moisture in the air—if there had been little or no moisture in the air heat would have been taken instead from the plants themselves, leading to frost damage by breakdown of their exploded cells.

To protect plants successfully from frost, the water should be applied continuously, especially during the time when thawing occurs. However the total amount of water used should always be kept to a minimum, both to minimise the ice load and to avoid waterlogging the soil. A balance is required … not too much, not too little!

Filtration
Problems with Clogging
Irrigation water that is a bit "soupy" (more than about 1000ppm TDS) should be filtered to some extent, unless surface or subsoil irrigation is used. Nozzles, water channels and pipes can and do become clogged with a range of foreign bodies, including organic matter, iron oxide, calcium carbonate, algae, fungi and microbial slimes.

Clogged nozzles reduce emission rates and cause uneven water distribution, as well as increasing the amount of water usage and wastage. The problem can become very serious and can only be solved by either cleaning/filtering the water or, sometimes, by using bigger nozzles, which in turn can create its own problems.

Filtration methods are many and varied: sand media, wire mesh screens, strainers, centrifugal separators and settlement tanks. In order of importance, it is increasingly desirable to clean or filter water for:

misting, fogging, drip, spray, sprinkler, bubbler and mobile sprays.

Saline Soils and Irrigation Water
A) *Problems with Saline Soils*
Saline soils are also discussed in **1Ca**.

If the irrigation water, or the soil, is saline, plants can only use some of the water because the osmotic pressure exerted by the salts in solution reduces the intake of water. As a result, salts accumulate within the plant, eventually reaching a level that becomes harmful to plant growth.

As a general "rule of thumb":-

"very saline" soils have more than 3 mmhos/cm^{-1} (or about 1920 ppm TDS);
"moderately saline" soils: between about 1 and 3 mmhos/cm^{-1} (640–1920 ppm TDS).
{1 mmho/cm^{-1} = 1 mg/litre = approx. 640 ppm TDS}

Solving the problem of saline soils generally involves *leaching* the soil, to wash away the accumulation of salts to below the bottom of the root zone. Obviously, the amount of water required to do the job depends on the quality of the water used to leach the soil, the salt content of the soil before it is watered and the soil type.

The land must be well drained, and many soils need to have a drainage system installed before the salts can be properly leached out.

The amount of water needed to leach out salts properly varies greatly, but roughly speaking between 5% and 35% extra water is needed in addition to the normal irrigation water used, as shown on the table below:

TDS (ppm)	Extra water needed
640	5%
900	15%
1280	25%
1800+	35%

Fertilisers and Leaching
When the soil is leached beneficial nutrients are also lost, together with the salts, so it is preferable not to apply fertiliser soon before leaching is in operation. If fertilisers are used during leaching periods, some fertilisers are available which are less soluble than others and so these should be used if possible.

When to Leach?
Leaching can be done at any time of the year but it is normally done in the cooler, wetter seasons—for good reasons. At that time plants need less water, rainfall is higher, and there is enough time to leave the irrigation equipment in one place to ensure thorough leaching. An additional reason is that water is often scarce and more expensive in the hotter summer months.

B) _Problem's with Saline Irrigation Water_
Even low levels of sodium or chlorine ions in irrigation water can accumulate in the soil and become harmful to some plants. Unfortunately the techniques used to remove salts from water are very expensive and use large amounts of energy/electricity. The techniques involve either reverse osmosis or electrolysis, and are generally only practical for irrigation schemes where the supply of energy is cheap.

The best way, indeed the only really effective way, to ensure that salts do not accumulate in the soil is to monitor the actual levels of salts on a regular basis. Due to the fact that the salt level rises and falls—sometimes dramatically, in rivers for example—the irrigation can be scheduled so that water is applied when the salt content is low. Alas, water is often most critically in demand at the very time that water sources are at their lowest, ie most salty, point.

Water that is used to irrigate plants can often become up to 3 or 4 times more concentrated when it enters the soil, especially when it is around the root zone. This fact is important to grasp, as crops which are tolerant to about 5000 ppm TDS should not be irrigated with water which contains more than about 2000 ppm TDS on sandy loam soils or 1000 ppm TDS on clay soils.

1O. STORAGE

The two critical factors for the safe storage of seed, and food, are **temperature** and **relative humidity**. Food also requires carefully controlled storage conditions, but even greater care is needed with seed because it is a live product that needs to be kept in certain specific conditions if it is to maintain its germination and vigour.

If the quality (germination rate and vigour) of seed is to be maintained for a long period of time, there is a useful guide:

Relative humidity (%) + Temperature ($^{\circ}$F) = 100 or less

This means, in theory at least, that in hot regions the relative humidity of the air should be reduced, and in humid regions the storage temperature should be reduced.

However, irrespective of the climate and storage conditions, **the higher the average temperature, the lower the moisture content of the stored seed should be**.

In practice it is usually difficult to adjust either the temperature or the relative humidity, so it can sometimes be a good idea to transport the seed to a cooler, drier place if it has to be stored for a long time.

All seed benefits and remains viable for longer periods if it is stored **cool and dry.** However the specific requirements for each crop can be very different, and the viability of seed depends on many other factors in addition to temperature, relative humidity and seed moisture content, as discussed in **1Fa. "Germination"**, page 46.

1. Seed Moisture Content

In general seed should not be stored for more than a few days if the seed moisture content is more than about 13%. In humid or very hot climates, 9–10% is advisable, while in cool, temperate climates seed can be stored safely for several months at up to 16–17%.

Metabolic activity in grain practically ceases below about 8% moisture.

Measurement of Seed Moisture Content

The moisture content of grain can be easily and quickly measured by using small hand-held meters. These meters can be used in the field if necessary, and measure to within 0.1% accuracy. They are operated either by batteries and/or mains/solar electricity.

Moisture meters measure either the electric conductivity or the dielectric constant of the sample. The moisture content is shown instantly on the meter, and is quite accurate enough for most field work. Even more accurate measurements can be made in a laboratory, but are rarely needed.

Mites in Stored Grain

Mites are often a problem in stored grain, normally when the temperature and humidity are high and when the moisture content of the grain is between about 12% and 18%. In general, the best conditions for most insect pests is a temperature around 30°C, with a relative humidity of 40 to 80 percent. Above 40°C, most species will eventually be killed, while reproduction ceases at temperatures below about 20°C, with dormancy induced at temperatures below 10°C. Reproduction is inhibited at less than about 40 percent humidity.

Mites (class *Arachnida*) of stored seed and grain are very small, 0.2–1 mm long, and unlike other insects have four pairs of legs (insects have three) and an apparently unsegmented body, with no wings or antennae. A hand lens is normally needed to see them. They are mainly whitish in colour, and they move rather slowly.

Some species of mites are beneficial, and eat the eggs of moths and other mites, although it is difficult to tell the difference between the species. As a *very rough guide*, smaller mites tend to be more harmful than larger ones.

Control of mites is possible with phosphine (online information available from http://phosphine.com) or with an acaricide that is proven to be both effective against mites and safe for use with stored food and/or seed. Lindane is highly effective but has wisely been banned, or has seriously restricted use, in most developed countries.

2. Storage of Grain Legumes for Seed

If the moisture content of legume seed is between about 9% and 14%, the two following guidelines can be applied independently of each other:

- *if the storage temperature is reduced by 5°C, the life of the seed (ie the length of time it remains viable) will be approximately doubled.*
- *if the moisture content of the seed is reduced by 1%, the life of the seed will similarly be about twice as long.*

3. Storage Fungi

The most important storage fungi are: *Alternaria, Aspergillus, Mucor, Penicillium, Streptomyces, Cladosporium* and *Sporendoma*. They can cause reduced germination, discolouration, heating, biochemical changes and loss in grain weight.

Seed which has become very "mouldy" ie infected with fungi, should be destroyed (burned or buried) and not used as either animal or human food.

Fungi in stored seed/grain is mainly controlled by reducing the temperature, humidity and seed moisture content; chemical control is very limited.

Clean, unbroken seed is less liable to be damaged by fungi than seed that is broken, damaged or dirty.

4. Storage Pests

Rodents, birds and insects (various beetles, weevils and mites) are the main pests. Damage by birds is mainly caused by spoiling the seed with their faeces, as well as by birds eating it.

The optimum temperature for most insects attacking stored seed is between about 28°C and 35°C. If the temperature rises much higher, insects normally either die or cannot reproduce, so very dry seed can be stored at about 28°C for short periods to minimise insect infestation. But even when the seed is dry insects can sometimes multiply very rapidly, especially if the seed is not well cleaned and contains many broken or damaged grains.

5. Storage Facilities

On a global scale post-harvest loss due to poor storage is estimated at around 5–10%, representing many millions of tons of lost food. To reduce losses in storage, the storage facilities should have as many of the features listed below as possible:

- an even, cool and dry internal atmosphere;
- protection against insects, rodents, birds, fungi and thieves;
- protection against rain and floods;
- constructions which are simple, cheap, preferably made from local materials, and most importantly, easy to thoroughly clean between harvests.

Types of Seed Store - seed stores can be either ventilated or non-ventilated:

1. Ventilated Stores

Exposure to air means that the crop can continue to dry while in storage. Heating is also reduced since moulds (fungi) are less likely to multiply. On the other hand, the seed is exposed to insect, bird and other damage.

Ventilated stores are most useful in humid regions, and are often used for storing maize cobs. As the climate or region becomes drier, less ventilation is necessary and also becomes increasingly inappropriate.

2. Non-Ventilated Stores

These are normally used in drier regions, and the seed must be at or below its safe moisture content before it is stored. Non-ventilated stores have solid walls and close-fitting lids, which ideally are designed to allow for fumigation of pests if this should become needed.

Seed is often also stored under the ground, where it is cooler.

Storing Seed (and food) in Sacks

There are four points to bear in mind:

1. The sacks should be well protected from rain, sun, insects, rodents etc., and—vitally—**OFF THE GROUND**. Pallets, and plastic (overlapped to avoid gaps) are ideal.
2. If possible use only **clean, undamaged sacks** to avoid losses from spillage and/or infestation by fungi and insects which may be present in the old sacks.
3. Observe all the normal rules applied to the storage of food in sacks, principally **regular inspection**, and fumigation if necessary. *Keep the store clean.*
4. The sacks should be clearly marked (seed type, lot number etc).

6. Seed Storage Life (Longevity)
The length of time during which seed remains viable, ie capable of germinating to produce a normal seedling, depends for each crop on a whole range of factors, such as the health of the parent crop, temperature, relative humidity, seed moisture content, handling and so on.

The germination rate of any sample of seed inevitably declines to some extent over time, and the rate of decline depends on a number of factors, discussed previously. As a result it is not possible to state for any particular species or variety the precise number of years and months that the seed will remain viable, though some broad guidelines for certain crops are outlined in the paragraph on germination, **1Fa**, page 46.

Some crops, such as finger millet, teff and cucumber have seed that can remain viable for several years. However the viability of seed at any point in time depends not only on its age, but also on its initial germination rate and vigour, the storage conditions and how often the seed has been either transported or handled roughly.

As a result it is a good idea to test for germination any seed that is coming from an unknown or dubious origin before it is distributed or planted. Germination, and testing for germination, is also discussed in **1Fa**, page 46.

In the United Kingdom, farmers can register their grain store through a scheme known as ACCS (Assured Combinable Crops Scheme), provided that the store is free, or nearly free, of vermin, pests, moisture etc. Further details are available on the ACCS website www.assuredcrops.co.uk.

SECTION 2

DESCRIPTION AND CHARACTERISTICS OF THE MAIN FOOD CROPS

(Page numbers in brackets)

For quick and easy reference, the food crops listed in Section 2 are all described in the same way, in the same order. For example, if you want to know about the soil requirements for barley, go to "Barley", then "Planting conditions/Soil":

- *Common English and scientific ("botanical" or "Latin") name.*

- *Local names, synonyms and names, where known, in French, German, Spanish, Portuguese, Italian, etc.*

- ***Introductory comments**, on plant description, origins, historical background, relevance as a major food crop or as a potential food crop, botanical classification/taxonomy, etc.*

- ***Planting conditions**, subdivided as appropriate into subheadings: propagation, soil, seed rate, germination, seed spacing, depth, inoculation, rotation, intercropping and weeding.*

- ***Growth conditions**, subdivided into subheadings: day length, growth period, temperature, rainfall, altitude, pests and diseases.*

- ***Yield**. Low, average and high yield expectancy. Lowest & highest national averages.*

- ***Utilisation**. The main uses for the plant, not only as food but also as a source of revenue, including its by-products, either on the farm, for sale (cash crop) or in industry.*

- ***Limitations**. Reasons for the crop being suitable to grow only in certain conditions, and any inherent dangers in the plant, seed or by-products.*

*

2A. CEREALS

Of all the agricultural crops, cereals are generally considered to be the most important. They are grown on about 75% of the world's cultivated area, and directly supply about two thirds of the energy and half the protein needs of the world as well as indirectly supplying large amounts of food when converted into meat, milk, eggs and so on.

The increasing global demand for certain cereals such as maize to produce meat and animal products may lead to a rapidly increasing demand for maize, possibly overtaking demand for rice and wheat in the next two decades. In the late 1990s the tonnage of maize and rice produced (but not the acreage) began to consistently exceed the tonnage of wheat produced for the first time in many thousands of years, and it is clear that wheat is slowly losing its place as the world's most popular cereal.

Genetic Modification (GM) has been very successful with maize, rice and soya, but much less so with wheat, mainly due to the fact that most wheat varieties are hexaploid—six copies of each gene—and are consequently more difficult to modify genetically. 200 million acres of GM crops were grown in 2004, with positive results in regard to yield, pesticide use, biodiversity and costs—and no negative effects on human health. Yet GM continues to face strong opposition from the environmental movement. The irony of this is that GM was invented in 1983 as a safer and more gentle approach than the existing methods of generating mutant, possibly useful, plant types. These methods include irradiation, x-rays, thermal neutrons and ethyl methane sulphonate to damage DNA and produce random mutations, while with GM scientists can add the specific attributes they want in a more or less predictable way—indeed, in a more "organic" method of manipulating genes to introduce desirable characteristics. The jury is out ...

As early as 1956 irradiation was used at the Atomic Energy Research Establishment at Harwell, Oxford, UK on a barley variety called Maythorpe. This gave rise to barley strains with shorter, stiffer straw but also with the same desirable characters of early maturity and malting quality, culminating finally in the release of Golden Promise.

The true cereals, which produce a grain type of fruit, are all members of the *Gramineae* family—the grasses. The "pseudo-cereals" are from other plant families, such as *Amaranthaceae* (Amaranths), *Chenopodiaceae* (Quinoa) and *Polygonaceae* (Buckwheat). They are often regarded as cereals because their seeds are similar nutritionally to those of the true cereals; their importance is slowly dwindling in comparison to the "true" cereals.

The great success of the cereal species is due to a number of factors: they adapt well to a wide range of soil types, climates and cultivation methods; they are relatively efficient in photosynthesis; they are all annuals; they are relatively hardy (tolerant of cold, and other factors) and they recover well from damage. Most important of all, their grain is contained in a neat package of stored energy which is convenient to harvest, and which is easy to handle, clean and store.

Cereal grain is an important source of carbohydrate, fibre (insoluble and soluble), some vitamins (B complex and E) and minerals. The fat content is about 2% (up to 7%, in dehusked oats), and highly unsaturated—maize (corn) oil is widely traded and consumed throughout the world. The iron content of Teff is an impressive 80–90 mg iron per 100 g. Some cereals provide useful amounts of calcium; finger millet grain for example contains about 350 mg/100 g.

On the negative side, cereal grains are deficient in a number of Vitamins. The dry grain contains no B_{12} nor C, and very little A, D, K or B_2 (Riboflavin). In general the quality of protein in cereal grain is also poor; all of the cereals are deficient primarily in lysine with a secondary deficiency in threonine or tryptophan. There has been some progress with breeding high lysine varieties of maize.

The most important cereal crops are **maize, wheat** and **rice**, which in global terms produce approximately 80% of the total production of grain cereals. Most of the other 20% is produced from the cereal crops described in the following pages: **barley, buckwheat, millets, oats, rye, sorghum** and **teff**.

Barley
Hordeum vulgare

Orge (French); Gerste (German); Cevada (Portuguese); Cebada (Spanish);
Orzo (Italian); Korn (Swedish); Sha'ir (Arabic); Sigem (Tigrinha),
Gebs (Amharic), Garbuu (Oromifa) in Ethiopia; Jau, Jar (Hindi and Dari);
Oorbashay (Urbashi)(Pashtu).

Cultivation of barley probably originated in highland Ethiopia and in Southeast Asia
where it has been cultivated for at least 2000 years. It was the main bread plant of
the Hebrews, Greeks and Romans. It is descended from wild barley (*Hordeum
spontaneum*), which still grows in the Middle East. Bread made from barley is
unleavened ("flatbread") due to its low gluten content.
 Nowadays barley is the most widely distributed of all the cereals, and grows in
almost all temperate regions as well as in hotter, drier areas such as those found in
North Africa and Ethiopia, and the highland tropics.
 The annual global production of barley in 2004 was 154 million MT, about half
of which is produced in Europe; it is the fourth most important cereal after wheat,
rice and maize.
 The plant is an annual grass (family *Poaceae* alt. *Gramineae*), 50–130 cm tall,
normally with many tillers and almost always with long (7.5–10 cm) awns, which
make the plants look like awned, or bearded , wheat.
 There are two main types of barley, which can be cross-pollinated by plant
breeders:

2-row (var. *distichum*)—only the central spikelet develops seed. Each head
has two rows of seeds, one opposite the other. This type of barley is the favourite for
making beer, though some 2-row barleys are feed types. The average protein content
is 11.5–13%.
 6-row (var. *hexastichum*)—both the central and lateral spikelets develop
seed, producing six rows of seed. This type is normally used for animal food, though
6-row barleys are also used for malting. The average protein content is 12–13.5%

A third type exists, **4-row** (*Hordeum tetrastichum*), but is not widely grown.
 There are both spring and winter varieties of both types. There are also
"intermediate winter-types", often called "facultative" varieties, such as Secret, that
can be planted either in the autumn or spring, but which are mainly less winter hardy
than the true winter types.
 A third type exists and is occasionally cultivated; sometimes called Abyssinian
intermediate or Ethiopian black barley, *Hordeum irregulare*, has fertile central
florets and varying proportions of fertile and sterile lateral florets.

To Distinguish Barley from Wheat
a) *Examine the Seedling (Young Plant):*
Barley has long smooth auricles, the leaves have no hairs. Wheat has hairy auricles,
the leaves have very small hairs. Oats do not have auricles:
("Big and Bare is Barley; Whiskery and Wee is Wheat; Oats have 0").

b) *Examine the Seed*:
Barley has the glume and palea (seed coverings) remaining attached to the seed after threshing. These can only be removed by grinding, to produce "pearl barley", unlike wheat which produces a "chaff" of glumes and paleas during threshing. In fact, there are a few "naked" varieties of barley, which are especially valued as human food.

The information below relates to feed barley types unless otherwise stated. Although malting barley types are grown in very similar (but normally better) conditions, farmers have to apply higher standards of crop husbandry to produce malt quality barley, which also can only be grown on certain soil types.

PLANTING

Propagation: by seed, occasionally hybrid. Almost 100% self-pollinated.
Soil: barley needs less fertile soil than wheat, and adapts to a wider range of soil types than wheat. More salt tolerant than most other cereals, though some Triticale and Durum wheat varieties are more tolerant. Soil must not be waterlogged, and should not be light or sandy or more acidic than pH6 (classified as "sensitive" to soil acidity).
Fertiliser—similar to wheat; a typical application is 200 kg/ha of 11:54:10 on poor soil, and 85 kg/ha on more fertile soil. Average UK application for winter barley is 140:75:100. Potash helps to reduce damage from mildew. Malt varieties normally receive little or no Nitrogen.
Seed rate: dryland 50–70 kg/ha, irrigated 70–120 kg/ha (maximum 200 kg/ha). If the seed rate is too high there may be too many heads, producing thin, shrivelled grain. Seed is often broadcast, when more seed is used than with "drilled" (machine planted) barley.
Seed spacing: 15–25 cm between rows. 28–30,000 seeds per kg.
Depth: 3–5 cm in temperate or humid regions, 5–8 cm in arid and semi-arid regions.

GROWTH CONDITIONS

Day length: long-day, but adapts to varying day length.
Growth period: in general, barley matures earlier than most of the other cereal crops. The moisture content of seed should not be much above 13% at harvest, if practical. Winter sown varieties need about 180 days or more to reach maturity, spring varieties need about 85–120 days.
Rainfall: barley is more drought resistant than wheat, and can grow with 500 mm a year. Nevertheless, hot and dry conditions lead to premature maturation, leaving thin seed, with high Nitrogen content. Not suited to warm, humid climates (fungi). Irrigation increases both seed size and yield, and decreases the Nitrogen content.
Temperature: a cool season crop, barley can grow well vegetatively in cold weather. Very hot weather also can be tolerated during and after heading provided that the air humidity is low. Optimum temperature for germination and emergence is 15–20°C—the minimum is 2°C.
Rotation: often grown as a monoculture on the same land for many years, in which case fertiliser is needed to maintain reasonable yields. Can be used as part of an arable silage mix, with oats, beans, vetches and/or grass.

Pests: mainly the same insects that attack wheat:
Aphids—especially on young plants, and after long periods of dry weather.
Barley Fly—not found on wheat. Larvae eat growing points of seedlings, producing dead central leaves. Controlled with seed dressings.
Paddy Bugs—attack seeds. Partially controlled by removing nearby wild grasses, which are alternate hosts.
Hessian Fly—less harmful on barley than on wheat.
Pentatomid Bugs—attack seed at the milky stage. Controlled with insecticides.
Chinch Bugs—worst on poor, weak crops.
Rice Weevil—attacks stored seed. Controlled by heat treatment.
Diseases: also mainly the same ones that attack wheat. Stem Rust and Leaf Rust (*Puccinia* spp.) are the most serious, causing loss of yields and thin seed. Also Net Blotch, Spot Blotch, Barley Stripe, Powdery Mildew, Barley Leaf Blotch (*Rhyncosporium secalis)*, Smuts (especially Loose Smut *Ustilago*), Scald and virus diseases such as BYDV[1], BaYMV[2] and BSMV (Barley Stripe Mosaic Virus).

(1)   Barley Yellow Dwarf Virus. Main control is by controlling the aphid vector with pyrethroid sprays, seed—dressing (eg Imidacloprid based) or destroying any any green bridge.
(2)   Barley Yellow Mosaic Virus. An increasingly serious problem in Europe and elsewhere. The vector is a soil—borne nematode. There are some tolerant barley varieties. BaYMV also affects some wheat varieties.

YIELD
FAO's estimate of the global average for barley in 2004 was 2.68 MT/ha, varying from 0.5–1 MT/ha in low input systems to 3–8 MT/ha in higher input systems. The Netherlands reported the highest average yield in 2004 at 7.87 MT/ha while the lowest was Lesotho at 250 kg/ha.

The maximum attainable yield is well over 8 MT/ha. Some 6-row varieties can attain 10 MT/ha under very intensive management.

UTILISATION
* About half of the barley grown is used as **animal food**, the grain normally being mixed with other foods to produce animal food concentrates. Due to its low gluten content, barley is not suitable for making leavened bread, though unleavened barley bread is quite tasty.
* The **growing crop** can be used as fodder, and can also be grazed, especially when irrigated. If grazing is stopped early enough a reasonable grain yield can result—or the crop can be cut for hay or silage.
* Barley is the most important grain used for **brewing beer**, for which special malting varieties are needed. In order to be acceptable for malting the grain sample should be of specific quality regarding its Nitrogen content, size, germination, moisture and *mealiness.*
* Malting quality seed is normally more valuable than food quality seed.

- **Pearl(ed) barley**, where the outer husk and part of the bran layer is ground off the seed (leaving only the "pearl"), and also **naked barley**, is eaten by humans. Barley flour can be made from pearl barley, to make flat (unleavened) bread.
- **Hulled barley (barley groats)** is the least processed form of barley, with only the outermost hull removed.
- **Barley water** is made by soaking pot (Scotch) or pearl barley, often flavoured with lemon or orange.
- The **straw** can be used as fuel, or fed to animals; it has a low food value and is often more suitable than wheat straw as animal bedding, being both softer and more absorbent.
- Barley can be very useful where the soils are saline, though the **tolerance to salinity** varies with the variety.

LIMITATIONS
- Barley grain is not easily prepared and eaten as human food.
- Rough-awned types used for hay or silage may damage the mouths of livestock.
- The growing crop needs either regular rainfall or irrigation.
- It does not tolerate heavy, poorly drained soils, nor very light sandy soils. It is less tolerant of acid soils than wheat, and soils should be no more acidic than pH6.
- Plants tend to lodge very readily. In some modern agricultural systems this problem is reduced by applying hormone growth regulators such as Ethepon, Terpal, Cerone, Meteor, Moddus, etc. which either reduce stem length and/or thicken the stem walls.
- In dry climates much of the grain can be lost due to shattering, especially malting varieties.

Buckwheat
Fagopyrum esculentum (Syn. F. sagittatum, F. vulgare)

Beechwheat, Brank, Fagopyrum, French Wheat, Garden Buckwheat, Saracen Corn, Sarrasin, Boekweit, Ch'lao Mai, Hua Ch'lao, Qamh Al Baqar, T'len Ch'lao, Wu'Mai. Blé Noir (French); Buchweizen (German); Trigo (Grano) Sarraceno, Trigo Negro, Alforfón (Spanish); Fagópiro, Trigo Sarraceno, Trigo Mouro, Trigo Prêto (Portuguese)

Buckwheat belongs to the *Polygonaceae* family (the docks) and is therefore, strictly speaking, not a true cereal (*Gramineae*). However the grain is very similar to the true cereals and Buckwheat is normally regarded as a cereal crop. It is normally consumed more or less locally, though there is an increasing international trade. It is thought to have originated in China.

On good soils, Buckwheat is less productive than other cereals, but it is well adapted to arid, hilly land and cool climates. It tolerates acidic, heavy and poor soils, but they must be well drained. It also tolerates dry and arid conditions.

The kernels of the triangular shaped seeds are about 6mm long and enclosed by a tough, dark brown, black or grey rind. Seeds contain about 6% of the essential amino acid lysine and are also a source of Vitamin B_6. The flour (96% extraction) contains about 11% protein (very rich in lysine) and 2% fat, as well as some Vitamin B_1 and B_2.

Three other species of buckwheat are also grown: Tartary or Mountain Buckwheat (*F. tartaricum*) which thrives and produces crops in very poor soils and in a short growing season, Winged Buckwheat (*F. emarginatum* syn. *Eriogonum alatum*) a perennial with edible roots and seeds, and Perennial Buckwheat (*F. cymosum* syn. *F. dibotrys*) with edible seeds, and leaves that are rich in rutin.

The plant is an annual, 50–160 cm tall, with a single stem and many branches. The root system is only about 3% of the weight of the plant, compared with 6–14% of the true cereals. The flowers are normally white, with a pleasant honey smell, and pollinated by bees, hoverflies and other insects.

Buckwheat grows best in cool, moist climates, though it is frost sensitive. It is sometimes grown in India and at high altitudes in the tropics.

The global production of buckwheat was 2.9 million MT in 2004, according to FAO. Russia is the main producer and it is also grown widely in France, Poland, Ukraine, Belarus, Austria, Germany, Romania, Canada, Brazil, USA, China and Japan.

PLANTING

Propagation: early sowings are for seed or leaf crops, later sowings are used mainly for leaf crops or green manure.

Soil: should be light and well drained. Buckwheat grows well on poor, sandy and acidic soils, and often does not respond well to fertilisers. However, phosphatic fertiliser may increase yields significantly in poor soil. Too much Nitrogen causes weak plants, and therefore exacerbates lodging. Good drainage is important. Tolerance to soil acidity similar to oats and potatoes.

Seed rate: 40–66 kg/ha. 40–50,000 seeds per kg. It self-seeds freely.

Seed spacing: buckwheat seed is very often broadcast. 7–15 cm between plants.

Depth: 1–3 cm. Germination takes about 5 days.

Rotation: it removes large amounts of nutrients from the soil, and should ideally follow a well fertilised crop, and be followed by a winter (ie autumn sown) crop which is well fertilised.

GROWTH CONDITIONS

Day length: there are both long-day and day-neutral varieties.

Growth period: about 100 days. A crop of leaves can grow in 8 weeks. The seed ripens irregularly, so it is difficult to harvest.

Temperature: buckwheat prefers a cool climate, though the plant at all stages is killed by frost. Seed set is poor at high temperatures. Seed will germinate between 7°C and 41°C, though the optimum is about 26°C.

Pests and diseases: these are rarely a problem. Leaf Spot (*Ramularia* spp.) and Root Rots (*Rhizoctonia*) can occur. Wireworms and aphids sometimes cause some damage. Pythium rot is especially virulent when the plants are in standing water, but can be cured with drainage, or fungicides such as Apron.

YIELD
Buckwheat yields globally in 2004 were estimated by FAO at 1.09 MT/ha. The maximum yield is about 5 MT/ha.

The highest and lowest average yields reported by FAO in 2004 were 3.5 MT/ha in France and 300 kg/ha in South Africa respectively.

UTILISATION
- **Buckwheat seed** has excellent quality protein and is normally ground into flour for cakes, biscuits, pancakes, soups, porridge, pasta and dumplings. The residue is suitable for animal food. The grain is also fed whole to poultry and game birds, but for other animals it is ground up. High quality protein content of 10–11%, 70" carbohydrate, 2% fat. It makes great beer.
- The **hulled kernels**, or "groats", are prepared like rice, called *kasha* in Eastern Europe and *sayraisin* in France. The flour does not make good bread but can be used—either alone or mixed with wheat flour or soybean flour—to make griddle cakes. Noodles called *soba* are made from it.
- The buckwheat crop makes a useful **green manure, smother crop** or **catch crop**. It is used to reclaim badly degraded soils and subsoils, and is said to reduce grass weed and winter wheat populations.
- It is a good source of **honey**; the flowers remain on the plant for 30 days or more. A brown dye is extracted from the flowers, and a blue dye from the stems.
- The **leaves and flowering stems** are widely used in medicine, normally in conjunction with vitamin C to aid absorption. They should be stored in the dark, and used with care as they have been known to cause light-sensitive dermatitis.
- **Rutin**, a flavanol glucoside used in vascular disorders associated with hypertension, is obtained from the leaves, stems, flowers and fruit. It dilates the blood vessels, reduces capillary permeability and lowers blood pressure. Rutin is also found in black tea and apple peels.

LIMITATIONS
- Buckwheat plants are "heavy feeders" ie they take up and remove large amounts of nutrients from the soil.
- Buckwheat crops may increase erosion as they can leave the soil more loose, and therefore more unstable, than other small grain crops.
- The plants are frost sensitive.
- The seeds do not develop uniformly, leading to harvesting problems, volunteer plants in the following crop, etc.
- An irritating skin disorder can appear on white or light-coloured skins if buckwheat is consumed in large quantities, especially when the skin is exposed to sunlight.
- Plant breeders have great difficulties in developing widely adapted improved varieties.

Maize
Zea mays

Corn, Mealies, Indian Corn, American Corn
Mais, Blé de Turquie, Turquet (French); Kukuruz, Turkischer, Weizen (German)
Milho (Portuguese); Mijo Turquesco, Maíz, Zara, Trigo de Turquía (Spanish)
Makka, Makai, Butta (Hindi); Mahindi (Kiswahili); Jawar{i} (Pashtu and Dari)
Ufun, Elbo (Tigrinha), Boqqoolloo (Oromifa), Bokkollo (Amharic); Epungu
(Angola)

Maize is one of the three major cereals in the world, together with rice and wheat. About half of the global crop is produced in North America; China is the second largest producer, then Brazil, Mexico and Argentina. Global production in 2004 was 721 million MT (FAO estimate).

In nutritional terms, maize grain is mainly useful as a source of carbohydrate and energy. 100 g of whole maize grain (at 12% mc) contains on average 362 calories, 71 g carbohydrate, 10% protein and 4.5% fat. Immature grain ("corn on the cob") is even less nutritious.

Although it is not a reliable crop for regions with limited or erratic rainfall, where sorghum or millet normally grows better, in good rainfall areas or under irrigation maize has a greater yield potential than any other cereal. Despite the unreliability of maize in low rainfall areas it is often planted by farmers who accept the risk of producing only a small yield, for two reasons: the green cobs help to fill the "hungry gap", and they prefer the taste and cooking qualities of maize to sorghum or millet (ie quality is rated higher than quantity).

Plant breeders have selected and developed maize varieties that can adapt to almost any environmental conditions. Some maize plants are 70 cm tall and mature in 50 days, others are 4 m tall and need more than a year to mature. This wide range of different plant and grain types is grouped into seven main types according to the nature of their endosperm; there are also several intermediate types, the "Semi-Flints", etc.

Dent (Horse-tooth) Maize
Zea mays var. *identata*. Large grains, normally yellow or white, with soft white starch which shrinks on drying to produce the characteristic "dent" at the end of each grain. This type has the biggest yield potential and is the most widely grown.

Flint Maize
Zea mays var. *indurata*. Compared to dent maize, they are usually earlier to mature, their grain is rounded and without the dent ie hard endosperm only. The growing plant is more likely to produce tillers, which is not a good habit because maize normally only produces grain cobs on the main central stem. On the other hand they are better adapted than dent types to growing in difficult conditions.

Soft (Flour) Maize
Zea mays var. *amylacea*. Grown in the drier parts of Western South America, North America and South Africa. The grain colour is very varied, and more rounded than dented. Soft endosperm.

Sweetcorn Maize
Zea mays var. *saccharata*. Less sugar (sucrose) is converted into starch than in other maize types, so the grain is sweeter. Not only that, it makes the best and the strongest beer. It is mainly grown in North America.

Popcorn Maize
Zea mays var. *everata*. There are two types, both mainly grown and eaten in North America; one, the rice popcorn, has pointed grains, while the pearl popcorn has very compact, rounded grains.

Podcorn Maize
Zea mays var. *tunicata*. This is the most primitive form of maize and is not grown commercially other than as an ornamental. It was this form of maize pollen found in Mexico that is at least 80,000 years old. Wild maize and also the earliest cultivated maize were podcorn types. The grain is enclosed in glumes, like other cereals.

Some South American Indians believe that podcorn maize has magical properties—as a result of a genetic throwback, plants of this type occasionally appear in a field of regular maize.

Waxy Maize
Zea mays var. *ceritina*. The starch of the grain is 100% amylopectin, while the other non-waxy types have a mixture of amylose and amylopectin. It is grown mainly in East Asia, and in some other areas for use as a substitute for starch.

POLLINATION OF MAIZE

The maize plant is one of the few in the plant kingdom in which the male and the female are situated on different parts of the plant. The male *tassel* appears on top of the plant, and produces an astonishing 2–5 million pollen grains—up to 20 million on a big tassel. These pollen grains drift off into the air until they make contact with a style, or *silk* from the female part of another maize plant. On the same plant, a few days later, the female inflorescence appears, somewhere in the middle of the stem. This female part consists of up to 1000 ovules, which can be fertilised by the pollen when it comes in contact with the silk. Normally only one or two cobs develop per plant, though some very prolific varieties, normally hybrids, can produce several cobs.

Maize is thus mainly cross-pollinated, as explained above, though because the plant is still producing some pollen when the styles appear some self-pollination does inevitably occur, but normally in no more than about 5% of cases.

Cross-pollination can occur between plants that are very widely separated, as the wind can carry pollen for great distances. Because of this habit it is difficult to maintain for any number of seasons a variety of maize which is genetically uniform. So farmers who "grow on" their own supply of seed every year have large and diverse collections of germplasm and plant types, known as composite varieties (**1Fd**). These farmers can be quite confident that these plants, or most of them, will succeed next year because during previous years they have selected seed from the best, most highly adapted, plants.

In this way man has selected for thousands of years maize (and other crops), which are best suited to grow in the particular area where they were farming.

However, locally produced varieties have a limited genetic resource, and introduced varieties, with years of international plant breeding effort behind them, can very often grow even better.

HYBRID MAIZE

Because of the maize plant's habit of bearing the male and female parts separately, the production of hybrid maize seed on a field scale is relatively simple. The tassels of the female parent of the hybrid are cut off and removed, so that they cannot self-pollinate, then pollen from the male parent is taken and introduced to the styles of the female parent.

The production of hybrid maize seed, using both this method and male-sterile plant material, is an enormous business that has been hugely successful in increasing maize yields all around the world. Chemical sterilisation to prevent pollen formation is not yet 100% effective.

However, as discussed in **1Fc**, page 52, *hybrid seed can bring its own problems. Note to the following:*

The agronomy of maize is an enormous subject, which has been very thoroughly discussed in hundreds of books, scientific papers, etc. The following is a brief and incomplete summary.

PLANTING

Propagation: by seed. In general, earlier planted crops perform better than later planted ones, provided that there is adequate moisture and the soil temperature has reached about 12°C. It is usually a good idea for food producers to plant at least some of their maize early—those plants may not only grow better than later planted ones, but they can be eaten sooner, and so "fill the hungry gap".

Germination: seedling emergence in moist soil of about 21–30°C is 4–5 days; in 12–16°C soils it can take 15 days or more. Maize seed remains viable for 3–5 years if stored carefully (**1O**).

Soil: maize adapts to most kinds of soil, from pH 5–8, though like most other crops it prefers well drained loams of pH 6–7. In alkaline soils there may be symptoms of iron or manganese deficiencies. It is a heavy feeder, and needs plenty of N, P and K, Potash probably being the most important major element. It responds well to direct placement of fertiliser (especially P) next to the seed. But although maize can and does respond well to fertiliser there are other less expensive and troublesome ways of increasing maize yields, such as using appropriate varieties, spacing the plants more sensibly and accurately in the field, and timely planting. Indeed, farmers are wasting their time, money and effort if they apply fertiliser to maize without first attending carefully to these factors.

Seed rate: from 10–45 kg/ha or more. In general, rates are higher for hybrids and early varieties, for early planting and for moist, fertile soils; rates are lower for later, tall varieties in poor dry soils, as discussed in **1Fb**, page 52. For forage maize, about 50,000 seeds/acre are planted.

Seed spacing: this is an important consideration because correct maize plant spacing can significantly increase yields. Unlike other cereals maize is not good at compensating for different plant populations, which can vary from between about 15,000 and 90,000 plants per hectare. In general the plant population should be higher in good growing conditions, but if there are too many plants they may *lodge* (bend and fall over), leading to lower yields, more harvesting problems and damage by termites, fungi, pests, etc. Plant Population is also discussed on page 4.

The seed company that supplies the seed should be able to advise on the appropriate spacings for their varieties in different growing conditions.

Two or more seeds are often planted per hole, or *station*, but it is very important that only the strongest plant is allowed to grow per hole. There are about 13–40,000 seeds per kg.

Depth: 3–5 cm in moist soil, 5–10 cm in drier soil. There are certain varieties that have been bred especially for growing in dry, sandy soils and which are sown very deep.

Intercropping: very common. Legumes are the obvious choice, and this was done thousands of years ago, in South America for example where maize is still traditionally grown with beans.

Rotation: maize is often cropped continuously, but this is only acceptable when large amounts of fertiliser and high level management is practised. Nevertheless, continuous cropping of maize has less negative effect on yield than continuous cropping of many other crops.

GROWTH CONDITIONS

Day length: short-day. Long days increase the duration of the vegetative stage, the plant size, and number of leaves.

Growth period: varies from about 50 to 365 and more ... Averages are 90–120 days at low altitudes, and 180–240 days at approximately 2500 m above sea level.

Temperature: even a light frost can kill maize plants, so it normally needs about 120 frost-free days. For germination, 18–21°C is optimum—it is very slow below 13°C, and does not germinate below about 10°C. The ideal temperature at tasselling is 21-30°C.

Rainfall: in temperate or subtropical regions 450–600 mm during the growth period is enough; in the tropics it needs 600–900 mm. A very dry spell just before or during tasselling is bad news for the plant, reducing yields. Maize responds very well to irrigation, if water is a limiting factor to plant growth. If only one irrigation is possible it should be at silking or tasselling. In arid regions, fast growing varieties such as Kalahari, Katumani or Kito can be grown.

Altitude: 0–3300 m (in Mexico and the Andes).

Pests: maize is attacked by more than 200 different insects, some of which are described below. Plant breeders have recently made good progress in developing varieties resistant to multiple species of insects.

Stem (Stalk) Borer—there are at least 4 spp.—*Busseola fusca* is an important one, found mainly in higher altitudes. The adult moths lay eggs on leaves, between the edges of the leaves and the stem; the larvae eat some of these leaves, leaving "shot-holes" or "windows" as the leaves open, then enter the stem and feed near the growing point. Later on they enter and eat the developing cobs. Control is very difficult if the larvae have already entered the stems, but some of the systemic insecticides give some control. Some varieties are said to have some resistance. Burn all infected plant material.

Sorghum Borer—very similar to Stem/Stalk Borer, but chemical control is very uncertain. Burn all infected plant material.

African Armyworm—the adult moths of *Spodoptera exempta* can fly for hundreds of kilometres and their larvae can cause devastating damage, normally at the end of the maize growth period. All of the leaves can be eaten, leaving only the midribs.

Lesser Armyworm—much less damaging, and easily controlled with Malathion.

Earworm(Bollworm)—eggs of *Heliothis armigera* are laid on the silks, and larvae feed on them and the developing grain.

Maize Leafhopper—transmits Mosaic Streak Virus (MSV) and other virus diseases. Some hybrid varieties have some resistance.

Cutworms—a serious soil pest, especially in fields which were weedy before land preparation. Seedlings are attacked, often bitten completely through. Some control is possible by preparing the seedbed six weeks or more before planting, and then keeping weeds well under control, or by using baits of wheat bran or maize meal.

Diseases: the global yield loss from maize diseases has been estimated at almost 10%, mainly caused by:

(White) Leaf Blight—*Helminthosporium*. Oval, grey papery lesions on the leaves, which may die. Control by using resistant varieties.

Maize Streak Virus—spread by a leafhopper, more serious on late planted crops.

Downy Mildews—*Sclerospora* spp. - pale yellow streaks on the upper leaves, which become brown and necrotic. Plants are stunted and may produce nothing. Control is difficult; early planting may help, some varieties have some resistance, and systemic seed dressings are also available.

Crazy Top—*Sclerophthora macrospora*, a form of Downy Mildew in which the tassel and/or ear develops a proliferation of bizarre, unproductive growth.

Common Maize Rust—symptoms appear at tasselling time, or on the seedling; leaves become covered with tiny brown spots, and then dry up. Not a serious problem.

Maize Smut—Galls form all over the plant, which then produce black spores. The best control is by cultivating resistant varieties.

Head Smut—ears and tassels are converted to shoot-like growths, and the plants are stunted. Controlled with seed dressings.

Eyespot—decimates the leaves; can be a problem in Europe and elsewhere.

YIELD

According to FAO the global average for 1988/90 was 0.6–4.9 MT/ha, for the bottom 10% and top 10% of producer countries respectively. In 2004 the global average was 4.9 MT/ha, Israel recording the highest (16.0 MT/ha) and Botswana the lowest (119 kg/ha).

The yield potential, using hybrids in optimum conditions, is well over 20 MT/ha.

UTILISATION

More than 500 products are obtained from maize, which is used in three main ways:

- **Human food**, especially in the tropics and in Africa. Maize grain lacks gliadin, one of the key proteins of gluten, and so cannot be made into leavened bread. Its protein is poor quality and is deficient in niacin. Green maize ("corn on the cob") is less nutritious than mature maize. So-called "baby corn" is harvested before fertilisation and contains about 90% water, and very little fat, protein or carbohydrate.
- **Animal food ("stover")**; maize supplies two thirds of the total trade in food grains. Compared with other cereal grains, maize has less fibre, less protein (and of lower value) and fewer minerals, but has a higher net energy content and is more easily digestible. The whole plant is often used as a forage crop in Northern Europe, where cool climates limit the efficient production of crops of maize for grain.
- **Raw material for industry**: both the grain and cobs are used, in adhesives, explosives, textile sizing, dyes, plastics, chemicals, paper and wallboard, paints, maize starch, for brewing and distilling, etc.

LIMITATIONS

- Maize will only produce good yields in relatively fertile soils and with careful management. The plant is less drought tolerant than most millet and sorghum varieties.
- Both the plant and grain are susceptible to insect and disease attack.
- The plant is sensitive to frost.
- The protein is of poor quality, with a very low content of lysine and tryptophan. It is deficient in niacin (also known as nicotinic acid), and thus diets in which maize predominates, and where there is little protein, often result in the niacin deficiency disease known as pellagra[1].
- The nutritional value of maize is inferior to most of the other cereals.
- The elastic protein (gluten) of maize is comparatively poor quality, and maize is not used to produce leavened bread.

[1] **Pellagra** is characterised by skin lesions and gastrointestinal and neurological disturbances, causing the so-called "3Ds" of pellagra: dermatitis, diarrhoea and dementia.

The disease is seldom a deficiency of niacin alone - recovery usually follows a treatment of multivitamins, or a well-balanced diet for mild or suspected cases of pellagra.

Maize grain is low in both niacin and tryptophan, an amino acid that is converted by the body into niacin. Certain foods such as milk and eggs protect the body from pellagra even though they are low in niacin itself, because they have a high proportion of tryptophan.

The native Americans in Mexico and Central America developed a way treating maize with lime when making tortillas which makes the niacin available.

Millets

The millets are various grass crops that are harvested for animal and human food. Sorghum is known as "millet" in parts of Africa and Asia.

Compared to other cereals, millets are mainly suited to less fertile soils and poorer growing conditions, such as intense heat and low rainfall. These poor soils may be deficient in one or more trace elements, in which case yields of millets will be reduced.

There are eight different types and many species of millet, cultivated in the warmer regions of the world. They are mainly used as a human food grain crop, but are also used for hay and forage (eg Guinea Grass, *Panicum maximum*) and for making beer, pombe and other drinks.

The FAO estimate of the global production of millets for the 2004 season was 29 million MT, ranking them as the 6th largest of the cereals, after maize, rice, wheat, barley and sorghum. Millets are widely cultivated in parts of Asia, Africa, China and Russia.

The eight most commonly grown species for human food are listed below; the first three millets are described in more detail later on. These eight species are all members of the *Poaceae* (alt.*Graminae)* family, and are all in the *Paniceae* sub-family except for Finger Millet which is in the *Chlorideae* sub-family:

> **Pearl** or **Bulrush Millet**—*Pennisetum typhoides* (Syn. *P. americanum, P. glaucum, P. spicatum*)
> **Foxtail Millet**—*Setaria italica* (Syn. *Panicum italicum, Chaetochloa italica)*
> **Finger Millet**—*Eleusine coracana*

Japanese (Barnyard) Millet—*Echinochloa frumentacea* (Syn. *E. crusgalli*). Also known as Sanwa Millet and Billion Dollar Grass. An awnless annual 60–120 cm tall, grown for forage, hay and grain. It is the fastest growing millet, some varieties maturing in six weeks. Grown mainly in SE Asia and the USA, with cultivation methods similar to those for Foxtail Millet—see later. Often grown in Egypt as a reclamation crop on land too saline for rice. Grain is light brown to purple. Jungle Rice *E.colonum* is a related species, usually considered to be a weed.

Browntop Millet—*Panicum ramosum* (Syn. *Brachiaria ramosa*). A native of India, it is a fast growing annual 60–120 cm tall, mainly grown for hay or forage, or for wild bird food. Its seed shatters readily and so can become a weed on arable land.

Common or **Proso Millet**—*Panicum miliaceum.* Also known as Panic Millet, Broomcorn Millet, Brown-corn Millet, Hog Millet, Hershey Millet, Russian Millet, India Millet and Cheena (Hindi). A very drought resistant fast growing (60–80 days) annual about 75 cm tall, grown as a human grain crop and also as fodder and bird seed. Good potential as a catch crop in hot dry areas with poor soils, but does not grow well on coarse, sandy soils. Reputed to need less water to grow than any other cereal. Mainly self fertilised, but some cross-pollination does occur.

Kodo or **Koda (Ditch) Millet**—*Paspalum scrobiculatum* var. *commersonii* and var. *scrobiculatum.* Grown on light soils, mainly in India for forage. Hardy and drought resistant, but low yields.

Little or **Kutki Millet**—*Panicum sumatrense* (Syn. *P. miliare*). Grown throughout India but not very important elsewhere. Can survive on poor soils that are unsuitable for any other cereal.

PEARL (BULRUSH) MILLET

Pennisetum typhoides (Syn. *P. americanum, P. glaucum, P. spicatum*)

Candle Millet, Spiked Millet, Cat-tail Millet, Dark Millet, Mands Forage Plant, Penicillaria.
Mil Perle, Petit Mil, Mil à Chandelles (French); Perl Hirse, Beger Hirse (German);
Panizo Negro, Panizo de Daimiel, Mijo Perla, Mijo Candella (Spanish);
Painço (Portuguese—grain & plant); (M)Assango, O'huwe (Angola); Bajra, Bajri, Cumbo, Sajje (India); Bultug (Tigray), Dehun, Zengada (Ethiopia), Sanio, Gero, Babala, Nyoloti, Dukkin, Souna (Africa).

Pearl Millet is the most widely grown of all millets, and is the staple food in many of the drier parts of Africa and India. It is more of a drought-avoiding crop than a drought-resistant one, in that it can grow very quickly and so needs little moisture. Although it cannot tolerate drought for long periods in the way that sorghum can, it is a dependable crop which generally produces at least some yield every year in areas where maize and even sorghum sometimes fail.

It is an erect annual grass, 0.5–5m tall, often producing many tillers especially when thinly spaced (low plant population) in fertile soil. The panicle (seed head) is 15–50 cm long, about 2.5 cm in diameter, very compact and light brown in colour. The seed is normally light brown, sometimes near white or even black. There are about 180–190,000 seeds per kg.

Pearl millet is often "dry-planted" ie before the rains begin, both to take advantage of the "flush" of Nitrogen that occurs with the first rains and also so as to utilise the entire rainy season, which may be very short.

In the driest and most infertile soils in India and Africa pearl millet is very often the most productive grain crop; if sorghum crops frequently fail, pearl millet can often be a good substitute.

Some semi-dwarf hybrid seed is produced, but mainly for forage varieties. Pearl millet is normally cross-pollinated.

PLANTING
Soil: sandy loams are best. Although it will grow in infertile soils that are too light for sorghum it does respond well to fertiliser. Nitrogenous fertilisers such as sulphate of ammonia are especially beneficial, applied at about 200 kg/ha, followed by a top dressing of about 100 kg/ha 4–6 weeks after planting.
Seed rate: 3–9 kg/ha for grain/seed production, 10–22 kg/ha for hay or forage. Low plant populations are acceptable in fertile soils as the plants compensate by producing many tillers.

Seed spacing: often interplanted with other cereals or legumes. Several seeds are often planted in stations 1.5–2 m apart, which should be thinned later. Sometimes a second crop is interplanted between plants from the first sowing.
Depth: 2–3 cm.

GROWTH CONDITIONS

Day length: there are both short day and day neutral varieties.
Growth period: harvest starts after about 90 days and continues for several weeks.
Temperature: pearl millet tolerates higher temperatures than sorghum—32°C is optimum. Heat is needed for growth, especially during and after flowering, though some varieties tolerate light frost.
Rainfall: it requires less water than most other grain crops, and can grow well with 400 mm a year. Even at 250 mm a year it can produce some yield if the rains are well distributed, or if soils are deep or retentive, and if evaporation is not too high.
Weeds: it can sometimes be grown in Striga-infested land (**1L**) which can no longer support maize or sorghum. The plants benefit from careful weeding in the early stages.
Rotation: ideally it follows groundnuts or other legumes.
Pests: pearl millet is often badly damaged by birds. The main insect pest is Stem Borer (*Coniesta ignefusalis*), and several caterpillars and grasshoppers also damage its leaves. Stored seed can be protected with insecticides. Bird damage, such as by *Quelea*, can be devastating.
Diseases: Downy Mildew (Green Ear)—probably the most serious disease, though some varieties have some resistance, Smut—the panicles become blackened, Ergot - the grains become purple and enlarged, and should not be eaten, by either man or animals, and Rust—leaves become infected, but little damage is caused. Rain at flowering may lead to Honeydew (Sugary) Disease.

YIELD

Pearl millet yields are very variable. The global average yield of pearl millet is about 850 kg/ha, though yields of 200–300 kg/ha are common.

3 MT/ha or more is possible in fertile soil with irrigation. More than 17 MT/ha of dry forage is possible.

UTILISATION

- **Grain:** normally eaten as a porridge, cake or unleavened bread. It is sometimes fed to animals, but is more commonly used to make beer. It stores well.
- **Stems:** useful for building material, fencing and fuel.
- **Whole plant:** grown for grazing and for hay, and sometimes for silage(eg in North America).

LIMITATIONS

- Yields of pearl (bulrush) millet are normally rather low.
- The seed heads are very susceptible to bird damage.
- The stems and leaves (haulm) is woody and not very digestible to animals.

FOXTAIL MILLET
Setaria italica (Syn. *Panicum italicum, Chaetochloa italica)*

Italian Millet, German Millet, Hungarian Millet, Siberian Millet
Sétaire d'Italie, Millet des Oiseaux (French); Borstenhirse (German);
Kakun (Hindi); Dana, Mijo Menor, Panizo Comun (Spanish);
Painço (Portuguese—grain and plant)

It is thought that this very ancient crop originated in China where it was grown in the north-west of the country 5000 years ago. It is now the most important millet in Japan.

It is an annual plant, 30–200 cm tall, with more slender and leafy stems than pearl millet. The panicle colours vary from creamy white, pale yellow, orange, reddish orange to dark purple—or mixtures of these colours. Unlike pearl millet the panicles often curve downwards; they are 7–25 cm long, 1–5 cm in diameter. The flowering period lasts for 10–15 or more days; they are mainly self-pollinated though cross-pollination also occurs.

The mature grain is up to about 2 mm long, with the husk (the lemma and palea) tightly held onto the testa. Grain colour varies from white, pale yellow to orange, red, brown or black.

Foxtail millet grows very quickly—it can mature in 10 weeks or so—and so can be very useful as a catch crop. The plants look very much like the 3 common weeds, the yellow, green and giant foxtails. It is mainly grown in India, Japan, China, southeast Europe, North Africa and America.

PLANTING
Soil: good drainage is important. Foxtail millet adapts to a very wide range of soil types including infertile ones.
Seed rate: 4–6 kg/ha when planted in rows, 11–17 kg/ha when broadcast on to clean land, 27–33 kg/ha when broadcast onto poor or weedy soil. There are about 450–500,000 seeds per kg.
Spacing: often intercropped with legumes or other cereals. As a monoculture, the plants should be closely spaced, 5–6 cm between plants and 20–30 cm between rows, to smother weeds.
Depth: foxtail millet seeds are small and should only be planted into moist soil; up to 3 cm deep if necessary, for example in sandy soil.
Rotation: best if grown after legumes or another small grain or maize. In dry regions some crops yield poorly after foxtail millet, unless irrigated and/or fertilised.
Intercropping: this is commonly done, with legumes, cotton, finger millet and other cereals.

GROWTH CONDITIONS
Growth period: the grain matures in 75–120 days. Hay can be made from some varieties in 55–65 days. This fast growth can make it useful as a late-planted catch crop (**1Gc**).

Temperature: warm weather is needed for plant development.

Rainfall: it is more of a "drought avoiding" species than a truly "drought resistant" one, by virtue of its rapid growth. It has a shallow root system, so it wilts readily, and also does not recover well from long periods of dry weather.

Altitude: up to 2000 metres ASL.

Weeds: the seedlings compete poorly with weeds, so weeds should be controlled in young crops.

Diseases: normally not a big problem as Foxtail Millet is normally grown in arid regions, but the following diseases can reduce yields to some extent: mildews, leaf spots, green ear, smut and bacterial blight. Seed dressings can give some control, but this is rarely done.

YIELD

The average yield of foxtail millet is between about 400 and 900 kg/ha of grain or seed, and between two and twelve MT/ha of hay.

UTILISATION

- **Grain**—used by humans, either cooked whole, ground into a flour or made into beer. It can also be fed to animals, after it is ground up thoroughly, or to caged birds.
- **Whole plant**—makes quite good hay and silage, suitable for cattle and sheep.

LIMITATIONS

- Yields of foxtail millet tend to be low.
- Weeds must be kept well under control.
- There is a shortage of improved varieties, and seed of any kind is often hard to source.
- The hay and silage is not popular with livestock.

FINGER MILLET
Eleusine coracana

African Millet, Birdsfoot Millet, Korakan or Coracan Millet, Indian Millet, Ragi (Raggee) Millet, Nagli;
Eleusine, Coracan, Millet de Yokohama (French); Korakan, Ragihirse, Afrikanische (German); Coracán, Ragi, Mijo Africano, Mijo Coracana, Mijo Digitado (Spanish); Milheiro (Portuguese—general term for "millets"); Wimbi (Kiswahili); Ragi, Koracan, Maruwa (India); Bulo (Uganda); Telebun (Sudan); Dagoosha (Tigrinha), Daguussa (Amharic and Oromifa); Oluko, Kaluku (Angola)

Finger millet is an important staple food in parts of Central and East Africa and India. In parts of Uganda it is the most important cereal. It is either prepared and eaten like most other millets or made into beer, for local consumption. It often produces some yield where other cereal crops would have failed due to drought.

The plant is a short, hardy annual, about 1–1.5 m tall, which is normally self-pollinated. It is less susceptible to bird damage than the other millets or sorghum.

In parts of Africa and India many different varieties are available, which are often grown together in the same field.

The seed, which is sometimes yellow or white but is more normally brown or red in colour, is 1–2 mm in diameter, with about 300,000–450,000 grains per kg.

One of the best attributes of finger millet is that its seed can remain viable for up to ten years if it is stored in a cool, dry place. In marginal conditions, finger millet is stored in this way for use in emergencies such as famine. The seed is a rich source of calcium (0.33–0.36%), as well as phosphorus and iron.

PLANTING
Soil: a wide range of soil types are suitable, though reasonably fertile sandy soils are best. Finger millet responds well to fertilisers (and good management); average applications are, in kg/ha, N 25–90, P 20–45, and K 0–45.
Seed rate: 20–30 kg/ha broadcast, 5–10 kg/ha sown in rows.
Seed spacing: there are 400–500 seeds per gram. Often mixed with legumes or sorghum. In a monoculture, the rows are about 45 cm apart. Five or six seeds are often planted in groups at 20 cm intervals along the row, and thinned later.
Depth: 1–2 cm (3 cm maximum). The land should be well prepared.
Intercropping: very common, with pigeon peas and other legumes, sorghum, Niger seed, etc.

GROWTH CONDITIONS
Growth period: there are 3 types of finger millet: early (95–100 days), medium (105–110) and late (115 or more days). Many older varieties mature in 125 days or more.
Rainfall: 600–1300 mm/a, which should be well distributed due to finger millet's root system, which although quite extensive is shallow. The plants respond well to irrigation, but do not like heavy rainfall.
Altitude: 0–2000 metres. In Africa, mainly above 1000m.
Pests: the plants can harbour the wheat curl mite, a vector of wheat streak mosaic virus, so wheat should not follow a finger millet crop.
Weeds: young finger millet plants (2.5–5 cm tall) should be well weeded. The wild species, *Eleusine africana*, is identical to finger millet before flowering and can become a big problem.
Pests & diseases: these are rarely any problem, but Helminthosporium diseases can cause leaf spots, seedling blight and head blight. Grain smut has been reported.

YIELD
The average for finger millet is about 1 MT/ha, though more than 4 MT/ha is quite possible. In Africa the whole head is usually cut off with a knife and stored until

needed. In India the whole plant is either pulled up or cut off at ground level, then stacked in the field until dry.

UTILISATION
- Finger millet is used in much the same way as pearl and foxtail millets. In parts of North and Central Zambia finger millet is the main cereal crop. Unlike sorghum and pearl millet, finger millet is not normally attacked by birds.
- It is frequently malted and made into beer.
- One great advantage of finger millet is that it can be stored, even in poor conditions, for up to about 10 years without serious deterioration or weevil damage, and so it can serve as a valuable famine reserve food.

LIMITATIONS
- There are very few problems with finger millet, though yields are often low. Threshing and cleaning the grain, seedbed preparation and weeding the young seedlings can be very labour intensive.
- The young seedlings are easily overcome by weeds or drought.
- The protein content is relatively low, about 5–6%.
- Very limited research work is done on this potentially very useful crop.
- Most of the improved varieties which are currently available are not adapted to grow well in marginal conditions.

Oats
Avena sativa

Dousar, Groats, Common Oats, Steel-cut Oats, Cat Grass, Rolled Oats
Avoine (French); Hafer, Saathafer (German); Avena (Spanish); Aveia (Portuguese);
Jaie (Hindi); Yulaf (Persian); Gandiala Ahelee (Pashtu);
Addja (Amharic, Ethiopia); Omborrii (Oromifa, Ethiopia)

Oats are grown throughout the world in the cool, damp parts of temperate regions, or at high altitudes in the tropics. They are one of the most important temperate cereals and can be grown in a wider ecological zone than wheat or barley. FAO estimated that the 2004 global production was 26 million MT, making oats the seventh most important of the cereals in terms of production. The main producers are the USA, Belarus, Russia, Kazakstan, Canada, France, Poland, Finland, Germany and Australia. The production of oats has decreased in many countries as a result of increasing mechanisation which is replacing horses, the main consumer of oats.

Oats are mainly used as animal food, both the grain and straw being eaten. They are commonly consumed on the farm, so a high proportion does not reach the marketplace.

The plants have large, open, spreading panicles (seed heads) with large spikelets, both awned and awnless (without awns) varieties They are all members of the Poaceae family and are 99% self-pollinated (cross-pollination is by wind).

The grain is high in carbohydrate and contains more protein than other cereals (up to almost 20%) and 5–9% of highly unsaturated fat. It is a source of calcium, iron, vitamin B_1 and nicotinic acid.

Two other cultivated species of oats are found:

- *Avena byzantina*—Red or Algerian Oats—of Asian origin, it is more resistant to heat and is grown in the warmer regions of North Africa, Argentina and southern North America.
- *Avena abyssinica*—Abyssinian Oats—grown mainly in Ethiopia. Normally intercropped with barley, the two crops being planted, harvested and eaten together. Plant breeders have crossbred Red Oats (*A. byzantina*) with Yellow or White Oats (*A. sativa*) to produce a number of new cultivars.

PLANTING

Soil: of all the cereals, oats are second only to rye in their ability to survive in poor soils. Best in neutral silt and clay loam, though oats can grow on moderately acid soil where wheat or barley would fail. Waterlogged and high N soils can lead to lodging, and oats are more susceptible to this than the other cereals. Moderate tolerance of saline soil. Fertiliser requirements similar to wheat and barley, but less N is required.

Rotation: oats should not be followed by another cereal, especially rye (or oats).

Seed rate: In the sub-tropics, 30–70 kg/ha on dryland, 50–80 kg/ha irrigated.
In temperate regions, 150–220 kg/ha for winter types, 190–250 kg/ha for spring types. 28–35,000 seeds per kg.

Seed spacing: 18 cm between rows.

Depth: 4–5 cm.

GROWTH CONDITIONS

Day length: long-day.

Growth period: 180–220 days. Harvest should be just before the seed is fully ripe; oats shatter easily, so a lot of seed can be lost on the ground.

Rainfall: about 750 mm per year is the minimum. Irrigation is most effective at flowering.

Temperature: oat plants can be damaged by frost in very cold winters. Hot dry weather before heading can cause seed loss.

Germination: storage conditions and seed viability for oats are similar to wheat.

Rotation: oats can be suitable as a pioneer crop, the first crop sown after breaking in new land. Best after a root crop and not after another cereal, especially rye. They should not be grown on the same land for too many years as eelworms can build up. The plants are more prone to lodge if grown after a legume or if the soil has a high Nitrogen content.

Pests: more resistant to insect attack than wheat or barley. Most problems are caused by: armyworms, grasshoppers, leaf hoppers, crickets, grain bugs and frit fly.

Diseases: seed diseases include loose and covered smut, and stripe, all controlled by seed dressings. Foliar diseases include rusts, especially *Puccinia* spp., controlled with resistant varieties. Powdery Mildew (*Erysiphe graminis*) can also be a problem.

YIELD

The global average yield of oats in 2004 was 2.2 MT/ha according to FAO, though 7 or 8 MT/ha is possible in good growing conditions.

The UK reported the highest average yield for that year (6.0 MT/ha) and Israel the lowest (363 kg/ha).

Harvesting should be done in good time (early) to minimise shattering.

UTILISATION

- Oats are mainly used for animal food; both the grain - rolled, and with husks - and the straw are eaten. The straw is highly absorbent, and so is good for bedding.
- The rest of the grain is dehusked and used for human food, as oatmeal, porridge, oat cakes, muesli, etc. Grain from so-called "naked oat" varieties, where the husk does not develop, is also used, mainly for breakfast cereals.
- Oatbran, and to a lesser extent oatmeal, are rich sources of inositol, part of the vitamin B complex and important in nerve transmission, the metabolism and movement of fats and for reducing blood cholesterol levels. Inositol is also found in a wide variety of other food, including beans, citrus fruit, meat, raisins, seeds and nuts, etc.
- In some cold regions the young plants of winter sown types are grazed in spring. If no top dressing is applied the grain yield may be reduced a little, though with crops which are too thick ("winter proud") the yield is often increased by grazing.
- The plants can make good hay and silage.
- In industry, oat hulls are sometimes used as a source of *furfural* (or *furfuraldehyde*) a chemical intermediate in the production of many industrial products such as solvents and resins.

LIMITATIONS

- Very few, although other cereals normally produce higher income per hectare.
- Oats and oat products can rapidly turn rancid due to the presence of free fatty acids which produce a soapy taste due to the activity of an enzyme called lipase, though this can be rendered harmless by steam treatment.
- Oat plants can and do lodge badly, especially in fertile soils.
- The seed shatters very readily when ripe.
- The seed:husk ratio is very low - the husk can weigh 25% of a grain sample. Also it can be more difficult to separate this husk (the "chaff") from the grain than wheat and barley.
- Breeding of improved varieties of oats with wide adaptation has only had limited success.

Rice
Oryza sativa

Riz (French); Reis (Germany); Arroz (Spanish & Portuguese); Ruz (Arabic and Amharic); Ruzii (Oromifa); Bariis (Somali); Vreejay, Oreeji, Wriji (Pashtu); Brenj, Shali (Dari); Dhan chaval (Hindi)

Rice provided 700 calories/day/person or more for approximately 2.9 billion people during the year 1997, according to FAO. Rice is the most important cereal in the tropics, and in global terms the tonnage of rice grown is similar to that of maize and wheat (FAO estimates for 2004 were 605, 721 and 627 million MT respectively).

There are more than 20 species of *Oryza*, but only *O. sativa,* Asian Rice, and *O. glaberrima*, African Rice, are widely cultivated. African Rice is very similar to Asian Rice; grown mainly in the flood plains of the Sahel, it has poor quality dark coloured grains; it is used in some West African rituals. The two species were first successfully crossed in 2002, with promising results for the hybrid.

One of the species, *Oryza rufipogon* and others, is Perennial Wild Rice (Brownbeard Rice, with many synonyms), an aggressive and serious weed in rice fields. Evidence suggests that *O.sativa* was originally derived from *O.rufipogon*; the two species hybidise readily, producing a highly variable range of weedy perennial wild rice types, including annual types, and much taxonomic confusion. The seed is red-brown colour, 5–7 mm long. It propagates by seed (which shatters prematurely) and by rhizomes.

The other Wild Rice, *Zizania aquatica* (family Poaceae), sometimes called American Wild Rice, grows in eastern North America and was once an important food of the Indians living in that area. It is now cultivated to produce a highly valued, nutritious food, the "caviar" of cereals. It is a rich source of niacin and riboflavin.

Improved varieties of rice have been developed so far mainly for irrigated and favourable lowland areas, but in some countries their yield advantages are more than offset by the higher prices of traditional rices such as the aromatic Basmati types.

Classification of Rice
The classification (taxonomy) of rice is not simple, and can be based on at least four different characteristics:
1. Cultivation Methods. On this basis there are three types:
- *Upland (Hill or Dryland Paddy)*—varieties which can be grown in regions with adequate rainfall for 3-4 months. Cultivation methods are similar to other cereals.
- *Paddy (Lowland* or *Swamp)*—grown on artificially flooded fields. Some varieties are adapted to grow under both paddy and upland conditions.
- *Floating*—very rapid plant growth keeps up with the rising level of water, up to 5m deep.

2. Grain Characteristics. On this basis there are two types, glutenous and non-glutenous. Most varieties are non-glutenous.
3. Grain Shape and Size. On this basis there are four types: long, slender grain; long grain; medium grain; short grain.

4. Growth Period. There are two main classifications, used in different parts of the world. In North America and other countries: very early maturing (96–117 days); early maturing (117–132 days); mid-season (132–150 days); late (150–180 days).

In India: very early (less than 110 days); early (110–140 days); late (150–170 days); very late (180 days +). Some varieties only mature in 260 days.

There are 3 sub-species of *Oryza sativa*:
1. ssp. *indica*—Tropical Rice, mainly grown between latitude 0° and 25° in the tropical monsoon zones. 1–3 month seed dormancy. Tall, leafy, many tillers, susceptible to lodging, low response to fertiliser, more hardy and resistant to diseases and and poor growing conditions, photoperiod sensitive.
2. ssp. *japonica*—Temperate or Pearl Rice, mainly grown in temperate regions. Short, stiff straw and therefore resistant to lodging, fewer leaves or tillers, early maturing, It is grown mainly in Japan, South Korea, North China, New South Wales, South Europe, California and South America.
3. ssp. *javanica*—Javanese Rice, mainly in equatorial regions. Low yield potential.

Varieties of Rice
Many thousands of varieties of rice are available—more than 5000 in India alone. As rice is mainly self-pollinated, farmers can save their own seed from year to year and so build up their own composite variety (discussed in **1Fd**, page 53) that is adapted to their area. Mixtures of varieties are also commonly grown together. Varieties often display different characteristics when grown in different environments.

The International Rice Research Institute (IRRI), is based in the Philippines and has had great success in breeding improved rice varieties, particularly hybrids. Varieties developed and released by IRRI are prefixed with "IR" eg IR8, IR20, etc. Development of perennial rice strains has not yet proved very successful.

Hybrid rice has been growing in popularity since the 1970s; in China in 1992-93 for example, 19 million hectares of hybrid rice was planted, around 65% of the country's rice area. Under irrigation, hybrids generally yield about one MT per hectare more than the semi-dwarf modern "conventional" varieties. Future hybrid rice seed could be based on *apomixis* (discussed in pages 2 and 3)—apomixis is asexual reproduction whereby farmers could retain apomictic hybrid seed from their own crops for many seasons.

PLANTING
Propagation: by seed. Although often planted first in nurseries and transplanted at 4–6 weeks, there is often no yield advantage to this system though there are other advantages. The seed can be planted directly into dry soil, and the field then submerged; or it can be broadcast into a water-submerged field; or sown into a wet field, in which case the seed is often pre-germinated ie soaked in water for 24–36 hours, then kept in the dark for a further 24–36 hours.
Germination: most *indica* varieties have a 1–3 month dormancy. *Japonica* varieties and others that are insensitive to photoperiod (**1Ej**) have no dormancy period. The

optimum temperature is 30–35°C. The seed can germinate under water, but more slowly than when planted into moist soil.

Soil: heavy soils, especially alluvial soils of river valleys and deltas, are better than light soils—they can be "puddled", and also they lose less water and nutrients through percolation. The optimum pH is 5.5–6.5 when dry, which becomes pH 7.0–7.2 on flooding. Some varieties are classified as "tolerant" to acidity; other varieties can grow in soils with pH 8–9, and these can be used to reclaim saline or sodic soils. Phosphate is often a major factor in limiting yields.

Seed rate: Paddy: 90–110 kg/ha drilled in 15–20 cm rows, 135–230 kg/ha broadcast, 160–170 kg/ha broadcast from a plane. Upland: 100–120 kg/ha in 20–25 cm rows, 30–50 kg/ha in 50 cm rows. There are about 30–35,000 seeds per kg.

Seed spacing: (between rows) Paddy: 20–30 cm for later varieties, 10–20 cm for earlier varieties. Upland: 20–50 cm. The plant population is normally about 100–120 per square metre.

Depth: Paddy: 5–6 cm in light loams, 2–3 cm in heavy clays. Upland: 1–3 cm.

Intercropping: not suitable for paddy rice. Upland rice is sometimes mixed with other crops, especially in shifting cultivation systems.

Rotation: rice is often grown in soils unsuitable for other crops, so monocropping is common. Green manure crops such as Berseem (*Trifolium alexandrinum)* are often planted between rice crops. The green manure should be ploughed in just before flooding. Or 2–3 years of rice can be rotated with 2–3 years of grazing, either with volunteer rice plants, or better still with pasture mixtures including legumes. There are often fish in paddy fields, which are sometimes introduced; even prawns and crayfish have been used. In these cases great care must be taken with pesticides.

GROWTH CONDITIONS

Day length: most rice types are short-day plants, but varieties differ in their response. There are two main groups of rice varieties:

1. *Season-limited (date fixed)*—they flower on a certain date whenever they are planted. For example, the growth period of the variety Siam 29 varies from 162 to 313 days depending on when it is planted. Many of the tropical *indica* rices are season-limited.

2. *Time-limited (period fixed)*—their growth period is more or less the same no matter when they are planted. Many of the temperate *japonica* rices are time-limited.

Most modern varieties are insensitive to photoperiod and so can be grown in a wide range of latitudes.

Growth period: see above. If early (ie short growth period) varieties which are insensitive to photoperiod are grown, 3 crops per year are possible.

Temperature: an average temperature of 20°C during the entire growth period is required, with 22–38°C being optimum during the main growth period, and a minimum of about 25°C during flowering. The plant requires long periods of sunshine, especially for the final six weeks or so. Irrigation water ideally should be between 21 and 30°C.

Rainfall: water is normally a more limiting factor than soil. Rice grows mainly in the humid tropics and is one of the few crops that can be grown in the evergreen forest belt.

Paddy rice—transplanted into a well-soaked field, with some standing water; the water level is slowly increased to 15–30 cm as the plants grow, then decreased at flowering until almost dry at harvest.

Upland rice—900 mm minimum during the growth period, or many light irrigations, 10–15 or more, to provide a total of between 400 and 600 m³/ha.

For both types of rice the two most critical periods for water supply are 10–12 days before tillering, and during flowering.

Altitude: the terms "upland" and "lowland" rice are misleading. Very often the so-called upland varieties are grown at much lower altitudes than lowland varieties, and vice versa.

Diseases: Leaf Blight and Leaf Streak are caused by bacteria. At least 11 fungi also attack rice, of which five are described below:

– Blast. The most widespread and devastating. Plants can be attacked by this fungus at all stages. Small blue flecks appear on the leaves, which become brown with grey centres; these spread until the whole leaf becomes brown. If the attack is early, the grains do not fill and the panicle falls over, hence the other name for this disease "rotten neck". Control is by resistant varieties and use of clean seed; copper fungicides and seed dressings may also help to some extent.

– Brown Spot. Damage occurs both in nurseries and fields, especially in cold weather. Control is by seed dressings and burning the stubble of infected crops.

– Narrow Brown Leaf Spot. Foliage dies off early. Controlled with resistant varieties.

– Gigantism (Foot Rot or **Bakanae Disease).** Caused by a *Gibberella* fungus. Common in Asia, especially in seedbeds. Plants become very tall and thin, with few tillers; the panicle emerges poorly, and the grain is shrivelled. Occurs in wet soils above about 20°C. Control is by seed dressings and destruction of diseased seedlings.

– Stem Rot (Sclerotial Disease). Excessive late tillers, and loss of grain. Normally only occurs when plants are in unfavourable conditions. Control is by burning infected stubble and by taking care with the irrigation water, which spreads the disease.

There are also nine **Virus Diseases**, including Hoja Blanca, Yellow Dwarf, Orange Dwarf, Dwarf, Tungro and Grassy Stunt.

Pests: extensive damage is frequent, both in fields and stores. The worst of the field pests are stem borers, leaf miners, armyworms, grasshoppers, locusts and various nematodes. Rice is also damaged by rats, crabs and birds. Storage pests include Rice Weevil, Lesser Grain-borer, Khapra Beetle, Saw-toothed Grain Beetle and the Angoumois Grain Moth.

YIELD

According to FAO, the global average for 1988/90 was 0.9–6.6 MT/ha for the bottom 10% and top 10% of countries respectively, being higher in Asia and lower in South America and Africa. The global average figure for paddy rice in 2004 was 4.0 MT/ha, ranging from 9.69 MT/ha in Egypt to 750 kg/ha in the Congo.

The International Rice Research Institute (IRRI) in the Philippines and other plant breeders have increased rice yields by two or three times since the 1960s. Modern hybrids and open pollinated varieties can yield around 10 MT/ha in ideal conditions.

The gap in yields between irrigated and non-irrigated areas has widened, and is expected to widen further. Yields of rice grown in the Northern, temperate zones are almost double those grown between the Equator and the Tropic of Cancer. This is mainly due to better management, increased use of fertiliser, more reliance on irrigation and less on rainfall, and the use of *japonica* types, which have a higher yield potential than the *indica* types normally grown in the tropics.

UTILISATION
- About half of the rice grown is consumed on the farms where it is grown. Apart from **human consumption** rice has few other uses. The husks left after pounding the "paddy" (this word applies both to the unhusked grain and to the growing crop) is not suitable as animal food nor as fertiliser; this bran is used in bedding, litter, fuel and building materials such as hardboard.
- The **rice bran** (or rice meal) left after pearling and polishing is a valuable source of animal and poultry food. Milling of rice greatly reduces its human food value.
- The **straw** can be fed to animals but it is inferior to other cereal straws. It is also used for strawboards, for thatching and brading, and for making hats, packing material, broom straws and mats. In Thailand and China the straw is used for the culture of mushrooms.
- So-called "rice" paper is made from the pith of the rice-paper tree (*Tetrapanax papyriferum*), a member of the *Araliaceae* or ginseng family, not from rice.

LIMITATIONS
- Rice can be a very **labour intensive** crop to grow. In some primitive systems more than 800 man-hours per hectare are needed to produce a crop.
- Monocropping is often inevitable as there may be few or no other crops suitable for growing in the paddies. This tends to lead to a **build up of diseases, insects and weeds**.
- Many varieties of rice, including hybrids and especially photosensitive *indica* types, are **adapted to grow well only in small, limited regions**. They sometimes also need to be planted within a specific and rather short period of time, and to be heavily fertilised, in order to produce an economic yield.
- The **protein content** of the grain of 6–8% is rather modest.
- When a high proportion of the diet is white rice, there may be a **risk of beriberi**[*], a disease arising from a deficiency of thiamin (Vitamin B_1 or aneurin), other B vitamins and minerals.

[*]**Beriberi**—also known as **Athiaminosis** and **Kakke Disease.**
A disease of the nerves of the arms and legs, commonly found in eastern and southern Asia. Symptoms include fatigue, diarrhoea, appetite and weight loss, disturbed nerve function (dry beriberi, or endemic neuritis) causing paralysis and wasting of limbs, water retention, swelling of body (edema) and heart failure (wet beriberi).

The name comes from the Sinhalese language, "I cannot, I cannot".

The risk of beriberi developing can be reduced by "parboiling", an ancient Indian technique, in which the rice is steeped in hot water, steamed and then dried prior to milling. This process causes the movement of vitamins and minerals from the hull and bran into the endosperm, so that the resulting milled, white rice is nutritionally more valuable than regularly cooked white rice.

Rye
Secale cereale

Seigle (French); (Saat)-Roggen (German); Segale, Centeno (Spanish); Centeio (Portuguese); Seigle/Segale Comune (Italian), Almindelig Rug (Danish & Norwegian); Råg (Swedish); Rúgur (Icelandic); Rogge (Dutch); Ruis (Finnish); Jau Dahr {"Bearded Barley"} (Pashtu and Dari); Zyto (Russian)

Rye is grown mainly in Europe (Germany, Poland, Belarus and the Czech Republic), Russia and the Ukraine, and in the US. Global production is slowly declining, and was estimated by FAO as being 13.8 million MT in 2004—of the "cereals" (so-called), only buckwheat produced less tonnage globally.

True rye bread is becoming less popular these days and a similar type of bread, retaining some of the original characteristics, is now made from a blend of rye and wheat flour. Ryebread is aerated by the use of a leaven (sourdough) rather than yeast.

The plant is a hardy annual, or biennial, 0.5–2.5 m tall, with long awns. The seed heads are 7.5–15 cm long, with 2-flowered spikelets. The seed looks very similar to wheat (about 20–40,000 seeds per kg). It can be successfully intercropped with wheat.

The main agronomic attributes of rye are its hardiness, its ability to grow on light, acidic soils and its high gluten content. There are both spring and winter types and varieties of rye. Spring varieties normally require 10–12 days vernalisation, winter varieties need 40–60 days.

Rye (family *Poaceae*) has been successfully hybridised by plant breeders. In 1875 a Scottish botanist was the first to cross rye with wheat, to form a "new" crop known as **Triticale**, which combines the high yield of wheat (**Triti**cum) with the winter-hardiness of rye (Se**cale**). This bigeneric hybrid contains one set of rye chromosomes and 3 sets of wheat chromosomes, and early Triticales were sterile and produced disappointing plants. Modern varieties can be intercrossed, and also crossed with wheat. The grain protein content is a respectable 14–20%, and it contains more thiamin and folic acid than either of its parents, but less niacin or Vitamin B6. Has enough gluten for bread flour.

The global production of Triticale in 2004 was 13.8 million MT, according to FAO, at an average yield of 4.1 MT/ha (with the highest average of 7 MT/ha in Belgium, and the lowest of 846 kg/ha in Portugal).

Rye can also be crossed with Durum Wheat to produce a sterile hybrid.

Rye is naturally cross-pollinated, by wind; self-pollination produces shrivelled, weak seed. As a result, seed should be bought in when possible (every year if feasible) as it is difficult for farmers to maintain their own pure seed.

PLANTING
Soil: rye grows well in a very wide range of soils, including poorly drained and infertile, sandy soils. Tolerant of acid soils. Relatively resistant to lodging, even under high Nitrogen conditions. Responds well to fertiliser; may need N and P in sandy soils.

Seed rate: 40–50 kg/ha on dryland, 55–70 kg/ha on irrigated land. Up to 200 kg/ha may be used, on fertile soils in Northern Europe for example. Higher rates are used for spring sown types.

Seed spacing: 17.5–38 cm between rows.

Depth: 1.2–5 cm (4–5 cm in dry, sandy soils).

Rotation: Ideally rye follows a summer fallow or a root crop. It should never follow an oat crop. Because the seed head shatters readily volunteer plants often appear in the next crop, which may be a problem if Ergot is present—see "Diseases", below.

GROWTH CONDITIONS
Day length: long-day.

Growth period: 180–220 days. Rye should be harvested when the seed is at the "wax-ripe" stage because much seed is lost from shattering if the seed heads are left to full maturity.

Temperature: winter sown types require vernalisation (**1Fg**). Rye is the most winter hardy of all the cereals, and needs only about 15–20°C for its seed to mature.

Rainfall: relatively drought resistant, though shattering increases if drought occurs when the seed heads are maturing. Optimum is 600–1000 mm a year, while it can grow with between 400 and 2000 mm. Often grown under irrigation, especially in South Africa and America; irrigation at flowering time is especially beneficial.

Altitude: up to 4000 m.

Pests: the same as for wheat and barley.

Diseases: rye is resistant to the Smut diseases. The most serious disease is **Ergot** (*Claviceps purpurea*) a fungus which produces large, misshapen purple growths in place of the seed. These ergots contain the alkaloid *ergotoxine*, poisonous to both man and animals. Control is by the use of clean seed. To remove ergots from a seed sample, the whole sample can be immersed in a 20% salt solution; the seed falls to the bottom, the ergots float to the surface and are skimmed off. The seed is then rinsed in clean water and dried.

YIELD
The global average for rye in 2004 was estimated by FAO at 2.5 MT/ha, ranging from an average of 6.7 MT/ha in Switzerland to 310 kg/ha in South Africa.

2–3000 kg/ha of straw is a good yield in optimum conditions. To reduce loss of seed from shattering, harvesting is often done at dawn and dusk.

Premature harvesting leads to low quality, low germination seed. Delayed harvesting can lead to large losses of seed from shattering.

UTILISATION
- The **seed** of rye contains a high proportion of gluten, and is used for making black bread (*schwartzbrot*), rye crispbreads (*knaeckebrot* in Scandanavia) and biscuits, whisky (in America and Canada), gin (in Holland), beer (in Russia) and rye starch. Pumpernickel is a dark brown bread made from unsifted rye flour, and for many centuries was a staple food in much of eastern and central Europe.
- Rye is often planted, either alone or in mixtures with other forage crops, and **grazed by livestock** when the plants are green and young.
- The **straw** can be fed to animals, but it is poor quality fodder and is more often useful for **animal bedding** or for **construction of buildings**.
- The dried sclerotia of **ergot** are sometimes used in pharmacy, as the source of certain alkaloids to produce *Ergotamine*; this causes constriction of the blood vessels and is used to treat migraine.

LIMITATIONS
- Rye is susceptible to Ergot, causing ergotism, the "St Anthony's Fire" of the Middle Ages—gangrene, abortion, hallucinations and other unpleasant symptoms.
- Timing of harvest is critical as the seed shatters very readily.
- The plants are intolerant of high temperatures.
- The straw is tough and fibrous and can normally be used only as bedding or litter.
- The crop takes a long time to grow ("long growth period");
- Expenses are incurred as new seed has to be frequently bought in by farmers, because rye normally cross-pollinates.

Sorghum
Sorghum bicolor (Syn. *S. vulgare*)

Milo, Milo Maize (America), Indian Millet, Sorgo (sweet sorghums);
Sorgo, Sorgho (French); Sorghum (German); Daza, Sorgo {comun} (Spanish);
Sorgo (Portuguese); Ovasa, Omassambala (Angola); Mtama (East Africa);
Guinea Corn (West Africa); Mechella (Tigrinha, Ethiopia/Eritrea), Mashella (Amharic, Ethiopia), Mishinga (Oromifa, Ethiopia); Durra (Sudan); Kaffir Corn (South Africa); Jola, Jawa, Jowar, Cholam, Great Millet (India); Bajra (Pashtu); Kaoliang (China)

Sorghum is the fifth most widely grown cereal crop in the world, after wheat, rice, maize and barley. The FAO estimate for global production in 2004 was 59 million MT. Sorghum plants prefer warm growing conditions, and the crop is mainly cultivated between about 40° North and South of the equator.

Hybrid sorghums are widely grown in South Asia and Latin America, and also have good yield potential in Africa. Yields of these hybrids often achieve more than 50% of the control varieties, but the production of hybrid seed is still not as well developed as other crops such as maize.

Some varieties are available which have some resistance to the parasitic Striga weed, as well as some lines of sorghum which display some tolerance to soil acidity and aluminium toxicity.

There are more than 30 different species of sorghum, and several thousand varieties, and there is some disagreement between botanists about their classification. Nevertheless, there are roughly speaking five main types; in between these five types there are many examples of plants with botanical characteristics of two or more types—not surprising considering that man has cultivated and selected sorghums for thousands of years, and they also cross-pollinate readily.

The five main types of sorghum are listed below:
1. Grain Sorghums:
Mainly grown for their large, relatively palatable seed. There are many different examples:

Milo—*S. subglabrescens.* Compact, goose-necked heads; many tillers.

Kafir Corn—*S. caffrorum.* Small, cylindrical heads.

Hegari—similar to Kafir, but thinner stems and more leaves.

Feterita—*S. caudatum.* Large white seed, which shatters readily.

Durra—*S. durra.* Bearded, fuzzy heads. Seed normally white. Grown mainly in the Sudan, North Africa and India.

Guinea Corn—*S. guineense.* Grown in West Africa.

Shallu—*S. roxburghii.* Grown in India.

Kaoliang—*S. nervosum.* Grown in China and the Far East.

Hybrids—the yield potential of sorghum has been doubled by the development of hybrids, which was made possible initially by the discovery 50 years ago of two plants which were male sterile.

2. Sorgos:
These are the so-called "sweet" or amber sorghums, which are mainly grown for animal food, hay and silage and for the production of syrup. The fresh, young stems are often eaten by humans.

3. Grass Sorghums:
These are also grown for animal fodder. The most well known is Sudan Grass (var. *sudanensis*).

4. Broomcorn (Broom sorghum):
Sorghum dochna (Syn. *S. bicolor* var. *technicum*). Grown for their stiff stems and heads suitable for making brooms and brushes. "Broomcorn" is also a frequently used name for Common, or Proso, Millet (*Panicum miliaceum*).

5. Special purpose sorghums:
Examples include Pop Sorghum (similar to Popcorn), and varieties developed for the manufacture of starch.

With this enormous range of plant types it may be misleading to make generalised statements about "sorghum". The information given below is only a brief guide and further reading is needed to fully understand the complexity of this crop.

Nevertheless there are some **characteristics of the sorghums** that are almost always found in common, listed below:

- **Drought resistance:** sorghum has been called the "camel of the plant kingdom", though in fact most millets are more drought resistant than most sorghums. Sorghum needs much less water than maize, though young maize plants can be equally drought resistant as sorghum plants of the same age. The plant has certain features that enable it to survive drought, such as its ability to remain dormant during dry periods, and then recover.
- **Adaptability:** sorghums can be found which will grow on a wide range of soils and rainfall.
- **Intolerance of cold:** virtually all of the sorghums are killed by frost.
- **Preparation for food:** sorghum grain is prepared for eating in many ways like maize, which it closely resembles both botanically and nutritionally.
- **Damage by birds:** in 1967 it was estimated that the loss of food that year alone in the Sudan area caused by the Weaver bird (*Quelea* spp.) was 4 million MT. These birds, and others, can make devastating attacks on sorghum, unlike maize, which protects its seed inside a cob.
- **Pollination:** sorghum is mainly self-pollinated, although about 6% cross-pollination can occur, depending on the variety, growing conditions and compactness of the inflorescence.

PLANTING

Soil: sorghum plants adapt to a wide range of soils; deep, fertile sandy loams are best, with pH 5–8.5. Tolerates slightly saline and alkaline soils; intermediate tolerance to acid soils. Better adapted to heavy soils than pearl (bulrush) millet.

Poor response to fertiliser in dry conditions, where it is usually water and not nutrients which is the main limiting factor to plant growth. In less arid regions fertiliser usually gives an economic response, Nitrogen being the most beneficial. Sensitive to iron deficiency, symptomised by chlorosis.

Seed rate: not very critical, as sorghum plants compensate quite well by producing fewer or more tillers, and so adjusting to different plant populations. A rough and ready rule of thumb is to plant 1 kg/ha for every 70 kg/ha of the "normal" yield expected.

Average figures are 2–5 kg/ha in dry conditions, 5–10 kg/ha in medium moisture conditions and 10–20 kg/ha in moist or irrigated soil. There are between about 25,000 and 70,000 seeds per kg.

Spacing: the seed is very often broadcast, but when planted in rows these are about 60-90 cm apart, with 20–60 cm between plants.

Depth: 2–3 cm in moist soil, 4–5 cm in dry soil.

GROWTH CONDITIONS

Day length: most of the sorghums are short-day plants, but there is wide variation in varietal response to photoperiodism (discussed in **1Ej**, page 42).

Growth period: a vast range, from between about 70 and 220 days. Sometimes the plants are cut down and harvested, then allowed to grow back up again as a *ratoon* crop. This itself is then harvested, producing a second, normally smaller, crop of grain and/or grazing.

Rainfall: some sorghum varieties are almost as drought resistant as pearl (bulrush) millet. The optimum is 350–800 mm a year, while some early, short varieties can produce a yield with no rain at all during the growth period if seed or seedlings are planted into a cool, moist, retentive soil. One way to grow sorghum during the dry season is to propagate seedlings in nurseries and transplant them into moist soil at the end of the rainy season. In other situations, tall, late maturing varieties that are growing in light soils may need 1000 mm or more during the growth period.

Temperature: 30°C is ideal for growth. Most sorghums are killed by frost. For planting, the soil should be at least 17°C at sunrise.

Rotation: sorghum is often intercropped, normally with legumes such as different types of beans, pigeon peas, etc. If it is monocropped, or rotated with maize or another *Graminae/Poaceae* species, Striga weed (page 89) may become a big problem.

Irrigation: sorghum responds well to irrigation, and can be grown under full irrigation without any rain. If only irrigated once, the best time is just before heading starts.

Pests: The small red-billed **weaver** or black-faced dioch (3 races of *Quelea quelea)* and other birds can cause devastating damage, so bad that sometimes farmers have to abandon sorghum and grow other crops.

   Bird resistant varieties such as Seredo/ Sereno (an improved Serena) are available, but they have a high tannin content and so tend to be unequally unpalatable to people and animals as to birds.

– Stalk Borer: may damage plants so much that whole seedheads break off, or the whole plants are killed. Chemical control is not very effective as the larvae are well protected within the stalk. All infected stalks should be destroyed.

– Corn Worms: young larvae eat the leaves and developing seed head. Not normally serious.

– Sorghum Midge: eggs are laid in the flower, and the larvae feed on developing seed. There are some resistant varieties available.

– Sorghum Flies: there are several species. The shoot fly is about half the size of a housefly; eggs are laid on young leaves, larvae eat the growing point. Not always a problem, as young plants can recover with new tillers, though these also may be attacked, resulting in bushy, stunted plants with few or no heads.

– Storage Pests: Rice Weevil is the worst, often found together with the Flour Beetle and the Grain Moth. Fumigation is often necessary. Seed should be stored at less than 10% moisture, either as threshed grain in dry areas, or "in the head" in more humid areas.

Diseases: these tend to be more serious in warmer, more humid conditions:

Anthracnose is very common. Stems and leaves become reddish or purple, the leaves dry up and stems may rot and fall over.

Leaf Blight, caused by *Helminthosporium*, attacks both seedlings and mature plants.

Covered Smut. The developing seed is replaced by grey sacs containing black spores. It is seed-borne, and can be very damaging, though it can be controlled with seed dressings.

Loose Smut is less damaging than covered smut. Black spores are released into the air. This smut can also be controlled with seed dressings.

Head Smut. The entire head becomes a mass of brown spores, but normally only a few heads are infected. Control: as the spores are carried over in the soil, seed dressing is not effective. All infected heads should be destroyed, by burning.

YIELD

Average yields of sorghum vary widely, from between 300 and 3000 kg/ha for rainfed crops, to 2–6 MT/ha for irrigated crops. According to FAO the average for the years 1988/90 was 0.4–3.4 MT/ha, for the bottom 10% and top 10% of producer countries respectively.

These figures are predicted to rise to 0.6–3.7 MT/ha by the year 2010.

The FAO estimated the average global yield for the year 2004 was 1.4 MT/ha, varying from the highest in Egypt (5.9 MT/ha) to the lowest in Eritrea (272 kg/ha).

Hybrid varieties in ideal conditions can yield 10 MT/ha plus.

In addition to this yield of seed or grain, the stalks are also often eaten by man and animals, and are also used as construction material. This fact means that in some circumstances lower yielding but very tall varieties may be just as useful as higher yielding dwarf varieties.

UTILISATION

- Sorghum is the staple food in many of the drier parts of Africa, India and China. The grain is mainly used in different types of leavened and unleavened flat bread (*kisra* in Sudan), in porridge (*ugali, kali, cuscur,* etc.) or low quality *enjera*. The entire grain can be boiled in soup or water, or made into beer (*soowa, t'alla, pombe*, etc). Sorghum flour is often added to bulk up other more valuable flours such as wheat, maize & teff.
- Sorgos are used for syrup manufacture, and their stems are also eaten by humans when they are young and fresh (the stems, not the humans).
- As animal food ("stover"). Sometimes the grain is fed, though this is a rare luxury in poor countries; more commonly it is the stems and dry leaves from the harvested plants that are fed to animals. If animals graze a harvested, ratooning sorghum field there is a possibility that the young plants may contain toxic levels of HCN (hydrocyanic, or prussic, acid).
- The stems of taller, more vigorous varieties are used for building material, and for making baskets, for cooking fuel and for heating.

LIMITATIONS

- Sorghum grain is susceptible to bird damage, and unfortunately the so-called "bird resistant" varieties such as Seredo (Sereno) are less palatable due to their high tannin (polyphenol) content, and so are not popular with farmers or their families. As a result, their commercial value is lower than other sorghum varieties.
- The crop is susceptible to infestation by Striga weed (**1L**, pages 89–90).
- The crop is also susceptible to insect damage, especially shoot-fly, stem borer and midges.
- Many varieties are slow to mature (though this may have its own advantages).
- The grain is generally less palatable than maize, and as a result sorghum seed is often less in demand than maize seed (ie quality rated more highly than quantity).

Teff

Eragrostis tef (formerly classified as *Eragrostis abyssinica* or *Poa abyssinica*)

Lovegrass, Toff, (Warm-season) Annual Bunch Grass, Ethiopian Millet
Tafi (Oromigna), Taf (Tigrigna)—Ethiopia

Teff, teff millet, or "t'ef", is the most important grain crop in Ethiopia, where about 2 million hectares are cultivated annually. Ethiopia is the only country in which it is widely grown as a grain crop apart from a few small areas in Kenya.

Virtually all of the teff cultivated as a cereal is made into the Ethiopian national food *enjera*. Often considered by Europeans to be "famine food" (*céréale de disette*), teff is in fact a luxury item, often associated with more affluent families, while others have to make do with sorghum or millet. White seed is especially highly valued.

In South Africa, Kenya and Australia teff is grown as a very palatable hay crop, and in India it is grown as a green fodder.

Teff is an annual grass, 40–120 cm tall, with very small seed (2500–3000 per gram), easily lost (*Teffa* means "lost" in Amharic). It is in the genus *Eragrostis*, tribe Festuceae.

There are two main types, differentiated by their seed colour:

White Teff (*"Tsa'da"*)—much more highly valued than brown/red teff, and suitable for deeper more fertile soils usually found below 2500 m ASL. Growth period is about 90–120 days.

Brown or Red Teff—suitable for less fertile, shallow soils. Less valuable seed, but considered to produce better animal fodder than white teff. The growth period is normally shorter than white teff, about 60–100 days.

The two types are often grown together in the same field, although there is a financial incentive for farmers to keep the more valuable white seeded types separate from the brown. Often the brown is for home consumption and the white is for sale.

Teff is a reliable crop that can generally be depended upon to produce at least some yield even when it is grown in poor conditions. However, even in good growing conditions teff does not normally yield as much as other cereal crops.

There is virtually no gluten in the grain, so teff cannot be made into leavened bread. The pancake-like enjera is made by fermenting the flour and water mix for a day or two, then baking the dough briefly in covered trays. Enjera is enriched with vitamins by the yeast which arises from the short fermentation of the dough.

The seed has a good balance of essential amino acids except lysine ,it is very rich in calcium (110 mg/100 g of edible portion) and iron (c. 90 mg/100 g), and contains about 9% protein. The high iron intake of people who eat a lot of teff is often associated with their resistance to hookworm anaemia.

Another great advantage of teff is that its seed can be stored for several years without losing much germination/viability if it is kept dry and well protected.

PLANTING

Soil: teff adapts to a wide range of soils, including badly water logged ones, but grows best on lighter, sandy soils. It is rarely fertilised, though Nitrogen and/or Phosphate will normally produce an economic response. Fertiliser should be applied at sowing time, but it can be top-dressed if this is not possible. Approximate applications are 130 kg/ha of DAP, plus about 35 kg/ha Urea on light soils and twice this amount on black soils. Several combinations of other N/P fertilisers can also be used.

Seed rate: 10–15 kg/ha should be enough, because the plants can produce many tillers and in this way compensate for low plant populations. In practice, rates of 25–55 kg/ha are more usual. The seed rate is lower for soils that have higher fertility and water retention.

Seed viability: teff can be stored much longer than other cereals, and even in basic traditional stores the seed can remain viable for at least two years.

Depth: very shallow, 20 mm max. Teff seed is normally broadcast by hand then covered gently by using light branches or by driving animals over the seeded area.

Timing: depends on soil type, altitude, moisture, variety, etc but in general the later (ie longer growing season) varieties are planted from May to early August, and the earlier varieties in August and early September.

GROWTH CONDITIONS

Day length: short-day; moderately sensitive.

Growth period: 90–120 days for white seed varieties, 60–100 days for brown/red varieties.

Rainfall: teff can survive with only 200–300 mm during its growing period if the soil is retentive and the temperatures are not too high. Some very fast growing varieties can produce some yield with only 250 mm.

Thus although it is more drought resistant than other cereals it is still vulnerable to severe drought due to its rather shallow root system. The plants do not thrive when rainfall is more than about 2500 mm.

Rotation: teff can be useful as a catch crop (**1Gc**), for example if a main crop fails. In theory it is preferable to rotate teff with legumes and/or oilseeds, but in practice many soils can and do produce teff continuously for many years. A typical 4 year rotation is: pulse/ teff/teff (or other cereal)/pulse.

Altitude: teff can grow from sea level up to about 2800 metres, but 1800–2200 m is optimum. Different varieties are adapted to grow at different altitudes. Normally only brown seeded varieties are grown above about 2500 m.

Pests: not normally a serious problem, although armyworms, locusts, grasshoppers and root-knot eelworms can cause some damage. The Welo (Wollo) bush-cricket (*Decticoides brevipennis* or *"Degeza"*) can cause great damage. Central shootfly (*Hylemya arambourgi*) is controlled with seed dressings. Red tefworm (*Mentaxya ignicollis*) is controlled with various sprays.

Diseases: Rust (*"Wag"*) is common and can be a problem, especially in humid conditions; late planted crops normally avoid rust, but run the danger of not having enough rainfall to grow well. Head Smudge (*Helminthosporium miyakei*, *"Aramo"*) can also be a problem in humid conditions; it is seed-borne, so infected crops should not be used for seed. Damping-off (*Drechslera poae*) can be severe, especially with high seed rates and/or early planting.

Other diseases include: **Stinking Smut**, which infects the whole seed head, **Leaf Blight** (*Septoria)*, **Phoma Leafspot** (on old leaves), **Soot** (*Alternaria cladosporium* and *Coniosporium* species) and **Anthracnose**.

YIELD

Anything from almost no yield up to 2–3 MT/ha is normal; 1 MT/ha is considered to be a good crop.

Teff is reliable; although other crops will provide more food in good years, teff will often produce at least something to eat in low rainfall years when other crops, even millet, produce nothing.

UTILISATION

- Teff is used almost exclusively for making **enjera** in Ethiopia, and by Ethiopians living abroad. The grain has very high levels of iron (80–90 mg/100 g) and calcium (100–110 mg/100 g), and about 9% protein. It is normally eaten with *wot*, a sauce made of meat and/or pulses; the wot supplements the lysine deficit in teff.
- The high fiber content of the grain means that it important in **preventing diabetes** and assisting with **blood sugar control**. An additional advantage is that teff is almost always grown and stored under **organic conditions**.
- The **plant** is a grass (*Poaceae* alt. *Graminae*), and can be used to make very palatable and highly valued hay for livestock fodder in countries such as Ethiopia, South Africa, Kenya and Australia.
- The **straw** after is a very highly valued animal food; it can also be used as reinforcement for mud plastering of tukuls & grain stores (*goteras).*
- In India the **young green plants** are often eaten by animals; sometimes the cut plants are used as a mulch.
- The **grain** is sometimes used to make alcoholic drinks (*tela* and *katikala*).
- Teff may also have applications for persons with Celiac Disease (gluten in-tolerance). Some gluten free food crops include maize, rice, buckwheat, arrowroot, chickpeas, quinoa, tapioca and potatoes.

LIMITATIONS (Teff)

- Teff seed is very small which makes it a very labour intensive crop. Every step, from land preparation, weeding and thinning, harvesting, threshing to the final cleaning and cooking is laborious. The seedbed should be fine, well prepared and free of weeds; the land is ploughed at least twice - up to an astonishing five times, a herculean task.
- The supply of well adapted, improved varieties is limited. Seed supply in general is not always regular, and the varieties most useful for certain areas are not always available.
- Brown and white seed types are often grown together which can cause problems with crops maturing unevenly and with adapting to soil type and altitude.
- Yields of teff grain are often low, though this is often more than compensated by high prices in comparison with other cereals.

Wheat
Triticum aestivum (Syn. T. sativum, T. vulgare)

Bread Wheat, Common Wheat; Blé, Froment (France); Weizen (Germany);
Trigo (Spanish & Portuguese); Gehun,Genhu (Hindi); Qamr (Arabic); Sernay, Shinray (Tigrinha, Ethiopia/Eritrea); Ajja/Addja* {*T. durum*}, Sinday (Amharic, Ethiopia), Qamaadii (Oromifa, Ethiopia); Otiliko (Angola); Ghaanum (Pashtu); Gandum/Gandom (Dari), Garma (Winter Wheat, in Dari)
*Also sometimes used as the word for oats.

Wheat is one of the three most important cereals in the world, in terms of both the area grown and production, together with rice and maize. According to the FAO, global production in 2004 was 627 million MT, grown mainly in China, Russia, USA, India, France, Canada, Australia, Turkey, Pakistan and Argentina.

Wheat was one of the earliest food crops domesticated, around 8000 BC, in the Fertile Crescent of southwest Asia, together with barley and some of the legumes. Einkorn (*T. monococcum*) and Emmer (*T. dicoccum*) were the early precursors of today's wheats.

Wheat is highly adaptable, and different varieties of wheat are adapted to grow at altitudes from sea level to 3500 m and between latitudes 60° North and South; it is grown in virtually every climatic zone except the lowland tropics.

Like the other true cereals wheat is an annual grass, a member of the *Poaceae (Graminae)* family, which grows about 30–120 cm tall. Plants can *compensate* for thinly sown crops by producing many tillers.

The panicle, or seed head, is formed into a compound spike which may be awned, or *bearded*, like barley, or, more commonly, without awns. It is mainly self-pollinated.

There are two types of wheat grain, hard and soft, & two colours, red and white:
Hard grain varieties (which have a vitreous endosperm) are normally grown in drier areas, have a protein content of about 11–15% ("strong") and are used mainly for bread.

Soft grain varieties (which have a mealy endosperm) are normally grown in more humid areas, have a lower protein content of about 8–10% (weak), and are used mainly for cakes, biscuits, pastry, etc.
Wheat grain can be classed into six groups:
- Hard, red winter wheat
- Hard, red spring wheat
- Soft, red winter wheat[1]
- White wheat[2]
- Durum (macaroni) wheat[3]
- Red durum wheat

[1] Normally grown in humid regions.
[2] Soft grain varieties which can grow with very little rainfall.
[3] *Triticum durum* (formerly classified as *T. turgidum* var. *durum*).

The grain is long and pointed, very hard and more amber in colour than red or white. It has a very high gluten content, but this does not retain carbon dioxide to the same extent as bread wheat gluten. The protein content is about 13%, and there are about 17–35,000 seeds per kg. Durum wheat is mainly used for macaroni, pasta, spaghetti, noodles, etc. Many types of Durum are grown in Ethiopia, mainly on the highland vertisols between 1800 and 2800 m asl. There are no fully winter hardy varieties of Durum wheat.

There are 3 main types of *Triticum aestivum* wheat, and many thousands of named varieties:

1. Spring Wheat
Mainly grown either in very cold regions where if wheat were to be planted in the autumn it would not survive the winter, or in warmer regions on the occasions when for one reason or another winter wheats cannot be grown. As their name implies they are normally planted in the spring, ideally not until the soil temperature has reached about 4°C; they are however also planted at other seasons, normally at the start of the rainy season.

Spring wheats need about 100 frost-free days, and do not need to be vernalised by low temperatures (**1Fg**, page 59) in order to flower. Flowering is induced when the day length becomes long enough.

2. Winter Wheat
Planted in the autumn in temperate regions. They need to be vernalised, sometimes for weeks at a time, in order for the plants to pass from the vegetative stage to the reproductive stage. In temperate regions, about 75% of all wheat grown is winter wheat.

3. Intermediate (Alternate or Facultative) Wheat

They differ from the winter wheats in that they do not need low temperatures in order to flower. They differ from spring wheats in that they do not normally develop during the short, cold winter days. Some varieties will only grow well when sown in the autumn.

PLANTING

Propagation: by seed. Seed size varies from about 20,000 and 53,000 seeds per kg.

Germination: Optimum temperature is 20–28°C, minimum 2–4°C. Seedlings normally emerge in 5–6 days. Dormancy is short, like other cereals, and is rarely a problem.

Soil: Wheat grows best in fertile, medium-heavy textured soils that are well drained and have a good lime content. The plants are prone to lodging; for this reason, and due to the declining value of wheat straw, plant breeders tend to produce "dwarf" (60–90 cm tall) or "semi-dwarf" (90–120 cm tall) types.

Fertiliser: This is a large and complicated subject. Very basically, in general Nitrogen should be applied at planting time in soils known to be deficient in this element, and an additional top dressing applied if adequate rainfall is likely and if the crop appears to be in need. Older, taller varieties can only utilise about 60 kg/ha of Nitrogen; modern, shorter varieties can utilise almost three times this amount. Applications of P and K should be based on soil tests. Winter wheat in the UK receives on average 200:75:120 NPK.

Seed rate: 25–250 kg/ha, average about 65 kg/ha. Not very critical as wheat plants compensate well, producing either more or less tillers according to the plant population. Rules of thumb: lower seed rates in low rainfall regions, in clean, weed-free fields, when planted early and for autumn sown winter wheats; higher seed rates in higher rainfall regions, in weedy fields, when planted later in the season and for spring wheats.

Seed spacing: a controversial subject. 15–20 cm between rows, to give a plant population of 200–400 plants/m^2. In good growing conditions, and when sown early, wheat plants compensate for low plant populations by producing many tillers, so 150–200 plants/m^2 may be enough. In poor growing conditions widely spaced plants are easily overcome by weeds.

Depth: 2.5–5 cm is normal. If sown into very dry soil it is often planted 8–10 cm deep. In theory this is a good idea, to ensure that the seed only germinates after adequate rainfall, but in practice trials have shown that 6–8 cm is the maximum for even the driest soils. Very deep sown wheat also produces fewer tillers.

Intercropping: wheat almost always grows better when in pure stand. Nevertheless, in India for example it is often intercropped with other crops such as barley, linseed, mung beans and mustard. Intercropping wheat with legumes may be justified in some soils low in Nitrogen and where fertiliser is unavailable or prohibitively expensive.

Weeds: wheat does not compete well with weeds, and also it is difficult to hand-pull or hoe weeds in the growing crop, so where herbicides are not available the seedbed should be cleared of all weeds and their seed as much as possible.

Rotation: a crop of wheat that is cultivated the season after the crops listed below may be expected to perform in the following ways:

Sorghum—wheat yields less than when it is in a highly managed wheat monoculture.

Maize and cowpeas—wheat should do well.

Cotton—wheat performs somewhere between the two above.

Green manure—wheat yields about the same as after a fallow, and green manuring is not worthwhile in areas with less than about 375 mm rainfall a year.

Sesame and cucurbits—wheat should yield well.

Sorghum and barley—wheat will not normally flourish.

Pulses and oilseeds—wheat will also often not flourish.

In low technology farming systems, wheat should not be grown consecutively on the same land for more than 3 or 4 years, or even less.

GROWTH CONDITIONS

Day length: most wheat species are long-day plants.

Growth period: anywhere between about 95 and 150 days, with a rough average of 130 days. The growth period is extended by about 15 days for every 300 metres increase in altitude. Sometimes it is possible to grow two wheat crops per year, but this is dangerous and can lead to a buildup of Rust and other diseases.

Temperature: the minimum for growth is about 3–4°C, and the optimum is 25–27°C. Varieties have been developed to grow well in a wide range of temperatures. Wheat can grow in very high temperatures, above 40°C, provided that the air humidity is not too high.

At the other extreme, some winter wheats grow in some of the coldest regions. High temperature may result in low yields of small grain, especially if there are also hot dry winds.

Rainfall: wheat is generally less tolerant of drought or irregular rainfall than barley, but can sometimes produce some crop with only 200–500 mm during its growth period. Most of the important wheat regions of the world have less than 700 mm a year. In the tropics, irrigation is usually necessary if there is less than 200 mm rainfall during the growth period.

Altitude: wheat is grown from sea level to about 3500 m above sea level. In the tropics it is normally grown between 1600 and 3500 m. In Ethiopia, mainly 1800–2500 m. Many tropical countries attempt to grow more wheat at high altitudes so as to reduce food imports. At lower altitudes in the tropics there is often too little rainfall, or it is too hot and humid, which increases the damage caused by diseases.

Pests:

– **Wheat Stem Maggot**—mainly in Asia, Europe and America. The maggots (larvae) are slender, glossy and pale green; they may kill one or more tillers of young plants. In older plants, the grain is shrivelled and the plants may die prematurely.

– **Aphids**—normally attack in the early growth stages and during droughts. Wheat aphids and the so-called Greenbug damage plants by sucking their sap and, more seriously, by transmitting virus diseases. Can be controlled with insecticides such as Parathion.

– **Wireworms**—these larvae are slender, hard, shiny and slow moving, with 3 pairs of short legs. They attack germinating grain and also roots and stem below ground, which usually causes death of the plant. Lindane soil treatment is possible but very expensive.

– **Cutworms**—more serious in arid and semi-arid regions. The larvae attack at or below the soil surface.

– **Suni Bug**—a yellow-brown stinkbug, which sucks sap from the plant and developing grain causing a reduction in seed yield and quality. Can be controlled with Parathion.

– **Sawflies**—black and yellow flies that look like wasps. Larvae enter stems, which break and fall over. Control is normally by natural predators, but crop rotations and efficient ploughing also help.

– **Hessian Fly**—adults have a dark coloured body about 3 mm long. Red, round eggs are laid on the upper leaf surfaces. The larvae are initially red, then become white; they can kill young plants, and cause older plants to lodge. Control is by using resistant varieties (not resistant to all strains of the fly), by destroying volunteer wheat plants, and by choosing an appropriate sowing date.

– **Grain Moth**—very serious in primitive farming systems in the Near East, where it is called *el Doodeh*—"the pest". Eggs are laid in the soil in the spring, larvae enter the soil and wait for good rains, then attack the leaves which turn yellow, killing many plants. Can be controlled by using heavy machinery to plough deeply and thoroughly.

– **Angoumois Moth**—eggs are laid on the wheat ears; larvae enter the developing grain and reappear as adults in the stored grain.

– **Rice Weevil**—adults are brown/black weevils about 3 mm long. They do some damage in the field, but main damage is in stored wheat. They can multiply very rapidly. Controlled by having clean storage containers and fumigating immediately they are observed.

– **Dusty Brown Beetle**—yellow larvae attack roots of seedlings. Control with seed dressing.

– **Shiny Cereal Weevil**—bronze coloured weevils which eat irregular patches from the leaf edges. Most damage is by the larvae, which attack the stem just below the soil surface. In Northern Europe, the frit fly, leatherjackets and slugs often cause major damage.

Diseases:

Stem (Black) Rust—*Puccinia graminis*. Can be very destructive. Red spots, which become black, on all parts of the plant but mainly the stem. Plants become yellow. Grain is shrivelled, or sometimes absent. Control is by using resistant varieties. Some species of Barberry (genus *Berberis*) are a host for the fungus and those near wheat fields should be destroyed.

Leaf (Brown) Rust—*Puccinia triticina*. Worse in humid and semi-humid regions, but losses are usually less than with Stem (Black) Rust. Orange-brown spots are mainly on the leaves and are smaller than those of Stem Rust; they also become black. Control is also by using resistant varieties. Durum wheats are mainly resistant to Leaf Rust.

Stripe (Yellow) Rust—*Puccinia striiformis*. Does not develop in hot dry weather. Small light yellow spots appear on all plant parts, which join together to from long yellow stripes. Spread by wind — the spores can travel enormous distances. Control is also by using resistant varieties.

Septoria Leaf Blotch—very similar to Leaf Blight. Brown, speckled areas appear on the leaves, which then die. Control: crop rotation, destroy volunteer plants and seed dressings.

Septoria Glume Blotch—very similar to Leaf Blotch.

Take-all—especially in wheat monoculture. Plants are yellow and wilted, with pale and often empty ears. Roots are short and thick, and poorly developed. Stems are black and shiny just above the soil level - the leaf sheaths must be removed to observe this. Control by crop rotation, seed dressing and cultural strategies eg preparation of a firm seedbed helps a little.

Loose Smut—*Ustilago nuda*. A mass of black spores appears in place of the seedhead. Control: use clean seed, either from clean fields or by anaerobic seed treatment. Covered Smut is often more serious in wheat, and Loose Smut is often more serious in barley.

Bunt—*Tilletia* spp. Worse in drier regions and seasons. Infected plants have smut balls instead of grain. A sample of infected grain smells of fish. Control is by seed dressing.

Pythium Root Rot—especially in moist soils and in wheat monocultures, and with heavy applications of Nitrogen. Control is by crop rotation, preferably with a legume.

Virus Diseases—there are several, such as Wheat Streak Mosaic seen in hard red wheats; yellow-green stripes appear on the leaves, which die, stunted plants, and shrivelled grain.

YIELD

According to FAO, the global average yield of wheat in 1988/90 was between 0.8 and 5.1 MT/ha, for the bottom 10% and top 10% of countries respectively.

The average projected yields for the top and bottom 10% of countries is estimated by FAO to be 1.2–6.4 MT/ha by the year 2010.

The FAO estimate for the global average yield in 2004 was 2.9 MT/ha, from a low in Eritrea of 302 kg/ha to a high in Belgium of 8.98 MT/ha.

Hybrid varieties grown in good conditions can yield around 10 MT/ha.

UTILISATION
- **Seed/grain:** wheat has become increasingly important in many tropical countries and has partly replaced other cereals in the diet in many areas. The amount and quality of gluten is the most important factor for wheat flour. The percentage of gluten increases with short ripening periods, and decreases with long ones. The protein content depends (apart from the variety) mainly on environmental conditions such as day length, temperature, soil moisture and nutrients.
- A small amount of wheat grain is used in industry for the production of starch, paste, malt, dextrose, gluten, alcohol and other products. Wheat germ oil is traded to a small extent; it is highly unsaturated and is a rich source of Vitamin E.
- **Whole plant:** one recent trend in mixed (crop and animal) farming systems is to cut wholecrop wheat at about 50% moisture and ensile the crop for high energy animal forage.
- **Straw:** used for thatching rooves, for fuel and for making baskets. Wheat straw is a good quality animal food, and also makes good bedding for animals (though barley straw is preferable).

LIMITATIONS
- Wheat is **susceptible to many diseases**, which can often rapidly produce new races and thereby overcome the plants resistance which has been bred into new varieties by plant breeders.
- Many varieties, especially older and taller ones, **tend to lodge** in fertile and/or windy, rainy conditions.
- The plants are **intolerant of poorly drained soils**.
- Wheat **responds to fertiliser applications** in various ways, which are sometimes difficult to predict; the correct usage of fertiliser on wheat often depends on a farmer's ability to predict rainfall.
- **Allergy to gluten** in the diet happens with some children and adults. Rye can also promote adverse reactions to gluten, while oats and barley have a variable, unproven effect. Maize and rice do not have this problem.

2B.   LEGUMES

The grain of food legumes—also known as grain legumes, or pulses—is second in importance only to cereals as a source of human and animal food. The greatest variety of legumes is found in the tropics and subtropics; in India and some other countries legumes provide the only high protein component of the normal diet.

There are more than 18,000 species of legumes, of which about 30 are important as food crops. Despite the enormous importance of legumes as protein sources, only soybeans and groundnuts have received much attention from plant breeders or research workers.

Many of the under-exploited crops with potential as human and animal food are members of the *Leguminosae* plant family—the Tepary Bean, Mat (Moth) Bean, Rice Bean, Winged (Four-angled) Bean, Lupin, Camel Thorn, Bambara Groundnut and Leucaena are just a few of the examples, some of which are described in Section **2G. "Under-exploited Crops"**, pages 266–287.

The average protein content of legume grain is about 26%, though some have up to 60%, and pulses have been described as "the poor man's meat". Although many of the legume seeds are deficient in certain essential amino-acids (cystine and methionine in particular), these are normally present in cereals which are often eaten together with legumes to form a balanced diet.

The seed of many legumes is poisonous, containing certain alkaloids that must be broken down by cooking before they are eaten. If the seed is sprouted, as in China where the practice is an ancient tradition, not only is this problem averted but also large quantities of vitamins are released which are not available from the dry seed.

Some of the legumes, such as soybeans, groundnuts and winged bean, not only have high protein content but are also rich in oil; these crops are the so-called "oilseed legumes".

The ability of legumes to produce, or "fix", Nitrogen in nodules on their roots is another reason for their major influence on food production. Cowpeas, for example, can fix 45 kg/ha of Nitrogen or more in the soil, equivalent to 112 kg/ha of urea or 225 kg/ha of ammonium sulphate. The topic of Nitrogen fixation is discussed in more detail in Section **2Fe**, page 54.

The Camel Thorn *Acacia albida* (syn. *Faidherbia albidia*)

Although *Acacia albida* is not strictly speaking a human food crop, no discussion of legumes would be complete without mention of this valuable leguminous tree (Mimosaceae family), the **Camel Thorn** (also known as Ana Tree, Apple-ring Acacia, Winter Thorn, Kertor, Grar, Gerbi, Derot, etc).

Farmers make use of this drought resistant tree in hot regions such as tropical and southern Africa, Cyprus, Israel and Lebanon. In West Africa and elsewhere it is intercropped with sorghum and millet.

The tree has several attributes which can improve the food production possibilities in the dry tropics and subtropics, the most interesting feature being that it tends to lose leaves just as the rainy season begins, with the following advantages:

- animal forage is available towards the end of the dry season, at a time when other trees and forage plants have few or no leaves;
- a single tree can produce more than 100 kg of 27% protein pods, which fall from the trees at the hungriest time of year, at the end of the dry season. In an emergency, the pods can be eaten by humans;
- the leaves shade the soil under the trees and so protect the soil from wind erosion;
- the leaves, together with manure from the animals grazing underneath, enrich the soil and so make it more suitable for crop production. Leaf fall is perfectly timed to provide nutrients when they are most needed;
- during the rainy season, sunlight can reach crops growing under the trees because there are no leaves.

Broad Bean (Horse Bean or Field Bean)
Vicia faba (Syn. *Faba vulgaris, Faba sativa, Vicia fava*)

Fava Bean, Faba Bean, Longpod, Pigeon Bean, Tick Bean, Tic Bean, Windsor Bean; Féverole, (Grosse) Fève, Fève de Marais (French); Ackerbohne, Saubohne, Puffbohne(German); Haricot Caballar, Haricot Común, Haba Común (Spanish); Bondbona (Sweden); Hestebonne (Norway); Pacae (Peru); Feijão de Cavalo, F. de Porco, F.Miúdo, F.Forrageiro (Portuguese); Atah-bahari, Abeeatah, Alkwhyee, Atah-barativeri, Bagila, Bakela, Baldenga, Baldunga (Ethiopia); Gutate (Tigray, Ethiopia); Foul Masri (Sudan); Boerboon (South Africa); Double Bean, Katjang Babi, Ontjet (Indonesia); Baakla (Hindi); Boqoli (Dari)

There are several different types of broad bean, with plants exhibiting great variation in growth period, yield, plant and seed size, colour of seed and utilisation.

In essence, there are four different groups of varieties of *Vicia faba,* although botanists do not always agree on the taxonomy of the broad/horse/field bean:

- var. *faba* or *major*—the broad bean;
- var. *equina*—the horse bean, or "longpod";
- var. *minor*—the tick (or "tic") bean;
- var. *paucijuga*—similar to var. *minor*, grown mainly in Central Asia; it is largely self-pollinating, unlike the other varieties.

The broad bean is an erect, hardy annual plant, normally 60–180 cm tall, although some dwarf types are only 30–45 cm tall. It is the most hardy of all the beans, and is commonly sown in autumn in temperate climates. The plant is easily recognised by its four-ribbed stem.

Although the seed has a relatively high protein content of 24–33%, in common with many of the grain legumes it is deficient in the essential amino-acids methionine and cystine, especially when the grain is baked. Nevertheless, the beans are a good source of energy (340 calories per 100 mg), fat (1.5%), carbohydrates (49–57%), fibre (4.5%), calcium (100 mg), iron (6 mg) and Vitamin A (25–100 I.U.). The seed is large, with about 1–5,000 seeds per kg.

Under certain circumstances the seed can also contain toxic substances which can cause ***Favism***, a disease characterised by haemolytic anaemia, a disorder of the blood in which the red blood cells break down. This condition is most commonly found in Mediterranean and North African countries and in the Middle East. It is also prevalent in China, and 100 million people are thought to be affected around the world. Favism can sometimes be caught just by inhaling pollen when walking through a field where the plants are flowering. Susceptibility to favism is inherited as a sex-linked trait; the disease is especially threatening to children. Symptoms include pallor, fatigue, breathlessness, nausea, abdominal and back pain, fever and chills. Jaundice and dark urine may develop in severe cases, and the disease can be fatal to children.

China is a major producer of Vicia faba beans, and it is also planted as a winter crop on the edge of the tropics, in countries such as Sudan, Ethiopia and Burma, and at high altitudes in the tropics, such as in Uganda.

PLANTING
Soil: Vicia faba plants like deep, heavy, well drained and fertile/well manured soil, with pH 6–7. They do not like acid soil, and saline or waterlogged soil also will not produce a worthwhile crop. The plants benefit from wood ash and/or comfrey.

Fertiliser: A 4 MT/ha crop of broad beans will remove 45 kg P_2O_5 (36 units) and 50 kg K_2O (40 units). The Nitrogen requirement is normally supplied by fixation. Manganese deficiency is quite common (also with other pulse crops, due to their high requirement).

Seed rate: very variable, from about 75 kg/ha in Sudan to 200 kg/ha in Egypt, or up to 300 kg/ha in the UK. Small-seeded types such as tic beans use about 200 kg/ha.

Seed spacing: compromise is required ... broad beans needs a high plant population, without too many gaps, to attain high yields as it does not compensate very well. But if the crop is too thick it self-competes, grows tall and lodges, and is also more prone to diseases. So if the post-winter plant population is too high, the plants should be actively thinned out, either manually or by deep harrowing. They can, and do, send out several "tillers" or "stems". The optimum plant density is about 24–35 plants/m^2 (ie plant about 35–40 seeds/ m^2).

Approximately 75–100 cm between rows, 15–25 cm between plants, according to thousand grain weight (TGW) which varies from about 400–1000grams/1000 seeds, depending on variety and growing conditions.

Depth: 7.5–10 cm minimum, but deep enough to avoid birds pecking out the seed itself. If birds only peck off a new stem, but the seed remains in the ground out of reach, it will often produce another stem below the broken off area. In the UK winter beans are often ploughed in, to a depth of 20 cm, and 15 cm for spring sown crops to avoid this problem (with rooks especially).

Inoculation: recommended for certain situations, especially if grown for the first time or if previous broad bean crop roots did not have active, pink nodules.

GROWTH CONDITIONS
Propagation: by seed. Most varieties are cross-pollinated. Bees are very fond of the flowers.

Growth period: 90–220 days, depending on variety and climatic conditions.

Temperature: they are the hardiest of all the beans. In temperate regions, both autumn-sown, frost-tolerant "winter" varieties and spring-sown varieties are grown. 18–27°C is optimum. Higher temperatures can cause the flowers to drop off and failure of seed to set. In the humid tropics the plants often fail to produce any seed.

Intercropping: broad beans are tall plants which make an excellent nurse crop for smaller crops such as lettuce, haricot beans or soybeans. They can be sown in rows 1.5–2 m wide early in the spring, with rows of more delicate plants sown later between the guard rows of broad beans.

Rainfall: the least drought resistant of all the grain legumes, they need about 650–1000 mm.

Altitude: in Ethiopia they are mainly grown between 1800 and 3000 m, especially in the so-called "cereal–pulse zone" between 2000 and 2500 m.

Rotation: broad beans can be a useful break crop between cereal crops, but only once every five years, to avoid build-up of Stem Rot (*Sclerotinia*) (see "Diseases", below).

Pests:

– Black Bean Aphid (Blackfly) *(Aphis fabae)*—these can devastate crops, by penetrating the plant tissue, transmitting virus diseases and encouraging fungus diseases. The aphids often congregate at the tops of stems, so with smaller plots of beans these can be cut off and destroyed. Pyrethrum effectively controls aphids but should only be sprayed at night to avoid killing bees, which habitually visit the bean flowers.

The **Pea Aphid** and **Vetch Aphid** also transmit virus diseases.

– Pea and Bean Weevil—the soil coloured adults live in the soil in the day and feed on the leaves at night, but it is the larvae which do the damage by feeding on the root nodules. Deterred by applying lime or soot on the dew soaked plants, or by spraying nicotine or quassia (a natural insecticide) on the plants and surrounding ground.

Diseases:

Chocolate Spot—the most serious disease, especially on young plants. Fungicides such as Bravo, MBC, Ronilan and Rovral, are all mainly contact/protectant chemicals and must be applied before the fungus gains entry to the leaf or stem. Once a leaf is infected, spraying has little or no effect. Some control by removing all the haulm from previous crops, deep ploughing and correcting any potassium deficiency in the soil with appropriate fertilisers or wood ash. Spring sown crops normally suffer less than autumn sown ones.

Downy Mildew seems to be increasing in significance. Mainly spread by wind-borne spores. The youngest leaves of the plant are the first to be attacked. Apply a foliar spray of metalaxyl plus chlorothalonil (Ridomil + Bravo) if the disease is spreading in the crop during flowering. A repeat spray 10–14 days later may be necessary if the disease pressure remains high. The disease occurs mainly in warm damp sheltered sites.

Leaf Spot (Ascochyta)—mainly a problem of winter-sown beans , and worse in hot and humid conditions, but rarely causes big yield loss. The disease is seed-borne so it is controlled with clean or dressed seed. No effective fungicide available. Destroy all volunteer beans in nearby fields before the new crop emerges.

– Bean Mildew—spray with Bordeaux mixture.

– Root Rots—occur mainly in the tropics and subtropics. Control with seed dressing eg Captan (NB Captan is very highly toxic to fish and most aquatic invertebrates).

– Powdery Mildews—can be very damaging, in Sudan for example. Spraying the crop two or three times with lime sulphur gives some control.

– Broad Bean Rust—a major problem in countries such as Peru and Egypt. However in northern Europe rust normally appears at the end of the season and is often welcomed as an aid to desiccating the crop prior to harvest.

– Stem Rot (*Sclerotinia sclerotiorum*)—although not common, beans may develop this disease. Sclerotia develop in the soil, and produce spores that infect bean stems (also linseed, oilseed rape and field peas). New sclerotia develop here and interfere with the plants water conduits, so that all plant parts above the infection die of "thirst".

– Virus Diseases—Mosaic, Leaf Roll and others. Transmitted by aphids.

YIELD
Highly variable, according to variety, climate, cultural practice, etc. The UK average is about 3 MT/ha. 7 or 8 MT/ha is feasible, with good conditions, while Argentina often produces record yields of 9 MT/ha.

The fresh, green pods that are eaten as a vegetable yield about 12 MT/ha.

UTILISATION
- Normally it is the dry mature beans of *Vicia faba* which are eaten. In the Middle East they are baked first, to produce "Ful midamis". In Ethiopia they are ground into porridge, and in India they are sometimes roasted and eaten like groundnuts. The beans are also suitable for sprouting.
- Modern food industries have produced meat and skim-milk substitutes from broad beans.
- The green immature pods are often used, either boiled as a vegetable, canned or frozen. The haulm can be fed to animals.
- The crop is sometimes used for green manure, or silage.

LIMITATIONS
- Yields of *Vicia faba* are very variable, often as a result of difficulties with pollination, including flower loss.
- The grain is deficient in sulphur amino acids methionine and cystine.
- Favism, especially serious when food products are fed to children.
- Susceptible to insect and disease attack, both in the field and in storage.
- Relatively high water requirement.
- The seed is quite fragile; germination rates can fall dramatically if it is subjected to excessive or sudden heat, or to mechanical damage such as rough handling.

Chickpea
Cicer arietinum

Bengal Gram, Calvance pea, Chestnut Bean, Chich (pea), Chick-pea, Dwarf Pea, Egyptian pea; Garbanzo(a) (Bean), Gram (pea), Indian Gram, Yellow Gram; Café Francais, Ceseron, Cicérolé, Ciche, Gairance, Garvance, Gairoutte, Pois Bécu, P.Blanc, P.Breton, P.Café, P.Chabot, P.Chiche, P.Citron, P.Cornu, P.de Brebis, P.Gris, P.Pointu, P.Tête de Belier (French); Kichererbse (German); Garbanzo, Gravancos, Sigró (Spanish); Grão de Bico, Chicaro, Ervanço (Portuguese); Pisello Cece, Pisello Cornuto, Ceci, Cesari, Cesco, Spizole (Italian); Kikart (Sweden); Blukkeert (Norway); Erevinthos (Greece); Yellow Gram (East Africa); Hummous, Kabkaza, Kebkabeik (Sudan); Dwergertjie (South Africa); Attah (Tigray); Shimbra (Ethiopia,Shumburaa in Oromifa)*; Chola, Chana, Chono, Chota But, Chunna, Kadale, Sangalu, Adas (India); Nakud, Nokhut (Iran); Nakhud (Pashtu and Dari)

* "Adunguaré" is also used, as a general name for beans, especially haricot beans.

The Chickpea is a bushy annual plant, normally about 45–60 cm tall, often bluish-green in colour and covered in sticky glandular hairs. The shape and size of the plants varies very much—some are semi-erect, with few branches, others are semi-spreading, with many branches. Chickpea flowers are small and either white or reddish.

There are four separate races and nine separate sub-races of the genus *Cicer*. A second species, the littleleaf chickpea *Cicer microphyllum* is a shrubby perennial that grows above about 2500m in the Himalayan area; the seed and young shoots can be eaten.

They are usually grown on a field scale, for their highly nutritious seed (c. 20% protein, 50–60% carbohydrate and c. 5% oils). Small seeded varieties are normally made into dhal or flour for poppadoms; larger seeded varieties are often roasted and eaten whole, or mashed with olive oil, lemon juice and sesame to make hummus.

The seeds are either white, yellow, red, brown or nearly black, and have a characteristic "beak". There are two main seed types: *kabuli* (European or garbanzo), with large beige coloured seed and lower fibre content and *desi* (or Asian), with coloured, small seed, angular and fibrous, mainly grown on the Indian subcontinent.

In global terms, chickpeas are one of the most important of the grain legumes; FAO estimated that 8.6 million MT were produced in 2004. They are normally cultivated in hot, dry climates from southern Europe and North Africa to China, and especially in India. They are also widely grown, at high altitudes, in Mexico. The major producers are India, Pakistan, Mexico and Ethiopia.

PLANTING
Propagation: by seed. Pollination is mainly by bees, resulting in very occasional cross-pollination (no more than about 1%).
Soil: chickpeas tolerate a wide range of soil types, if they are not waterlogged. Sandy soil is preferred. Optimum pH is 7–9, chickpeas are classified as "sensitive" to acidity and "moderately tolerant" of salinity.
Seed rate: very variable, from about 30 kg/ha in parts of India, up to 120 kg/ha in Greece. Averages are about 100 kg/ha for larger seeded varieties, and 35 kg/ha for smaller seeded varieties. Average weight of 100 seeds varies from 13 to 83g depending on the type and variety.
Seed spacing: in southeast Asia, 10–30 cm between plants and 25–30 cm between rows; in Mexico, 10–15 cm between plants, up to 1.4m between rows.
Depth: 5–10 cm
Intercropping: sometimes happens, usually with a cereal such as wheat or barley.
Inoculation: in ideal circumstances inoculation of the seed has been shown to increase yields by between 20 and 62%.
Viability: some varieties can remain viable for 2 or 3 years, while others—often the white seeded types—lose viability after a year or less.

GROWTH CONDITIONS
Day length: long photoperiods (short nights) of 16 hours or more favour high yields. In general, chickpeas are moderately sensitive to photoperiod; long days tend to

shorten the vegetative stage, but short days do not prevent flowering. Most varieties are long-day types.

Growth period: most modern varieties mature in 115–125 days; older varieties, 185 days or more.

Temperature: for germination, 15°C is optimum, 5°C is minimum. For growth, 18–29°C is optimum though they can tolerate much higher temperatures. Frost resistance varies according to the variety, ranging from very resistant to very susceptible.

Rainfall: good resistance to drought, 600–750 mm per year normally being adequate. Chickpeas do not grow well with more than about 1000mm rainfall per year.

Altitude: up to 1200 m in Kenya and 2200 m in Ethiopia.

Rotation: often planted following a cereal crop such as wheat, barley, teff or rice.

Pests: insects are not normally a big problem, though the gram caterpillar or gram pod borer *Heliothis armigera* {Helicoverpa spp.} can cause problems. There is some control by using resistant varieties. Storage insect pests can also be a problem, often from Bruchid (pulse) beetles, and others.

Diseases: these are also not usually too serious, but the following can appear:

– Ascochyta (Gram) Blight (*Mycosphaerella (Ascochyta) rabiei)*—seed-borne. The leaves, stems, pods and seeds become covered with brown lesions. Control: resistant varieties, disease-free seed, mixed cropping, burn all infected plants.

– Fusarium Wilt—especially in hot weather and acid soil. Control: resistant varieties. Rust *Uromyces ciceris arietini.* Leaf Spot (*Alternaria* sp.), *Ascochyta pisi*, Grey Mould (*Botrytis cinera*), Powdery Mildew *(Leviellula taurica), Pythium debar-yanum, P. ultimum*, Dry Root Rot *(Rhizoctonia bataticola), R. solani*, Foot Rot *(Sclerotium rolfsii), Sclerotinia sclerotiorum* and Wilt *(Verticillium albo-atrum)*.

– Viruses isolated from chickpea include Alfalfa Mosaic, Bean Yellow Mosaic, Cucumber Mosaic, Pea Enation Mosaic, Pea Leaf Roll and Pea Streak.

– Deficiency Diseases—if the soil is low in Phosphorus or Manganese the leaves may turn yellow or brown, with stunted plants which may die.

YIELD

In general, brown seeded chickpea varieties yield more than green seeded ones. According to the FAO the global average yield in 2004 was 769 kg/ha, varying from a high of 5.0 MT/ha in China to a low of 346 kg/ha average yield in Kenya.

UTILISATION

*   The dry grain is the main product, normally for human consumption (eg falafel). in some countries such as Mexico chickpeas are mainly used as animal fodder.
*   The grain may be eaten when green, eg. when no other green vegetable is available.
*   Flour ("meal") can be made from the grain, often mixed with cereal flour before use.
*   Hummus is made from ground up and puréed chickpeas, olive oil and lemon juice.
*   Chickpeas produce good and tasty sprouts, with a high Vitamin C content.
*   The broken grain and residue from dhal production can be used as animal food.
*   In Asia the fresh young plant shoots and green pods are eaten as a vegetable.
*   The straw after harvest is valuable animal food, containing about 13% crude protein.

LIMITATIONS
- Chickpea yields are normally rather low, and are very variable.
- The flavour is bland and insipid, sometimes branded as "the poor man's food".
- The grain is deficient in the sulphur containing essential amino-acids and tryptophan.
- The grain is very susceptible to insect infestation in storage.

Cowpea
Vigna unguiculata and other *Vigna* sp.

Asparagus Pea or Bean, Black-eye (d) Pea or Bean, Bodi Bean, Catjang, Cowgram, China Pea or Bean, Crowder, Indian Pea*, Kaffir Pea or Bean, Lubia, Marble Pea, Southern Pea or Bean, Snake Bean, Tonkin Bean, Yardlong Bean; Bannette, Dolique de Chine, Haricot à Oeil Noir, Haricot Dolique, Niébé, Pois de Brazil, P. de Canne, P. Poona, P. Vache (French); Kakjangbohne (*V. cylindrica)*, Spargelbohne (*V. sesquipedalis*), Augenbohne (*V. unguiculata)* (German); Feijao de Ojo Negro, Caupi, Frijol (Spanish); Feijão Macunde. F. Frade/Fradinho, F.da China, F.Nhemba (Portuguese); Sebereh**, Lamattarh, Engwoyeh (Tigray); Adonguari, Digir, Eka-wohe, Degera, Fasolea-dima, Gaisa, Nguno, Nori, Nyoari, Wuch (Ethiopia); Batong, Kibal, Karkala, Otong, Paayap (Philippines); Acundeh, Ohalé, Caupí, (Angola); Feijão Brabham, F.deCorda, Ervilha de Vaca (Brazil); Dau Den, D. Trang, D.Tua, D.Xa (Vietnam); Enkoole, Enkoore, Imare, Laputu, Liboshi, Likote, Likotini, Loputa, Omugobe, Osu (Uganda); Lubia Beida, L. Helu, L. Kordofani, L.Tayiba (Sudan); Nyemba Bean (Zimbabwe); Tonkin Pea, Tua Dam (Thailand); Lobia (Hindi); Loobia (Pashtu & Dari).
*More commonly used for *Lathyrus sativa,* the (Chickling)Vetch or Grass Pea.
**Also used for *Lathyrus sativa.*

Cowpeas are mainly grown in India and West Africa, and also in the warmer parts of South and North America, and are found throughout the tropics and subtropics.

All cowpeas are annuals, but their plant forms vary considerably, from great long trailing stems several metres long to climbing, bushy and erect forms. Mankind has grown and selected cowpeas for thousands of years, leading to this extravagant range of plant characteristics.

As a result, the classification of cowpeas is confused, and botanists are unable to agree on a common method. One system which is fairly well recognised is to divide cowpeas into three distinct subspecies:
1. *Vigna unguiculata* (Syn. *V. cylindrica, V. sinensis* var. *cylindricus, V. catjang, Dolichos unguiculatus, D. catjang, Phaseolus cylindricus)*—the **Catjang Cowpea**, grown mainly in Asia and to some extent in Africa.
2. *Vigna sinensis* (Syn. *Dolichus sinensis)*—the **Common Cowpea**, grown mainly in Africa.

3. *Vigna sesquipedalis (Syn. V. sinensis* var. *sesquipedalis, Dolichus sesquipedalis) —* the **Long Bean, Snake Bean, Asparagus Bean or Pea, Yardlong Bean, Bodi Bean, Pea Bean, Snake Bean** (Australia)**, Dolique Geánt,** etc, grown mainly for its green pods.

Cowpeas were probably first domesticated in Ethiopia 5–6000 years ago and to this day the largest range of different wild and cultivated cowpea types are still found there. From here it was taken in the earliest days to West Africa (some argue that in fact the cowpea was first domesticated in West Africa), and to North and West India, from where it developed various forms as it was selected for its adaptation to different growing conditions and the food needs of man.

Cowpeas are now widely distributed throughout the tropics and subtropics, grown both for dry seed and green pods. They are a valuable protein source (c. 22%) for subsistence farmers in many semi-arid parts of Africa and Asia. They enrich the soil by fixing Nitrogen at up to 245 kg per hectare.

Cowpeas have enormous potential as a food legume in the semi-arid to sub-humid tropics, if disease and insect attack can be controlled.

The vast majority of cowpeas are grown in Africa, and some are also grown in Asia, Australia, the Caribbean, southern North America and the lowlands and coastal areas of South and Central America. Global production in 2004 was estimated by FAO at about 3.9 million MT.

PLANTING

Propagation: by seed. Germination is rapid above about 65°F. Seed weight is about 60lb/bushel. Can be stored at 12% for short term, but 8–9% is recommended for longer term, or for warm and/or humid storage conditions.

Soil: a wide range is tolerated if they are well drained. Saline soils are not tolerated. Optimum pH is 5.5–6.5, though it can tolerate even greater acidity.

Seed rate: for seed, 17–28 kg/ha when monocropped, 22–33 kg/ha when inter-cropped. For forage, 50–100 kg/ha.

The number of cowpea seeds per kilogram varies between about 4000 and 10,000. 3000–4000 per pound is normal.

Seed spacing: very variable. Very often 2 or 3 seeds are planted on hills, about 50 cm apart for early, erect varieties and wider for late or spreading varieties. Cowpeas are not normally planted in pure stand but are intercropped with cereals or other crops.

Since the earliest days man has observed the beneficial effects of growing cowpeas (and other legumes) intercropped together with sorghum, millets and so on.

Depth: 2–5 cm

Inoculation: this can be worthwhile if well-nodulated cowpeas have not grown recently (**1Fe**).

GROWTH CONDITIONS
Day length: short-day, long-day and day-neutral types exist.
Growth period: 90–240 days for grain is normal, though 60 day varieties are available, 50–100 days for green pods, depending on the variety and growth conditions. Most cowpeas are indeterminate (**1Ei**) so their pods and grain mature unevenly.
Temperature: ideally the soil should be no less than about 20°C at planting. The optimum for growth is about 20-35°C. The plants are frost sensitive, and young plants are weakened below about 10°C.
Rainfall: 600 mm per year is enough for some early, determinate types. High rainfall or humidity tends to reduce yields due to an increase in fungus diseases; asparagus beans tolerate high rainfall better than common cowpeas.
Altitude: in East Africa they are grown up to about 1500 m.
Pests: probably the most serious problem with cowpeas, which are attacked by more than 100 different insect species, such as pod borers (*Maruca testulalis)*, blossom beetles (*Coryna* spp.), thrips, root-knot nematodes and a pod-sucking insect (*Acanthomia horrida*).
Diseases: also a big problem. The most important are rust, bacterial canker, cowpea (fusarium) wilt, mildew, charcoal rot, anthracnose and several virus diseases. In Africa, leaf and pod spot (*Ascochyta*) is often a problem; although fungicides can be effective, the best control is by using resistant varieties, crop rotation, and by destroying all diseased plant material in the field.

YIELD
Although the potential yield of improved varieties of cowpeas with good management is more than 4 MT/ha, the average grain yield for African subsistence farmers is about 300–600 kg/ha (though pure stands (monocropping) are rarely found in subsistence farming systems).

FAO estimated the global average yield in 2004 was 388 kg/ha, from a high in Croatia of 4 MT/ha to a low in Niger of 157 kg/ha.

When grown for hay, 5 MT/ha is an average yield.

UTILISATION
- Cowpeas are an important food legume, especially in Africa; the dried seed/grain is almost always the part that is eaten. The objective of much of the current research in Asia and Africa is more towards increasing the utilisation of cowpeas in the human diet than to improve the agronomy of the crop.
- Cowpea seed is highly nutritious and palatable, containing about 22% protein (up to 35%), 1.3–2% fat and 60–67% carbohydrate. The energy value is a respectable 340 calories/100g of edible portion. The proportion of protein, carbohydrate and Vitamin B varies considerably according to the variety and origin of the seed. Like most legumes, the amino acid profile complements cereal grains.
- The green grain, pods or young shoots can be eaten as a vegetable; or the leaves are boiled, then either eaten or dried and stored for use in the dry season.

- Cowpea haulm is a useful animal food, containing 11–12% protein. When grown for hay, which is common in southern USA, the protein content is about 18%.
- Cowpea plants can be used as green manure (p. 70) and as a cover crop (p. 63).
- Sometimes used as a shade crop (p. 64), for example to protect other crops from the sun loving Chinch Bugs (*Blissus leucopterus*).

LIMITATIONS
- Cowpea yields are normally low, as a result of insect damage, poor management and the use of unimproved varieties.
- The pods shatter quite readily when mature, losing seed and sometimes causing problems with volunteer plants in the following crop.
- Harvesting of indeterminate types can be a problem since the pods must be removed every few days over a period of several weeks.
- Storage losses can be high, often as a result of attack by Bruchid insects.

Field Pea
Pisum sativum (Syn. *P.arvense*—see below)

English Pea, Garden Pea
Pois des Champs, Pois Fourrager (Fourrages), Pois Gris, Pois Capucin (French); Erbse, Felderbse, Futtererbse, Ackererbse, Kapuzinererbse (German); Ervilha (Portuguese); Guisante Gris, G. de Campo, G. Forrajeiro, Arvejas, Bisalto (Spanish); Basilla (Sudan); Onjolovilha (Angola); Amashaza, Obushaza (Uganda); Atnattarrh (Tigray); Attarrh, Danguleh (Ethiopia); Ertjie (South Africa); Mar, Matar-mar (India); Mashung/Moshong (Dari); Polong (Indonesia); Citzaro (Philippines)

Field peas (*Pisum sativum* var. *arvense*) are only slightly different from Garden peas (*Pisum sativum* var. *sativum*):

Field Peas	Garden Peas
• More hardy • Small pods and seeds • Normally grown for dried seed, as unsupported plants • Flowers are normally reddish/purple; many new varieties have white flowers	• Less hardy • Large pods and seeds • Normally grown for green peas (or pods). On sticks or other supports • Flowers are normally white

These two botanical varieties are completely cross-fertile; botanically they are very similar, and are often regarded as the same variety. So-called "garden peas" are often grown on a large scale as a field crop.

A third type, *Pisum arvense* var. *abyssinicum*, is grown in northern Ethiopia and Eritrea. It has very sweet seeds, with a black hilum, and very small reddish/purple flowers.

All three types are entirely self-pollinated. Varieties with edible pods are known as Mange Touts or Sugar Peas. Many of the more recent varieties are bred to be leafless or semi-leafless.

Peas are nutritionally valuable, the dried seed containing about 25% protein, 1% fat, and 57% carbohydrate, as well as 337 calories and 100 (10–200) I.U. Vitamin A potency per 100 mg edible portion.

There are a very large number of regional types of peas, and a large range of named varieties, all variously adapted to a wide range of growing conditions. They are grown in temperate regions throughout the world, and as a cool season crop in the subtropics and at high altitudes in the tropics.

The major producers of dried field peas are China, Russia, India, America and Zaire.

PLANTING
Soil: peas tolerate a wide range of soil types, provided they are not waterlogged, though in clays and very sandy soil they do not grow well. The optimum pH is 5.5–6.5, though some varieties tolerate a pH of 6.9–7.5.

In general, but not always, they give an economic response to fertilisers, though less than most other legumes. Maximum Nitrogen is c.60 kg/ha—nitrogenous fertilisers may sometimes even reduce yields. In the UK pea crops receive about 80 kg/ha P and K.
Seed rate: 65–100 kg/ha for smaller seeded varieties, 130–170 kg/ha for larger seeded ones. 100 seeds weigh about 15–25 g. Seed benefits very much from fungicide/insecticide dressings.
Seed spacing: 30–60 cm between rows, 6–18 cm between plants. In trials in India, the best yields were obtained with 7.5 × 7.5 cm square grids.
Depth: 2.5–6 cm. In light, dry soil they can be planted even deeper.

GROWTH CONDITIONS
Growth period: 90–160 days for dried seed, 56–84 days for green seed or pods, depending on variety, climatic conditions and planting date.
Temperature: peas are essentially a cool weather crop. For germination, $4°C$ minimum, $37°C$ maximum. High temperatures stimulate the plant to flower too early. Frost is tolerated in the vegetative stages, but at flowering can cause heavy pod loss, and at pod set can cause deformed and discoloured seed.
Rainfall: the optimum is 800–1000 mm per year evenly distributed, with dry weather at harvest. They can grow with as little as 300 mm per year if the soil is deep and retentive, as the taproot can reach down more than one metre in search of water.
Altitude: in the tropics, about 1200 m is the lowest they will survive. In Kenya the best altitude is between 2100 and 2700 m. In Ethiopia the best altitude is between 1500 and 2200 m with rainfall less than 600 mm, and 2200–2300 m with rainfall more than 600 mm per year.

Rotation: in general, peas should not be grown on the same land more often than every 3–5 years, to minimise the increase of soil-borne pests and diseases.

Pests: these can cause big problems, both in the field and in storage:

– **Pea Aphid**—causes stunting and also transmits more than 20 different virus diseases. Control: some varieties have some resistance, or spray with organo-phosphorus insecticides, nicotine sulphate or rotenone.

– **Pea Cyst Eelworm**—plants are stunted, turn yellow and may die. Control is difficult, and a crop rotation with a 10–12 year break is the only practical method.

– **Pea Weevil**—larvae feed on the seed. Control: rotenone dust works well and is safe, though parathion may also work.

– **Pod Borer** and **Leaf-eating Caterpillar (Lesser Armyworm)**—both of these can cause big problems, and can be controlled with sevin or malathion.

Diseases: these can also cause large losses in yield:

– **Powdery Mildew**—this is the most widespread and serious. A white dust appears on the leaves and sometimes on the stems and pods. Control: dust regularly with sulphur, and crop rotation.

– **Ascochyta Blight (Leaf Spot, Pod Spot)**—a seed-borne disease which is a widespread and major problem. Control: use clean seed and/or 3–4 year crop rotation.

– **Pea Wilt**—plants are stunted, with yellow leaves. Control: resistant varieties.

– **Bacterial Blight**—all plant parts above the ground show water-soaked lesions, which can kill young plants, especially in humid conditions. No known control.

Stem Rot (*Sclerotinia sclerotiorum*)—although not common, field peas may develop this disease. Sclerotia develop in the soil, and produce spores which infect bean stems (and linseed, oilseed rape and spring beans). New sclerotia develop here and interfere with the plants water conduits, so that all plant parts above the infection die of "thirst".

– **Virus diseases**—there are more than 20 of these, some of which can be controlled with resistant varieties; failing this, the aphid vectors must be destroyed.

– **Deficiency diseases**—Molybdenum is essential for nodulation of the roots. Manganese deficiency, occurring mainly in alkaline soils, causes the seed centre to darken (**Marsh Spot**).

YIELD

The yield of field peas varies enormously:

Dried Peas—2 MT/ha is considered satisfactory; 4–5 MT/ha is common in northern Europe. Global average yield quoted by FAO for 2004 was 1.9 MT/ha, from a low of 268 kg/ha in Croatia to a high of 5.8 MT in the Netherlands.

Green Peas—a good average yield of green peas in the pod is 6.5–7 MT/ha, and 4.5 MT/ha of shelled green peas.

UTILISATION

- **Dried grain (seed)**—this is a valuable and nutritious food, which is often ground into flour or made into soup. There is a huge industry for canning (both green and dried peas), dehydrating and freezing peas. The grain is sometimes also used for animal food, when its price is similar to cereals, either whole, split or as flour. This practice is becoming more common in recent years, with GM soybeans being so widespread. Russia and China together produce nearly 80% of the world production.

- **Green, immature peas and pods**—these are also very nutritious, though they rapidly become tough and lose their flavour after harvest due to loss of sugars. Compared to dried grain they have less protein, fat and carbohydrate. USA and the UK are the largest producers of green peas.
- **Leaves** are used as a potherb in Burma and parts of Africa.
- **Animal food**—the pods and haulm are a valuable animal food, either dried or ensilaged. When grown for hay, peas are often grown mixed together with a cereal crop.

LIMITATIONS
- The main problem with peas is their susceptibility to **diseases**
- The seed/grain can be heavily attacked by **insects during storage.**
- The crop can only be safely grown on the same land **every 3–5 years.**
- The **food value quickly deteriorates** when the grain is stored, processed or cooked.
- **Yields** are often rather low.
- The crop can be **difficult to harvest**.

Groundnut
Arachis hypogaea

Peanut, Mani, Monkeynut, Earth Nut, Goober Pea, Pindar, Valencia/Spanish Peanut; Arachide (French); Erdnub, Erdnuss (German); Cacahué, Cachuate, Cacahuete, Maní (Spanish); Amendoim (Portuguese); Jingooba, Olongupa, Olongumba (Angola); Foul Sudani (Sudan); Foul (Tigray); Ochioloni and Loozii (Ethiopia); Mungphali (Hindi, Urdu, Pashtu and Dari).

There are two main types of the tropical legume crop groundnut: Bunch and Spreading (or Runner). The genus *Arachis* has literally scores of species, all of which are generally considered to be native to South America, particularly Matto Grosso State in Brazil.

They are annual plants, but can survive as short-lived perennials in frost-free regions. Despite its various names the fruit is a legume, or pod, and not a true nut.

After the (normally yellow) flowers are fertilised, almost always by self-pollination, the stalk which bears the ovary turns down towards the soil surface, and is then known as a "peg" or *carpophore*. The peg enters the soil and the ovary develops there into a seed pod containing 1–4 seeds (the small Valencia varieties may have 5).

The "nut" has formidable nutritional value; weight for weight, it has more protein, minerals and vitamins than beef liver, more fat than heavy cream and more calories than sugar.

The seed contains 35–55% oil, and groundnuts are often classified as an oilseed crop (like soybean) even though they belong to the *Leguminosae* family. In fact groundnuts are mainly cultivated for their oil. The oil is high in oleic acid and fairly high in linoleic acid, but low in other acids. It is "non-drying" ie it remains liquid, without a surface film, when exposed to air.

Groundnuts are grown in almost all tropical and subtropical countries, up to $45°N$ in Asia and Canada, where the summers are hot. The main producers are China (10–12 million MT pa), India, Nigeria, Senegal, Sudan, America, Indonesia and Brazil.

With an annual global production of about 30 million MT, groundnuts rank in the top 25 of the world's food crops.

PLANTING

Soil: should be deep, well drained and sandy to encourage development of the long tap root, up to 1.5 m long, and to allow the pegs to enter the soil surface. Soils should be high in phosphate, calcium and sulphur.

Groundnuts tolerate a wide range of pH, though they have some preference for slightly acidic soil—the optimum pH for light, very sandy soils is 5.5, and for sandy loams is 6.0, up to about 6.5, root nodules often forming even in these acidic conditions.

Fertiliser is very often not applied, although about 50–100 kg/ha of superphosphate is normally beneficial. Requirement for phosphate is high—see *Rotation*, below.

The plants survive in high aluminium soils that would be toxic to other food crops.

Seed rate: 50–80 kg/ha for Bunch types, 35–40 kg/ha for Spreading (Runner) types. Minimum viable seeds to be planted per hectare: 80,000 on ridges, 110,000 on the flat. **Depth:** 4–5 cm.

Seed spacing: 45–75 cm between rows, 10 cm between plants, either on the flat or ridges. High plant populations, up to about 250,000 plants per hectare, are needed to produce high yields.

Inoculation: this is advisable, unless it is known that well nodulated groundnuts, cowpeas or velvet beans have been recently grown on that land.

Germination: the seed is very fragile and must always be handled carefully. The seed should be left inside the shell for as long as possible before planting in order to maintain its viability. Sometimes the whole pod is planted, but germination is then slower and more uneven.

Viability is often good for 3–6 years if the seed is stored dry and cool. Medium sized seed gives the best results. The seed of some varieties can remain dormant for up to two years.

Rotation: often best to follow a well fertilised crop such as potatoes, maize or cotton and apply no fertiliser to the groundnuts, which efficiently uses the residual fertiliser.

They should not be planted after tobacco, soybeans or sweet potatoes, to reduce nematode and stem rot damage. In parts of India groundnuts are rotated with rice.

GROWTH CONDITIONS (Groundnuts)

Day length: day-neutral.

Growth period: from planting to dry pods, about 120–160 days for slower growing Runner type varieties, 90–110 days for faster growing erect Bunch varieties. The plants are ready to harvest when the leaves turn yellow and the insides of the pods show dark markings.

It is important to harvest in a dry spell and at the correct stage, since the nuts, even on a single plant, do not mature simultaneously. Fresh pod moisture of 30–40% must be dried to around 8–10% for storage.

Temperature: high temperatures favour high yields, around 15°C average is needed, and there should be no frost during the growing season.

Rainfall: about 600 mm per year is the minimum, with at least 500 mm during the growing period. Most varieties are grown where annual precipitation is 1000–1200 mm. Some varieties can produce some yield with only 300 mm per year. In Asia and elsewhere it is grown under irrigation.

Pests: rarely a problem, though the following insects may cause some damage: thrips, termites, ants, nematodes, leafhoppers, velvet bean caterpillars, corn earworms, cutworms & armyworms.

Diseases: at least five diseases can reduce yields:

– **Rosette**—a viral disease, spread by aphids, where leaves turn yellow, shoots become distorted and plants are stunted. Control: plant early at high plant densities, spray aphicides and use resistant varieties.

– **Leaf Spot**—a fungal disease, causing dark brown or black spots on both sides of the leaves. Control: plant early, dress/treat seed, destroy infected plants; fungicides are generally not cost effective.

– **Aflatoxins** are naturally occurring mycotoxins produced by species of Aspergillus, notably *Aspergillus flavus* and *A. parasiticus,* which can develop rapidly in humid conditions. They are carcinogenic to animals and humans, affecting mainly the liver. They are most dangerous with young animals and poultry, and are thought to be the cause of high rates of liver cancer in parts of Africa and Asia. The only control is to rapidly dry the seed to a minimum of 10% moisture content.

– **Groundnut Blight**—plants become wilted, and stems and even the whole plant may die, especially in humid conditions. Can be controlled with seed dressings.

– **Bacterial Wilt**—caused by the bacteria *Pseudomonas solanacearum.*

YIELD

Groundnuts can yield well in the warmer arid and semi-arid regions as well as in its more traditional homeland of the humid tropics.

The global average in 2004 was 1.45 MT/ha, according to FAO statistics.

The lowest average yield reported was 336 kg/ha in Zambia, and the highest was 6.7 MT/ha in Israel (Figures are for groundnuts in the shell).

UTILISATION

- Groundnut oil is used in margarine and cooking oil, and for lubricants; there are more than 300 by-products, including flour, soap, shaving cream and plastics.
- Whole groundnuts for human consumption, roasted, peanut butter, etc, and for fattening pigs.
- Seedcake or oilcake, the residue left after the oil has been extracted, contains about 50% protein and is used for cattle cake, pig and poultry food and fertiliser.
- Hay, about 7% protein, can be valuable if plenty of leaves remain on the plant. Sometimes the whole plant, including pods, is fed to livestock.
- Recent research in Sweden indicates that the amino acid arginine (or "arg"), which is present in large quantities in groundnut seed, could help to combat Mycobacterium tuberculosis—Tuberculosis. Arginine boosts nitric oxide, which helps the body to produce macrophages that combat some liver diseases.

LIMITATIONS

- Groundnut seeds mature unevenly, they are delicate and must be stored and handled very carefully.
- Danger of Aflatoxins developing, causing liver damage and cancer—see "Diseases".
- Growing a crop requires high labour input unless systems are mechanised.
- Some groundnut varieties have a long seed dormancy.
- Germination falls rapidly if the seed is stored out of the pod or in hot and humid conditions.

Haricot (French) bean
Phaseolus vulgaris

Black Bean, Green Bean, French Bean, Common Bean, Field Bean[1], Pole Bean[2], (Red) Kidney Bean[3], Navy Bean, Pea Bean[4], Pinto Bean, Salad Bean, String Bean, Snap Bean. Flageolet, Frijoles; Haricot Commun, Haricot Vert, Haricot Pain, Haricot à Couper (French); (Garten)bohne, Fisole (German); Frijol(es), Judía Verde, Judía Comun, Alubia Tierna, Frijol Verde, Poroto Comun (Latin America), Ejote (Mexico); Feijão da Índia, F. Ervilha, F. Colubrino (Portuguese); Bollokeh, Zada-adagonna, Adagura (Tigray); Adenguaré, Bolooqee(Ethiopia); Edihimba, Ojoo, Teiko (Uganda); Bush Bean (Zimbabwe); Feijão Ervilha, Ochipokeh[5] (Angola); Fasulia (Sudan); Tua Phum, Tua Kack (Thailand); Jungli Sem (Hindi); Mula, Michigan Pea Bean (Philippines)

[1] Broad (Horse or Field) Beans are also called Field Beans.
[2] Lima Beans, Runner Beans and Hyacinth Beans are also called Pole Beans
[3] Kidney Beans in the USA are a specific type, with a definite kidney shape and seeds which are either red, dark red or white.
[4] Pea Bean is the name often used in the commercial grain trade.
[5] Ochipokeh is a general name for "beans" ("Chitanga" is for all types of climbing beans)

Haricot beans are grown either for their green pods, when they are commonly known as French beans, or for their dried grain, when they are commonly known as Haricot beans, or Haricots.

Several hundred varieties of *Phaseolus vulgaris* are available, exhibiting a wide range of different plant types, which vary from the tall climbing or "pole" types with twining stems which may grow up to 5 m long, to the dwarf or bush types no more than 30 or 40 cm tall.

They are the most important grain legume, or "pulse", in Latin America and some parts of Africa, where they provide a valuable, protein rich supplement to carbohydrate staple food crops such as cassava, plantains and rice.

Haricots, as they are often known, are grown widely throughout the world, but are of relatively minor importance in India and most of tropical Asia where other legumes, which are indigenous to these areas, are more popular.

The annual global production is about 10 million MT, but is difficult to estimate since FAO includes haricots with many other "dry edible beans" such as mung bean, lima bean, etc.

In addition to *Phaseolus vulgaris* there are three important species of the genus *Phaseolus*:

- *P. lunatus (*Syn. *P. limensis, P. inamoenus)*—Lima Bean (large, usually white seed), Sieva Bean (small, coloured seed) and Butter or Madagascar Bean (**2B**)
- *P. aureus (*Syn. *Vigna radiata)*—Mung Bean, (Green or Golden Gram) (**2B**)
- *P. mungo (Vigna mungo)*—Black Gram, Mash, Urd or Woolly Pyrol

There are also a few other less important but useful and productive Phaseolus species, such as *P. acutifolius* var. *latifolius*—Tepary Bean (**2G**) and *P. coccineus (P. multiflorus)*—(Scarlet) Runner Bean

The **Moth Bean**, also called Mat bean or Turkish gram, was formerly classified as *Phaseolus aconitifolius*, but is now generally classified as *Vigna aconitifolas*.

The **Adzuki Bean** was formerly classified as *Phaseolus angularis* (or *Dolichos angularis*) but is now generally classified as *Vigna angularis*.

Plant breeders have successfully produced some hybrids between some of these species, such as *P. vulgaris* X *P. coccineus*, but so far none have proved to be very useful.

PLANTING
Propagation: by seed. Haricots are almost 100% self-pollinated.
Soil: haricots grow in most soil types. Optimum pH is 6.0–6.8, minimum 5.2, maximum 7.0.—"intermediate tolerance" to acidity. Sensitive to high concentrations of manganese, aluminium and boron. Rainfed crops need about 60–100 cm of topsoil to produce strong, healthy plants.
Intercropping: very common, for good reasons, with crops such as maize, sweet potatoes, coffee, cotton, etc. Earliest food growers appreciated the value of a haricot/cereal partnership.

Seed rate: this varies greatly, according mainly to variety. Some examples, in kg/ha: Canadian Wonder 55–70, pole snap beans 22–34, bush snap beans 56–170. There are approximately 2000–5000 seeds per kg.
Seed spacing: 5–23 cm between plants, 52–90 cm between rows.
Depth: 2.5–5.0 cm in heavier soil, 5–10 cm in light soil.
Inoculation: generally not necessary. The common "Cowpea Group" of rhizobia are often ineffective on *Phaseolus* species.

GROWTH CONDITIONS

Day length: French bean plants respond to day length according to species and sub-species. Most climbing types are either long-day or short-day, most bush types are day-neutral.
Growth period: 60–150 days for mature, dry seed; 45–75 days for green pods.
Temperature: optimum is 16–24°C. Growth stops altogether at c.10°C, and the plant is killed by frost. Maximum is about 30°C.
Rainfall: haricots can produce a reasonable crop with 400 mm rainfall per year if the rain is well distributed. They benefit from some rain at flowering and seed-set, and dry weather for harvest of dry beans. Yields start to decrease when rainfall exceeds 1500 mm, due to both flower drop and increased disease damage.
Altitude: approximate range, in msl: tropics 600–1950, Kenya 900–1500, Ethiopia 1650–1950.
Pests: in the tropics, pests are usually the main limiting factor to growing haricot beans:
– **Aphids**—cause damage by both sucking the plant sap and by transmitting virus diseases. Black aphids are often the worst, causing yellowing and distortion of the leaves.
– **Mexican Bean Beetle**—these eat the leaves, and are easily controlled with insecticides.
– **Bean Leaf Beetle**—the larvae attack the roots, the adults attack the leaves and stems.
– **Bean Flies**—very common in Africa. Stems crack and distort at the base (hypocotyl). Control is best by seed dressings, and also with early planting, rotation and the removal or destruction of volunteers and infected plants.
– **White Flies**—found mainly in Central and South America, they also transmit viral diseases.
– **Bean Pod Weevil**—the seed is attacked in the pod.
– **Potato Leaf Hopper** and **Green Leaf Hopper**—plants are stunted. Some resistance in some varieties, otherwise insecticides may become necessary.
– **American (Cotton) Bollworm**—mainly in Africa. Round holes are bored into the seed pods.
– **Spotted Borer**—the larvae, which are olive-green and hairy and have rows of dark spots, eat the seed inside the pod. Difficult to control.
– **Spiny Bugs**—cause damage by sucking plant sap, and also transmit a fungal disease.

Other insects which have a taste for haricot beans include the corn seed maggot, green stink bug, spider mites, cutworms, armyworms, root-knot nematodes and several storage insects.

Diseases: very common, especially fungal and bacterial diseases in lowland humid tropics and subtropics, and viral diseases in drier climates:

– **Anthracnose**—attacks the leaves, stems and pods. Seed-borne, and it also survives for at least two years in the soil. Control: resistant varieties, rotation and clean seed.

– **Bean Rust**—the leaves develop small white spots, which become rust coloured, the leaves then turn yellow and dry up. Control: burn infected plants and crop rotation; some varieties have some resistance to some of the races of Bean Rust.

– **Ashy Stem Blight**—seed-borne. Black marks on very young seedlings. Control: clean seed.

– **Angular (Grey) Leaf Spot**—grey spots on the underside of leaves, which become brown, and brown spots on upper leaf surfaces. Control: very difficult, though some varieties have some resistance; also the use of clean seed and crop rotations, and the destruction of infected plants.

– **Powdery Mildew**—all plant parts except the roots are infected. Control: various fungicides and resistant varieties.

– **Root Rots**—controlled with crop rotations.

– **Sclerotinia Wilt (White Mould, Water Soft Rot)**—usually after warm, humid weather. Control is not easy, but widely spaced plants and crop rotations, with cereals for example, or fungicides can limit the spread of this form of Wilt.

– **Southern Blight (Southern Wilt, Crown Rot)**—mainly in hot climates, the lower leaves become yellow and fall off. Very difficult to control.

– **Bacterial Diseases**—such as Common Blight, Halo Blight and Bean Wilt. Halo Blight is different from the other bacterial diseases in that it is more common in cool, wet areas, while Common and Fuscous Bacterial Blights (*Xanthomonas*) are more important in warm, wet areas.

Symptoms: irregular dark spots on leaves or pods, each spot surrounded by a yellow band or "halo". Control: clean seed and resistant varieties.

– **Virus Diseases**—at least five viruses attack haricot beans: Bean Common Mosaic (BCM, or Bean Virus 1) is the most common.

Also Bean Yellow Mosaic (BYMV, or Bean Virus 2), Curly Top, Golden Mosaic and Mottle Dwarf Virus Diseases.

YIELD

The global average yield of dry grain is approximately 500–1200 kg/ha, and in fact the haricot bean generally has a bad reputation for low yields even though its yield potential is at least 3.5 MT/ha. The average yield in Ethiopia is about 800 kg/ha.

Yields of the green pods are more stable and reliable, and vary from about 3 MT/ha in India and Africa to about 7 MT/ha in Europe.

UTILISATION

- **Dried beans** of haricots provide the bulk of the protein intake of huge numbers of people in many parts of South America, and some tropical parts of Asia and Africa. The beans can also be processed into protein concentrates such as milk substitutes. Although the nutritive value of haricot beans can be very high, it can also be very variable according to the variety, growing conditions and storage conditions. Fortunately, *Phaseolus* species (and other legumes) have a high content of the essential amino acid lysine, and so can supplement cereal proteins that are generally low in lysine.

- **Green pods** are the other main use for haricot beans. In Europe, America and other temperate zones they are more normally grown for these immature pods, either eaten as a vegetable or canned, dehydrated or frozen.
- **Green beans** are sometimes shelled from their pods and eaten as a vegetable.
- **Leaves** are sometimes used as a salad or potherb, or as famine food, in Asia and Africa.
- **Stems, dried leaves** and other parts not suitable for human consumption make useful livestock or poultry food, having a protein content of about 6%.

LIMITATIONS
- **Yields** of *Phaseolus vulgaris* are generally low, for a number of reasons, the most common being damage caused by diseases and/or insects. Inoculation of seed can sometimes be beneficial, but the yield response is unpredictable partly due to the uncertainty regarding appropriate Rhizobia strains for specific varieties and growing conditions. This topic is discussed in **1Fe**, page 54.
- **Cooking time** for the dry beans is relatively long, and must be vigorous to destroy lectins.
- **Deficiency** of the essential amino acids methionine and cystine (and tryptophan).
- **In storage** the grain is liable to develop moulds (fungi), hard shells and a bad taste.
- **Frost**—the plants are extremely sensitive to sub-zero temperatures.

Lentil
*Lens culinaris (*Syn. *L. esculenta, Ervum lens)*

(Red) Dahl/Dal, Split Pea; Gram
Lentille (French); Linse (German); Lenteja (Spanish); Lentilha (Portuguese); Adesi (East Africa); Ads Masri (Sudan); Missar (Amharic), Bersem, Bursun, Birsin (Ethiopia); Masur (Hindi); Sharkhal (Dari); Chaunangi, Chirisanagalu, Misurpappu, Thulukkappayar (India)

The lentil is one of the most ancient food crops, and has been cultivated since at least 6700 BC in the eastern Mediterranean. The red pottage of lentils for which Esau sold his birthright was probably made from the red Egyptian lentil. A number of different types of lentils are now grown in large areas in warm temperate and subtropical regions, and in the tropics as a cool season crop or at high altitudes.

The major producers are India, Pakistan, Turkey, Syria, Russia, Spain and Ethiopia. Lentils were introduced successfully into the New World and are now grown in Argentina, Chile and Washington State, North America. Global production estimated by FAO in 2004 was about 3.8 million MT.

The lentil plant is an annual, with slender stems and many branches, 15–60 cm tall, with small pale mauve, blue, pink or white flowers 6–8 mm long. The pods are small, about 3–9 mm across and no longer than 1.3cm in modern varieties. The pods contain 1–3 (normally 2) seeds which vary in colour from yellow, green, orange, red or grey to dark brown, sometimes mottled or speckled.

Grain protein content ranges from 22 to 35%, but the nutritional value is low because lentil is deficient in the amino acids methionine and cystine. Lentil is an excellent supplement to cereal grain diets because of its good protein/carbohydrate content, used in soups, stews, casseroles and salad dishes. Sometimes difficult to cook because of the hard seed coat that results from excessively dry production conditions.

In unmechanized ("traditional") farming systems the entire plants are normally pulled up and hung to dry, then threshed when required. The International Centre for Agricultural Research in the Dry Areas (ICARDA) in Syria is developing tall upright lentil varieties adapted to mechanical harvesting

The genus *Lens* is closely related to the *Lathyrus* and *Vicia* genera, and from a farmer's point of view lentils can be propagated and cared for as if they were field or garden peas (*Pisum sativum*)—except that, in general, lentils tolerate higher temperatures and drier soil than peas. The plants should be protected from wind whenever possible.

Lentils can be classified into two main groups:

- sub-species *macrosperma*—large-seeded varieties, with seeds 6–9 mm in diameter. Grown mainly in Africa, Asia Minor and the Mediterranean.
- sub-species *microsperma*—small-seeded varieties, with seeds 3–6 mm in diameter. Grown mainly in southwestern and western Asia.

PLANTING

Propagation: by seed, which is 2–7mm diameter. 15,600–100,000 seeds per pound. Usually self-pollinated, though some cross-pollination can occur.

Soil: lentils can thrive on fairly poor dry soil, but they prefer well-drained light to medium soils; they do not tolerate waterlogging. Moderately tolerant of salinity and acidity. Molybdenum is essential, and can be applied as molybdated gypsum at about 50–60 kg/ha or as a foliar spray.

Seed rate: when intercropped 10 or 15 kg/ha is enough. As a pure stand anywhere from 25 to 90 kg/ha or more are needed. 100 of the larger seeds weigh about 2g (18–20,000 seeds/kg).

Spacing: 15–30 cm between plants, 60–90 cm between rows. Sometimes planted in an almost square grid approximately 22 × 30 cm.

Depth: 1–6.5 cm according to seed size, soil type and moisture at planting.

Inoculation: seed should be inoculated with *Rhizobium leguminosarum* just prior to planting (within 24 hours).

GROWTH CONDITIONS

Day length: most varieties are long-day plants, some are day-neutral.

Growth period: 80–100 days for early varieties, 125–180 days for later maturing varieties.

Temperature: for germination the optimum is 18–21°C, minimum c.15°C. For growth, the optimum is about 24°C though varies according to variety. Maximum tolerated is about 30°C.

Rainfall: the plants are moderately tolerant of drought, especially ssp. *microsperma*, and survive long, hot summers. About 700 mm per year is usually enough, and pre-harvest and harvest periods should be dry. Lentils are not well suited to the hot, wet tropics.

Altitude: up to about 3500 m.

Pests: the most important are the gram caterpillar, white ants, gram cutworm and weevils (which are also a storage insect problem with lentils).

Diseases: there are two important fungal diseases:

– **Rust**—the leaves and stems turn purple, and may die, especially in humid conditions.

– **Wilt**—the leaves curl and the root system develops poorly, especially in light, dry soil.

Both diseases can be controlled with crop rotation, seed dressing, use of resistant varieties and destroying the diseased haulm.

YIELD

As part of a mixed cropping system, lentils may only provide a few hundred kilos per hectare, while pure stands even in relatively simple systems should yield at least one metric tonne(MT)/ha. With adequate rainfall and proper cultivation techniques a modern "improved" variety should give at least two MT/ha provided that disease and insect attacks are kept under control.

FAO reported that the global average in 2004 was 966 kg/ha, from a high in China of 2.5 MT/ha to a low of 100 kg/ha in Uzbekistan.

UTILISATION

- Lentils are not only highly nutritious, containing about 25% protein, 1% fat and 56% carbohydrate, they are also more easily digested than most of the other legumes, and so can be invaluable in emergency feeding programmes. The split seed, known as *dhal*, is normally eaten in soups and porridges. Lentils are also a good source of Vitamin A (up to 200 IU/100 g), Vitamin B_1 and B_2, iron (7 mg/100 g) and phosphorus.
- The grain is a source of commercial starch, used in textiles and printing.
- The flour can be mixed with cereal flours to make cake or bread, or invalid and baby food.
- The young pods are sometimes eaten as a vegetable, in India for example.
- The crop can be grown as a green manure, enriching the soil with Nitrogen and organic matter.
- The straw, or haulm, is a very nutritious animal food, being richer in proteins and lower in fibre than the other legumes.

LIMITATIONS

* Lentil plants are very susceptible to diseases.
* There is a lack of well adapted varieties, even modern ones being unsatisfactory with regard to both disease resistance and yield.
* As human food the lentil has a certain notoriety regarding flatulence.

Lima Bean

Phaseolus lunatus (Syn. *P. limensis, P. inamoenus)*

Sieva Bean (small, coloured seeds), Madagascar Bean, Butter Bean, Burma Bean, Curry Bean, Rangoon Bean, Sugar Bean, Towe, Potato Limas; Pois Amer, Pois de Sept Ans, Haricot de Lima, Haricot du Cap, Haricot du Kissi (French); Limabohne, Mondbohne (German); Judión, Judía de Lima, Poroto de Manteca (Spanish); Feijão de Lima (Portuguese); Abangbang, Chuku (Uganda); Feijão Espadinho*, Chitanga ** (Angola); Roaj (Sudan); Habichuela (also for Haricot Bean), Sibatse Simaron, Tagalo Patani, Zabache (Philippines); Tua Rachamat (Thailand); Lobyan (Pashto)
Feijão Espada (also known as *Feijão de Cobra, Feijão Holandês* and *Feijão de Porco)* are Portuguese names for the common jackbean (*Canavalia ensiformis).*
The Lima Bean is also described as *Feijão de Porco* in Brazil.
**Chitanga* is often used in Angola to describe any type of climbing bean

The Lima Bean is widely grown in Central and South America, North America and Canada and many parts of Asia, especially Burma. Lima beans are one of the major food legumes in the humid rain forests of Africa and Madagascar. It is generally a tropical and lowland species, and needs a hot growing season.
    There are two main types:
– **Pole (climbing) varieties**—twining, perennial herbs which either trail along the ground or are grown up supports. Can grow 2–4 metres in length. Long growth period.
– **Bush varieties**—developed by selection of appropriate mutations during cultivation to be annuals 30–90 cm tall. Can mature in 60–70 days.
    Botanists sometimes divide the Lima bean into two separate species, *Phaseolus limensis*—the Lima bean, mainly perennial but cultivated as annuals, both pole and bush types, with large normally white seed, and *Phaseolus lunatus*, the Sieva bean, an annual with smaller, coloured seed, and which is more tolerant of drought and high temperatures.
    However these two species (so-called) cross pollinate readily, and both display all forms of growth habit, seed size and colour, etc., so the Lima bean is best considered as one species, *Phaseolus lunatus*.

PLANTING

Propagation: by seed. Mainly self-pollinated, with up to about 18% cross-pollination.

Soil: adapts to a wide range, including acidic soil, provided they are well drained and aerated, and not saline. Optimum pH is 6–7; Lima beans (and cowpeas) often will grow well in soils that are too acid for other beans. Mulching often improves the growth and yield of Lima beans.

Seed rate: from 10 kg/ha for pole types in India to 130–170 kg/ha for large-seeded bush types.

Seed spacing: Bush types: 5–20 cm between plants; 10–15 cm between rows for large seed, 7–12 cm for small seed. Climbing types: 15–30 cm between plants, 75 cm between rows—or on hills, 120 × 120 cm, with 3–4 plants per hill. Seed size varies from 500–1100 seeds per kg.

Depth: 2.5–5 cm in moist, heavy soil, up to 10 cm in lighter, drier soil.

GROWTH CONDITIONS

Day length: the wild types and some of the varieties from the tropics are short-day plants, while most other types are day-neutral.

Growth period: the fastest is about 65 days, up to about 100 days, for the earliest (ie fastest growing) bush varieties. Large-seeded white types can take 200–270 days, producing pods for several months ie indeterminate growth habit. In the tropics lima beans can be grown as a perennial.

Temperature: Limas need more heat than Haricots to germinate and grow well, and growth is slow below $13°C$. Frost sensitive. Optimum is $16–27°C$.

For planting, soil should be at least $18°C$.

Rainfall: tolerates higher rainfall than Haricots, and does best in humid and sub-humid tropics with 900–1500 mm annual rainfall or more. Can grow with only 500 mm during the growth period, but then needs some top-up irrigation and also at least about 70% relative humidity at flowering and pod set.

Rotation: ideally Lima Beans should follow a well-manured crop such as potatoes, though they are often grown on the same land for many years.

Altitude: the optimum is 900–1800 m in the tropics and 900–1200 m in temperate zones.

Pests: insect attacks can be serious, though it is a tough plant and more resistant than other legumes of the humid tropics. The plant is attacked by the same insects that damage other *Phaseolus* species, and the Lima Bean Pod Borer is also a widespread problem. The Cowpea Weevil attacks both in the field and in storage, and the Bean Bruchid can be a serious problem in stored grain.

Diseases: like most beans, the Lima is susceptible to a number of diseases, including:

– **Root Rots**—controlled with seed dressing.

– **Pod Blight**—seed-borne, so can be controlled by only using clean seed.

– **Downy Mildew**—worse in warm and humid conditions.

– **Bacterial Blights**—there are three types: Common, Halo Blight and Bacterial Spot. All can be controlled with clean seed and rotation.

– **Yeast Spot**—the seeds are damaged, "pitted", without visible damage to the pod.

– **Anthracnose**—controlled with clean seed and crop rotation.

– Virus Diseases—there are several, such as Bean Golden Yellow Mosaic, which prevents the plant from producing seed.

YIELD

In unmechanized agriculture yields of lima beans can be up to about 1.5 MT/ha, but about 300–600 kg/ha is more common. In good conditions lima beans can produce more than 3 MT/ha, and in North America yields of 4.5 MT/ha are common.

These high yields indicate the strong potential of this crop, though it is not always easy to find good seed of varieties that are adapted to marginal conditions.

UTILISATION

- **Dry beans**—Lima beans are normally grown to produce the dry, mature bean, often called "butterbeans". They are normally boiled, fried in oil or baked. Care should be taken as some older varieties, mainly the ones with dark seed colour, contain dangerous levels of poisonous cyanogenetic glucosides such as phaseolunatin and linamarin which can produce toxic HCN, hydrocyanic (prussic) acid. The dry beans contain adequate but not exceptional levels of protein (20%), fat (1.5%) and carbohydrate (60%). They are also used to produce protein-rich bean flour.
- **Green pods**—these are normally "picked and picked again". In the tropics, and elsewhere, they are sometimes cooked and eaten as a green vegetable. In North America they are canned and also sometimes frozen, and the immature seed is canned or frozen.
- **Leaves and stems**—can be fed to animals direct, or made into hay or silage.
- **Cover crop** (page 63) and **green manure crop** (page 69).

LIMITATIONS

- The yield and cooking quality of lima beans can be disappointing.
- Modern, improved varieties are in short supply.
- cooking time can be long for dark seed colour types, to destroy the poisons (see "Utilisation", above). The beans should be either roasted or well boiled in two changes of water.
- Temperature requirements for growing a crop are quite specific, and the plants are also frost sensitive.
- Indeterminate growth habit of the climbing types and older varieties (though this is often more of a benefit than a limitation for subsistence farmers).

Mung Bean
Phaseolus aureus (Syn. Vigna radiata)

Green Gram, Golden Gram, Jerusalem Pea, Oorud Bean;
Haricot Doré, Haricot Mungo*, Ambérique* (de Madagascar), Boubour (French);
Mungbohne (German); Judia Mung, Poroto Mung, Frijol de Oro (Spanish);
Chickasano, (Lubia)Chiroko* (Africa); Batong-hidjao*, Mongo, Mungo
(Philippines); Tientsin Green Bean (Asia); Tua Kiew, Tua Tawng, Tua Tong
(Thailand);
Mai (Pashtu)

*these are also names for Black Gram (a.k.a. Urd (Bean), Mash, Woolly Pyrol)
Phaseolus mungo Syn.*Vigna mungo*

There are more than 2000 types of this fast growing legume, which display a large
range of plant type, form and adaptation. The majority are 30–90 cm tall, while other
types are more than a metre high.

Two main types of mung bean are recognised: **green gram**, grown mainly for
human food, either cooked or as sprouts, and **golden** or **yellow gram** which is
mainly grown for hay and other animal food, or for a green manure or cover crop.

The mung bean is very similar to the black gram (see above); the two can be
distinguished as follows:

	Black Gram	Mung Bean
Pods	Erect or sub-erect, with long hairs. 4-10 seeds	Spreading or reflexed, with short hairs. 10-15 seeds
Seed	Larger, oblong, smooth, mainly black, sometimes olive green, with "beak"	Smaller, round, mainly green, sometimes yellow or blackish, without "beak"
Hilum	Concave	Flat, round, white

Mung beans are probably the most important bean crop in Asia and in large areas
East of Pakistan. They are also widely grown in India, Thailand, Indonesia,
Bangladesh, the Philippines, Africa, Australia and the Americas.

PLANTING
Propagation: by seed. Almost always self-pollinated.
Soil: must be well drained. Mung beans tolerate both alkaline and saline soils, but
are sensitive to acidic ones. Responds well to phosphatic fertilisers, in both granular
and foliar form.
Seed rate: 3–4 kg/ha when broadcast or intercropped, 5–22 kg/ha when grown in a
pure stand. 100 seeds weigh 2–4 g.
Seed spacing: very often broadcast. If in rows, 4–5 cm between plants, 25–88 cm
between rows.
Depth: 3–4 cm.

Rotation: often grown after rice or another cereal. Rootknot nematodes can cause serious problems if mung beans are grown too often on the same land.

Intercropping: mung beans are often grown as a subordinate crop, mixed with cereals such as sorghum, maize or millet, and even sugarcane.

GROWTH CONDITIONS

Day length: most varieties of mung beans are short-day, though long-day and day-neutral varieties also exist.

Growth period: 80–120 days, up to 150 days, on average, though some new varieties mature in 65 days. The green pods can be eaten 50–70 days after sowing.

Temperature: frost sensitive, and requires warm conditions, 30–36°C being optimum.

Rainfall: 650 mm minimum, 750–900 mm optimum. Poor seed-set if it rains during flowering, and ideally the plants should mature during the dry season or a dry period.

Altitude: 0–1800 m.

Pests: in Africa mung beans appear to be less susceptible to pests and diseases than haricots (French beans) or cowpeas, while in southeast Asia it is attacked by the same pests and diseases as the other legumes, such as:

– **Bean Fly**—the most serious pest in southeast Asia. Control: granular carbofuran at planting.

– **Root-knot Nematodes**—controlled with crop rotations. Some mung bean varieties also act as hosts of the soybean cyst nematode.

– **Aphids, Cutworms, Pod Borers and Red Spider Mites**—can be troublesome, such as in the Philippines, and may need chemical control.

– **Storage Pests**—such as the Cowpea Weevil, which also attacks the plants in the field, and the Storage Weevil, a problem in Thailand, which affects some mung bean varieties more than others.

YIELD

Average yields of mung bean are about 250–700 kg/ha (1 MT/ha in North America), though modern improved varieties can yield more than 2 MT/ha in good conditions.

Individual pods can be removed as they ripen, or the whole plant can be uprooted and either sun-dried or dried indoors.

UTILISATION

- **Dry beans**—mung beans are mainly grown for human food, the dry beans being either boiled and eaten whole, or split and made into dhal. They are highly nutritious, with a protein content of 22–31% (average 25%) with modern varieties, and are more digestible than most other legumes. The beans are also made into flour, or used as a source of starch, or made into sprouts; the sprouts are also canned and quick-frozen. Dry beans are sometimes used for animal food, mainly poultry, when they are either roasted or boiled.
- **Green, immature pods** are sometimes eaten as a vegetable.
- **Leaves and stems** after harvest can be used as fodder; protein content is c. 10%.
- **Cover crop** and **green manure crop.**

LIMITATIONS
- Traditional mung bean varieties have relatively low yield and protein content.
- Modern, improved varieties are in short supply.
- Shattering of the pods can be a problem.
- Pests and diseases can devastate crops.
- The plants are frost sensitive and need warm growing conditions.
- Weeds can easily overcome the growing crop, especially in the first few weeks.

Pigeon Pea
Cajanus cajan (Syn. *C. indicus, C.bicolor, C.flavus, Cytisus cajan*))

Red Gram, Yellow Dahl, Angola Pea, Congo Pea or Bean, Gungo Pea, No-eye Pea, Cajan (Catjang) Pea, Puspo (Sacha) poroto, Shantouken, Tuvaram, Toovar; Embrevade, Pois Cajan, Pois Nain, Pois d'Angola, Pois de Congo (French); Straucherbse, Taubenerbse, Strauchbohne (German); Frijol Guandul, F.de la India, Guisante de Paloma, G.Enano (Spanish); Guando, Guandu, (Feijão)Andu, Jinjonji, Ervilha (Feijão) de Congo (Angola), Feijao Boer (Mozambique); Ervilha de Angola, Ervilha de Sete Anos, Feijão de Árvore, F. Guandu, F. Andu, Guandeiro, Guando, Guandu, Andu, (Portuguese); Yewof-attah, Ohota-farengota (Ethiopia); Ads Sudani, Lubia Adassi (Sudan); Burusa, Apena, Lopena (Uganda); Arhar, Tur, Adhaki, Arahar, Ihora, Kandalu, Cror (India); Cadios (Philippines); Togare (Thailand); Frijol de Palo (Salvador); Guando/u (Latin America); Gandul (Central America)

The pigeon pea is an erect, hardy, woody shrub 0.5–5 m tall which adapts to a wide range of growing conditions. About 90% of global production is grown in India, other main producers being Uganda, Malawi, Tanzania, Puerto Rico, Dominica, Burma and southeast Asia. The annual global production in 2004 was estimated by FAO as 3.3 million MT, though very little enters international trade; it is the world's fifth most important pulse crop.

The plant is a short-lived perennial which can be harvested for five years or more, though yields are normally lower after the first year; it is therefore often grown only for one or two years.

It is a useful crop to grow in poor soils and where fertiliser is unavailable or expensive, and is a suitable crop for "no-till" (ie no ploughing) agriculture. It can nodulate in most soils, producing more nitrogen than most other legumes.

The grain is a good source of protein (18–30%), carbohydrates and minerals. There are two distinct botanical varieties, or sub-species:

var. *flavus*—earlier maturing, semi-dwarf, with yellow flowers and green seed pods 4–10 cm long with 2 or 3 light-coloured seeds. Known as "*Tur*" types in India, or No-eye Pea in Jamaica.

var. *bicolor*—later maturing, larger perennial types, with red and purple streaks on the flowers and hairy, purple seed pods 4–5 cm long with 1–5 darker coloured, spotted and coarser seeds. Known as "*Arhair*" or "*Arhar*" types in India, or Congo Pea in Jamaica.

There is also a wide range of intermediate types, with great variation in height, growth period and photoperiodic response. Many named varieties are available, more than 100 in India alone, and there is a major plant breeding effort underway to improve yield and disease resistance.

PLANTING
Propagation: by seed, but cuttings can also be used. They are mainly self-pollinated, though up to 40% cross-pollination can occur. Inoculation can greatly increase yields.

Soil: pigeon peas grow well in infertile, but not waterlogged, soils (ridge planting is recommended in soils liable to waterlogging). Moderate tolerance to saline soils. Optimum pH is 5–7. Yields are reduced when soils are deficient in Manganese or Phosphorus. Rarely responds well to fertilisers, though Phosphorus often increases the efficiency of root Rhizobia, which can fix up to 100 kg/ha of Nitrogen.

Intercropping: very common, with maize, sorghum, finger millet, pearl millet, cassava, sesame and groundnuts.

Weeds: weeds can quickly overcome a pigeon pea crop and should be well controlled for the first few weeks.

Depth: 2.5–8 cm.

Seed spacing: not very critical as the plants are good at compensating. Average is 60 cm between plants, 120 cm between rows. When intercropped, one row of pigeon peas is often planted for every 3–5 rows of the primary crop.

Seed rate: 1–6 kg/ha intercropped, 9–30 kg/ha in pure stand. 17–19,000 seeds/kg.

GROWTH CONDITIONS
Day length: all three types are found, though most modern varieties are short-day, in which case the sowing date is critical as this affects the growth period, height and yield.

Growth period: very variable, from 80 days to more than a year for the first grain harvest, depending on variety, planting date, growing conditions and location.

Pruning of the branches after the first harvest encourages branching and helps maintain yield of the second harvest. Both determinate and indeterminate types exist.

Temperature: pigeon peas are sun-loving plants and most varieties are frost sensitive at all stages. The optimum temperature is 18–29°C, minimum is 10°C and maximum about 35°C.

Rainfall: drought resistant; ideal growing conditions are 600–1000 mm/a., with heavy rainfall for the first weeks of growth and a dry period for flowering and harvest. The minimum is about 350 mm/a., maximum is about 2500 mm/a. Wet weather tends to produce excessive vegetative growth at the expense of grain production.

Altitude: in Venezuela pigeon peas are grown up to 3000 m.

Pests: not often a big problem; the worst are the Gram Caterpillar, Red Gram Plume Moth, Gram Pod Fly, American Bollworm and Spotted Pod Borer.

In storage Pulse Beetles and Bruchids can cause serious damage if they are not kept under control with fumigation.

Diseases: Fusarium Wilt is usually the most serious; it is a soil-borne fungus, controlled with a 3–5 year crop rotation and/or resistant varieties. There are also a number of diseases caused by viruses, such as Pigeon Pea Mosaic.

YIELD

Commonly achieved yields of dried pigeon pea seed are from about 250 to 900 kg/ha when intercropped, and 1.7–3.5 MT/ha when in pure stand.

FAO estimated the global average yield for 2004 was 721 kg/ha, with a high in Trinidad and Tobago of 2.7 MT/ha and a low of 479 kg/ha in Panama.

Green pods harvested from pure stands should yield 1–5 MT/ha.

UTILISATION

Pigeon peas are potentially immensely useful and are a fine example of a **multi-purpose crop**, which can be used to satisfy a vast range of needs, listed below:

- **Mature, dry grain** contains 15–32% protein (24% on average) and is also a good source of Vitamin B, methionine and cystine. Normally used in the form of "split peas" soaked in water for several hours, then either boiled, or pounded and then fried or steamed. Often eaten with maize, it is said to be more nutritious when eaten with rice. The seed can be germinated to produce sprouts; small-seeded varieties or crushed seed can be used as poultry food.
- **Fresh, green seed** is eaten raw as a vegetable in many countries and can be canned.
- **Green pods** can be boiled and eaten as a vegetable or in curries.
- **Animal fodder**—either fresh plants can be fed or grazed, or dried plants can be ground into a meal for winter or dry season fodder.
- **Green manure, support crop, windbreak, shade crop** and **erosion control.**
- **Dry stalks/stems**—used for firewood, thatching, basket-making and charcoal.
- **Traditional medicines** and raising **silkworms.**

LIMITATIONS

- Pigeon pea plants are sensitive to frost, acidic soil and waterlogging.
- The germination rate falls rapidly in humid conditions.
- The plants are often slow to reach maturity, particularly the older, traditional varieties.
- Humid conditions reduce yields, while dry conditions can cause shattering.

Soybean
Glycine max (Syn. G. soja, G. hispida, Soja max)

Soya, Soy, Soy Bean, Soya Bean, Soyabean;
Soya, Soja (French); Sojabohne (Germany); Soja (Spanish & Portuguese)

Soybeans are highly efficient producers of protein and oil for both human and animal food. They are also widely used in industry, and are the most important of all the grain legumes in the world in economic terms.. Global production in 2004 was estimated by FAO at 204 million MT.

The main nutritional attributes of soya are, per 100 g of edible portion of the seed, 380 kcals energy, 35–40 g protein and 18–24 g oil. In common with groundnuts, soybeans can be considered to be an oil producing crop (they are sometimes referred to as "oilseed legumes") because of their high oil content. The oil contains useful amounts of phosphorus, and is normally considered to be a "drying" type, and sometimes as "semi-drying" ie it normally forms an elastic film when in contact with air. The glycerides are of the unsaturated type.

The amino acid distribution of soybean protein is more similar to animal protein than the protein from most vegetable sources, containing for example 5.4% lysine. It is thought that the protein in soya may lower blood cholesterol, and its isoflavines may reduce the risk of some cancers.

They were first cultivated in China around the eleventh century BC, the earliest soya plants being climbing types. Soybeans need climates with hot, humid summers to grow well and are rarely satisfactory as a European crop. Plant breeding work is being done to produce varieties suitable for cultivation in the UK by selecting fast growing plants able to mature in the short North European growing season.

The plant is an erect, annual legume, 30–170 cm tall, which sheds its leaves at maturity. The flowers are small, and either white or lilac/purple. The pods contain 1–5 seeds, normally 2 or 3, which are highly variable in size and colour between varieties. The seed is normally either yellow, green, brown or black.

PLANTING
Propagation: by seed. For northern temperate regions, modern quick maturing varieties are adapted to flower and fruit in the shortening days of July and August. Traditional varieties only flower during the lengthening days of early summer and so cannot produce any seed in the cool, short northern summers.
Soil: soya will grow well in slightly acid soil, the optimum being about pH 6–6.5. On more acid soils lime should be applied 2 or 3 months before planting. Heavy soils will grow soya if they are well drained, and light soils also if they have sufficient moisture, but the best soils are rich, sandy loams. Adequate amounts of phosphate, potash and calcium are essential, and about 25 kg/ha of Nitrogen is normally beneficial even for soya growing in good soils. Most soya varieties are moderately tolerant to salinity, and some varieties are tolerant.

Seed rate: 40–90 kg/ha. Seed size varies from 5000 to 40,000 per kilogram, the average for modern varieties being about 10,000–20,000.

Seed spacing: a rule of thumb is to have between 25 and 40 plants per metre along the row, with 60–100 cm between rows, to arrive at an optimum plant population of about 400,000 per hectare.

Depth: 2–4.5 cm.

Intercropping: generally not recommended for soybean except when grown for hay or forage, when it can be mixed with maize, sorghum or Sudan grass.

Inoculation: the seed should be inoculated with the appropriate strain of Rhizobia when planted in soils which have either not grown soyabeans for two or three years or which have never grown soya. In most cases, and if carried out properly, this procedure will increase both the yield of seed and the amount of Nitrogen fixed in the soil (**1Fe**).

GROWTH CONDITIONS

Day length: understanding the ways in which soybeans respond to the changing length of day—the latitude, in other words—is very useful when deciding to introduce new varieties into an area. Most varieties of soybeans are very sensitive to photoperiod—the relative lengths of day and night—and are said to be "photoperiod sensitive". They tend to flower earlier as the day length becomes shorter. As a result any one particular variety is well adapted to grow properly only in a narrow latitude band, approximately 160–250 km from North to South. As an example: in the Northern hemisphere, if a soybean variety adapted to one latitude band is grown north of this band it will flower, and therefore mature, later than normal, perhaps too late before the winter. If that same variety is taken South of this latitude band, it will flower and mature earlier than normal, producing small, underdeveloped plants—or it may not flower at all.

Most soybean varieties are short-day plants   they normally start to flower soon after the day length begins to shorten ie 21st June in the northern hemisphere and 21st December in the southern hemisphere. This topic is also discussed in **1Ej**, page 42.

Growth period: for mature beans, anywhere from 80 to 175 days, depending on the variety, location and planting date.

Temperature: soil temperature at planting should be at least 10°C; 15°C is even better. The ideal temperature range for growth for most varieties is 24–32°C. Both young and mature plants of most varieties can withstand a little touch of frost, but not too much.

Growth habit: most varieties of soybean are *determinate*, meaning that the plant only increases in height very little after it has started to flower, and also the flowering period is relatively short so that the grains tend to mature at more or less the same time. Other varieties, including many of the older, traditional ones, are *indeterminate*, meaning that their height, or length, increases from two to four times after it has started to flower. These types continue to flower over a much longer period, so that the pods and seed also mature over a much longer period, as discussed in **2Ei**, page 42.

Rainfall: to produce really good crops of soya, a regular supply of rain, or irrigation, is needed. Water stress is particularly damaging during the period from the flowering stage to just before pod maturity. On the other hand, soybeans need less water than maize.

Altitude: soya cultivation is restricted to altitudes which have about 100 days of growing period with average temperatures above about 20°C. Some varieties, such as Chippewa, are adapted to the lowland tropics.

Pests: insects can devastate fields of soya. The most serious are the **Semi-looper caterpillars, Bean Fly, Cutworms, Aphids** and **Snout Beetles.**

Diseases: altogether at least 50 diseases attack soybeans; all three types of disease occur:
– **Bacterial**—Stem Blight, Leaf Pustule, Bacterial Blight
– **Fungal**—Downy Mildew, Wild Fire, Frog-eye Spot
– **Viral**—Soybean Mosaic, Yellow Bean Mosaic. Most of these diseases of soybean can be controlled to some extent by using a combination of resistant or tolerant varieties with seed dressings and crop rotations. Aphid numbers may have to be kept under control with insecticides to reduce Mosaic and other viral diseases, but spread of the other diseases should be possible without recourse to pesticides.

Nutrient deficiencies: the symptoms of major, minor and trace element deficiencies are described below; though these symptoms are often partly masked by diseases. The most common are:
– **Phosphorus**—plants are stunted and have blue-green leaves.
– **Potassium**—leaf edges become yellow.
– **Iron**—yellowing between the leaf veins.
– **Manganese**—also yellowing between the leaf veins.
– **Molybdenum**—plants are stunted; more common in acid soils.
– **Zinc**—leaves & stems smaller than normal, yellow between leaf veins, especially lower leaves.

Trace element toxicities:
– **Manganese**—especially in acid soils; leaves, especially those at the top of the plant, are crinkled with downward turned margins.
– **Boron**—crinkled leaves, which may fall off, and which die from the leaf margins inwards.

For more information on Nutrient Deficiency, Trace Element Toxicity, etc., see **1Cd, "Trace Elements"** pages 23–29.

YIELD

The global average yields of soybeans have been steadily increasing for more than 20 years, from around 1400 kg/ha in the mid-1970s, to more than 1700 kg/ha in the mid-1980s to around 2 MT/ha in the late 1990s (2.23 MT/ha in 2004, according to FAO). In good growing conditions, modern soybean varieties can consistently produce more than 5 MT/ha.

Ideally soya seed should be stored at 10% moisture, and at no more than 12% moisture if it is to be kept for any length of time.

All seed stocks should be **handled and stored with great care** in order to avoid falls in germination rates that can be dramatic.

UTILISATION
- Soybeans are widely used in a vast range of industries—adhesives, nutrients, fertiliser, textiles and fibres—and in the food industry for both humans and animals. The grain contains no starch and so it is a good protein source for diabetics. Soya is the main source of lecithin, used in food processing such as an emulsifier in margarine.
- Cheese-like products such as tempeh, tofu and miso are made from soya
- In many countries the high protein content of soybeans is consumed as soya "milk". The soya grains are lightly toasted and ground into a rough flour which is then boiled with water and drunk as a kind of soup.
- A few varieties such as Butterbean, Frostbeater and Envy are eaten as green beans, cooked when the seed is full size but still green and tender.
- Soybeans can be used as a nitrogen enriching green manure crop, sometimes planted together with cowpeas.
- The growing crop can be fed to animals, either grazed or as forage, but care should be taken with immature seed which can cause abortion due to the oestrogens present in it.

LIMITATIONS
- Most varieties of soybean are only adapted to grow well in a relatively small geographical area, mainly in temperate regions, and to perform to their full potential in technically advanced agricultural systems, including high fertiliser input. There are relatively few varieties adapted for the tropics or subtropics or for cultivation by subsistence farmers with modest inputs.
- No varieties have yet been developed with true frost resistance, and they require a longish growing season, two reasons that they are rarely grown in northern Europe.
- Yields are low in acidic and/or infertile soil, and when there are insufficient accumulated heat units during the growing season;
- The beans are not very palatable to humans until they have been elaborately prepared and cooked.
- The seed may have to be inoculated (**1Fe**, page 54), which can be logistically challenging.
- The seed is delicate, so the germination rate falls rapidly in storage when the temperature or humidity is high, or if the seed is transported and/or handled roughly.

Vetch (Grass Pea or Chickling)
Lathyrus sativus

Chickling Pea, Chickling Vetch, Blue Vetchling, Lathyrus (Pea), Manila Bean, Indian Vetch;
Gesse Blanche, Gesse Chiche, Gesse Commune, Gesette, Lentille d'Espagne, Vesce {Gesse} (French); Saat Platterbse (German); Pisello Bretonne, Pisello Cicerchia (Italian); Almorta (Spanish); Sebbere, Sabberi* (North Ethiopia), Guaya (Amharic and Oromifa); Batura, Chural, Kansari, Kisari, Latri, Santal, Teora (India); Matri (Pakistan); Kalool, Patak (Dari)
*also used for Cowpea.

The Vetch or Grass Pea is a hardy and very drought resistant crop that is grown mainly in India, and in other dry areas with poor soil and low rainfall such as Eritrea and Ethiopia, especially in times of famine. The plant is a native of southern Europe and western Asia but nowadays it is not commonly grown outside India, Ethiopia and Eritrea. It is in fact a fodder crop, but the seeds are often eaten by poor people in emergencies such as famine. It was one of the earliest crops grown and was farmed in the Middle East more than 5000 years ago.

It will grow in very infertile soils, if they are not too acidic. The plant has a climbing or straggling growth habit, with many branches, which can grow up to 9 m long; it looks similar to *Pisum sativum* the common or garden pea but has much narrower and more elongated leaflets. The flowers are usually blue, but are pink, red, violet or white in some varieties. The pods are small and flat, 2.5–4 cm long, and contain 3-5 seeds which are either white, black, brownish-grey, yellow or spotted. Large, white seed is the most popular for human food. The smaller seeded types (called *Lakhori* in India) have a weight of about 5–7 g per 100 seeds, while the larger seeded types (*Lakh*) have a weight of 7–18 g per 100 seeds. The seeds contain about 25–28% protein, 0.6–1% fat, 45–61% carbohydrate and 3% mineral matter. It is the cheapest available pulse crop in most of India.

Maps showing areas where Vetch is grown in Ethiopia are available as gif files on http://www.general.uwa.edu.au/u/enneking/lathyrus/ It is estimated that 100,000–140,000 Ha is currently cultivated in Ethiopia.

Lathyrism
The seed of vetch also contains a non-protein, neurotoxic amino acid (beta-oxalyl-diamino-propionic acid) which can cause paralysis of the lower limbs in children, men, horses and cattle, known as lathyrism, if the grain is eaten for an extended period of time. It is said that a diet with 30–50% of grass pea eaten for 3–6 months can cause lathyrism, which is sudden, acute and irreversible ie there is no cure. In some very poor parts of India and Africa lathyrism is a public health problem (and the crop is sometimes banned).

Vetch is also found in some parts of North Africa, including Eritrea and North Ethiopia, where its local name of *T'Sebbere* means "the breaker". It is thought that lathyrism sometimes occurs due to admixture with seed of the common vetch (*Vicia sativa*). The toxin is greatly reduced if the seed is soaked in plenty of water for some hours, then sun-dried, or if the grain is partially boiled in the same way as rice. Unfortunately this causes the B Vitamins to be lost. The toxin is not destroyed during the normal methods of preparing chapattis. There is much research work currently in progress to breed varieties with lower levels of the amino acid.

People who regularly eat vetch (and HIV patients) may appear to be well nourished but they should be included in food relief and supplementary feeding programmes since malnutrition, particularly Sulphur amino acid, Zinc and Manganese malnutrition during famine caused by drought is a key factor leading to lathyrism. Diets with N-acetyl-cysteine are likely to be beneficial, but a nutritionist should be consulted for formulation together with other essential nutrients and safe dosage of this very acidic Sulphur amino acid precursor.

(Source of information: Dr Dirk Enneking, http://barley.ipk-gatersleben.de)

PLANTING
Soil: vetch/grass pea adapts to a wide range, including poor, infertile soils and heavy clays. The plants tolerate waterlogging better than other food legumes but they do not grow well in acidic soils. Fertiliser is not normally applied to the crop.
Intercropping: very common, for example with rice, barley, linseed and chickpeas.
Seed spacing: normally broadcast. When drilled, 2.5 cm between plants, 30–50 cm between rows.
Seed rate: 45–55 kg/ha pure stand, 20–40 kg/ha when intercropped.
Inoculation: not normally necessary, nor worthwhile.

GROWTH CONDITIONS
Growth period: 120–180 days. The plants are either cut with a sickle or uprooted, and then dried, as soon as the pods turn yellow - delayed harvest leads to a big seed loss from shattering.
Temperature: the optimum is 10–25°C—in India, for example, it is considered to be a "cold weather crop".
Rainfall: very drought resistant. 380–650 mm a year is usually enough, though it also tolerates heavy rainfall.
Altitude: in India vetch/grass pea is grown from sea level up to about 1200 m.
Pests: not normally a problem. Red-legged mites and aphids sometimes cause some damage.
Diseases: in India, mildews, rust and wilt are reported to cause some damage. Some wilt resistant varieties are available.

YIELD
The average seed yield of the grass pea in India is about 250–450 kg/ha.

In a pure stand, and in good conditions and with efficient cultivation it can yield one MT/ha or more.

Yields of forage ("hay") in a pure stand also average just over one MT/ha.

UTILISATION
- **Seed**—mainly for human food, especially in times of famine. It is either boiled and eaten as a pulse, or split and used as dhal, or ground into flour to make chapattis, paste balls and curries. It can be mixed with other more valuable pulses such as chickpeas or pigeon peas. The seed is also used as animal and poultry food and in homeopathic medicine.
- The **leaves** and **immature pods** can be boiled and eaten as a vegetable.
- **Whole plant**—either grazed, or cut and carried to animals. Hay can be made from it, but not silage. Fresh young plants are harmful to horses, but not to sheep, cattle or rabbits.
- **Green manure**—at a seed rate of about 65 kg/ha; about the same amount of Nitrogen is sometimes applied.
- **Catch crop**—between rice crops, both as a grain crop and for fodder.

LIMITATIONS
- Yields of vetch/grass pea are normally low.
- The seed is very deficient in methionine and tryptophan - see the first two pages of this section on the vetch/grass pea.
- The long, straggling plants can become a weed, in barley for example.
- Lathyrism (paralysis of the lower limbs) is a potential danger when the grain is eaten for prolonged periods—see the first two pages of this section on the vetch/grass pea.

2C. OILSEED CROPS

About 90% of the oil derived from plants comes from only ten or twelve crop species, though about two hundred other crops contain usable amounts of oil and are used locally. Crops such as soybean, maize and cotton are important sources of vegetable oil, though because their oil is a by-product they are often not regarded as being true oilseed crops. The term "oilseed" is used to describe what are in fact either fruits (such as olive and sunflower) or seeds (such as rape and soybean). The importance of oilseed crops is likely to increase as non-renewable sources of mineral oils become exhausted, unavailable and/or increase in price.

The seed or grain of oilseed crops contains about 20–50% oil, up to about 70%, which is extracted either by physical pressure or by solvent extraction—normally using hexane—or a combination of both methods. The oil is mainly used for making cooking fats and margarines, though linseed and castor oil are not edible and so are mainly used in industry.

Most vegetable oils are unstable, and the glycerides in the oil break down after some time, making the oil rancid. Three types of oil are extracted from plants:

- **Non-drying**, where the oil is mainly glycerides of saturated acids, and which remains a liquid and does not form a surface film when exposed to air. Examples include *castor, grape, groundnut, olive* and *oil palm*.
- **Drying**, which can form an elastic film when in contact with air, and which have glycerides of the unsaturated type. Examples include *linseed, safflower, soybean* and *tung (Vernicia fordii). Safflower* and *soybean* are sometimes classified as "semi-drying".
- **Semi-drying**, which together with drying oils are important in industry, for the production of paint, varnish, soap, detergent, etc. Examples include *cottonseed, Niger seed, maize* and *sunflower*.

Average Oil Yields
Yields of oil vary enormously, but palm-oil is generally considered to be the highest yielder, and sesame the lowest. Approximate average yields, in kg/ha, are as follows: palm-oil 1000, safflower 500, groundnut 270, sunflower 225, olive and soybean 200, sesame 150.

Minor Oil Crops
Safflower (*Carthamus tinctorius*). Widely grown in parts of North and South Africa, India, China, North America, Canada and Australia. Almost 70% of global production is grown in North America. Produces a drying or semi-drying oil; some types are rich in linoleic (polyunsaturated) acid, others are rich in oleic acid (monounsaturated).

Tung/Mu-Tree/Noix d'abrasin (*Aleurites montana*). A perennial tree grown in China, Paraguay, Argentina and Malawi. Produces a drying oil.

Buffalo Gourd (*Cucurbita foetidissima*)—described in **2G**, page 273.

Jojoba/Goat Nut (*Simmondsia chinensis*). A drought resistant evergreen shrub 0.5–1.5 m tall that produces beans with about 40% oil that is highly valued in industry, similar to sperm whale oil, which withstands extreme pressure.

Other oilseed crops include **rape, cotton, olive** and **coconut**.

Oilcake

Also known as *presscake, pomace* or *oilmeal*, oilcake is the material that remains after the oil has been extracted from the seed. With physical pressing systems, up to 10% of oil may remain in the oilcake. Oilcake is often used as animal food, and is a valuable source of oil, protein (up to 50%) and other nutrients.

Groundnut oilcake may contain aflatoxins and so is potentially toxic, and rape seed oilcake may contain glucosinolates (responsible for the bitter or sharp taste of many common foods such as mustard, cabbage, horseradish and brussel sprouts).

Oilcake can be used as fertiliser, though this normally has little influence on soil condition due to the small quantities applied.

Castor

Ricinus communis

Mole Bean, Castor Bean, Castor Oil Plant, Palma Christi;
Ricin (French); Rizinus (German); Rícino, Mamona, Mamoneira(o) (Portuguese);
Palma Christic, Ricino, Tartago, Higuerilla (Spanish); Khirwa (Arabic); Arand, Rehri (Hindi); Gulo (Amharic), Kobo (Oromifa)—Ethiopia; Olomolo, Mono (Angola)

Castor oil, a member of the spurge family *Euphorbiaceae*, was used by the ancient Egyptians as an illuminant. It was popular in medicine in the modern world up to the 20th century, principally as a purgative, and nowadays it is used mainly in industry and to some extent in the home. The oil is non-drying and has unique physical and chemical properties, such as its viscosity which does not change with temperature.

The wild plants and older varieties are short-lived perennial trees up to 12 m tall, while most varieties cultivated these days are annual herbs 1–7 m tall. The majority of these modern varieties, including hybrids, are dwarf types no more than 3m tall.

These days castor is mainly grown for the oil, used in the plastic and synthetic fibre industry, and for production of synthetic aircraft lubricants for high speed engines. It is also used as a solvent, in paints and varnishes, lino, oilcloth, ink and in the treatment of leather and textile dyes. The traditional method of oil extraction is to boil the crushed seed in water and then skim off the oil that floats to the surface.

The seeds are poisonous due to their highly toxic blood coagulant called *ricin* and a powerful allergen, a protein polysaccharide. Ricin is a lectin, and is used in certain experimental cancer therapies. Eating just one seed can make you feel nauseous; eating several can be fatal. The ricin and the allergen are removed from the oil if it is extracted properly, but they remain in the oilcake or residue. As a result this oilcake, the *castor pomace,* is normally used as fertiliser, though there are methods of detoxicating the meal so that it can be fed to livestock. The leaves and stems contain smaller amounts of ricin and should not be fed to animals.

The crop needs plenty of heat and is grown in the hot regions of Asia, India, Brazil, Africa, North America, the Mediterranean, China and Russia. About one third is produced in Latin America and more than half in Brazil and India. Global production in 2004 was 1.3 million MT, according to FAO.

PLANTING

Propagation: by seed, normally planted directly into the final growing place.

Soil: should be deep and well drained and without compact layers or hard pans as castor is intolerant of waterlogging. Moderately susceptible to salinity, the giant varieties being more tolerant than the dwarf types.

The plant is a heavy feeder, and needs relatively high fertiliser use, typical applications being, in kg/ha 25–50N, 50–70P and 40–50K. Too much Nitrogen causes an overproduction of vegetative growth and lower yields of seed.

Seed rate: 10–20 kg/ha. Seed dressing is often used to combat damping off and insects.

Seed spacing: very variable, but modern varieties normally have 70–100 cm between rows and 25–30 cm between plants. Dwarf varieties are planted at about 40,000 plants/ha. Often 2–4 seeds are planted at each station and thinned to single plants when about 30 cm tall. Seed size varies from about 1000 to 11,000 seeds per kg.

Depth: 3–8 cm

Germination: castor seed is quite delicate and should be handled carefully. Undamaged seed can remain viable for 2–3 years. It will germinate at a lower soil temperature than maize, but only slowly, and emergence may take ten days or more.

GROWTH CONDITIONS

Day length: castor is a mainly day-neutral species, with some long-day varieties.

Growth period: the average is 140–180 days for most tall varieties, minimum about 120 days, for dwarf varieties. Basically it is a short lived perennial.

Temperature: castor is killed by frost. Optimum growing temperature is 20–30°C. Seed set can be reduced above about 40°C.

Rainfall: castor is moderately drought resistant and can grow with only 300–500 mm a year, although it does need more water than maize in the early growth stages. Perennial varieties need less water than annuals. It does not tolerate heavy rainfall during flowering, nor waterlogging. It is grown under irrigation in North America.

Altitude: 0–2100 m

Weeds: should be well controlled until the plants are about 80 cm tall, taking care not to damage the plants root system which is near to the surface.

Rotation: it is often rotated with finger millet (*Eleusine coracana*). If forage crops are grown after castor, care should be taken to remove any volunteer castor plants. The root system improves soil tilth and in this way is beneficial to the crops that follow castor in the rotation, though the plant does remove large amounts of nutrients from the soil (See "**Soil**" above).

Intercropping: commonly grown mixed with legumes, maize, sorghum, cotton, sesame, groundnuts and cassava.

Pests: a large number of insects attack castor, which is not toxic to them, but damage is not normally severe. Some varieties have some resistance (tolerance) to some insects.

Diseases: also not normally a big problem, though Alternaria Leaf Spot may cause the leaves to fall off in humid conditions. Other diseases include Bacterial Leaf Spot, Cotton Root Rot (in cold, wet soils), Rust and Grey Mould (in warm, humid conditions). Modern varieties have partial resistance to some diseases.

YIELD

The average for annual varieties of castor in dryland conditions varies from about 500 kg/ha in India and Africa to 1700 kg/ha in France. Irrigated castor should yield 2.5–3 MT/ha; some hybrid dwarf varieties can produce 5 MT/ha and more.

The FAO estimate for the global average seed yield in 2004 was 1.07 MT/ha, from a maximum of 1.24 MT/ha in India to a minimum of 231 kg/ha in Kenya.

Perennial castor varieties are not normally grown in pure stands but are widely spaced all over the farm, either alone or as part of a mixed cropping system.

UTILISATION

- The **seed** or **"bean"** contains 35–60% of a highly valued non-drying oil. Traditionally this has been used as a medicine, for lighting and for the curing of hides and skins.
- The **seed** can be crushed and used to spread on *enjera* plates before cooking, in Ethiopia and places where Ethiopians eat together.
- The **castor pomace** presscake is used mainly as a fertiliser, or as animal food.
- an **insecticide** is extracted from the leaves.
- The plants can be used as a **shade crop**, to protect coffee bushes for example.
- They are also used in **landscaping** for their attractive giant, twelve-lobed, fanlike leaves.

LIMITATIONS

- The castor plant is a **heavy feeder** which depletes the soil of large amounts of nutrients.
- Ricin and an allergen are **toxic substances present in the seed, leaves and stems** and can be dangerous to animals (and humans) which eat them. These substances must also be removed from the oilcake before it is safe to be eaten by animals.
- The plant **requires hot conditions** and it is **sensitive to frost**.
- The **seed clusters mature unevenly**, so that two or more harvests must be made if it is harvested manually. With mechanical harvesting, some of the seed is too green and some will have shattered from mature pods, creating unwanted volunteers the next season.
- **Low yields** combined with **low prices** often discourage farmers to cultivate castor.

Linseed
Linum usitatissimum

Flax, Flaxseed, Oil-flax;
Graine de Lin (French); Flachs, Saatlein, Leinsamen (German);
Lino—*flax*, Linaza—*linseed* (Spanish); Linho, Linhaça (Portuguese); Lino usuale (Italian);
Alsi (Hindi); Sharkhal (Dari); Entateh (Tigrinha); Talba (Ethiopia); Kathân, Malsag (Arabic); Yama (Chinese); Tisii (Nepalese); Aliviraaii / Alivira (Tamil);
Keten (Turkish)

Linseed, or *flaxseed*, is a dual-purpose crop which is grown both for its oil and its fibre (or "flax"). Traditionally, oil has been the normal product in hot, dry regions and fibre in the more humid, temperate regions. However, in more recent years large areas of oil varieties are planted in temperate areas such as Argentina, Canada and northern Europe. Varieties are developed to produce either oil or fibre, but not both.

It is one of man's earliest crops and was grown for fibre in southern Asia and the Mediterranean by about 1000 BC. Linseed oil was used in ancient rituals in India.

The plant is an annual, 30–120 cm tall; oilseed varieties have shorter plants with many branches and seeds, fibre varieties are taller with few or no branches and fewer seeds.

The flowers are white, blue or rose-coloured and are mainly self-pollinated, though some varieties have a high rate of cross-pollination. Most temperate, oil-producing varieties are blue flowered, and fibre varieties are often white flowered.

Linseed's main advantages are the low input requirement, the short growing season and the fact that it makes a very good entry crop for wheat.

The oil is classified as "drying" because it thickens and becomes hard on exposure to air. It is slightly more viscous than other vegetable oils and is good source of dietary fibre and omega-3 fatty acid. Like most vegetable oils, linseed oil contains linoleic acid, an essential fatty acid needed for survival. But unlike most oils, it also contains significant amounts of another essential fatty acid, alpha linolenic acid (ALA).

The major linseed oil producers are India, North America, Argentina, Canada, Russia and Uruguay. Global production in 2004 was 1.9 million MT (FAO estimate).

Up to the turn of this century, linseed oil production was highly subsidised in northern Europe, stimulating increased levels of production. Fibre production from linseed is mainly in Russia, Poland, Belgium, France and Holland.

PLANTING
Rotation: linseed is considered to be a very good entry crop for wheat.
Soil: medium to heavy soils are best. The linseed plant has a short root system and so needs good soil moisture in the upper soil horizon. It is moderately susceptible to salinity. Fertiliser is not always used as the plant either does not respond or has an unpredictable response.

Seed rate: average figures are 40 kg/ha for oil varieties, 120 kg/ha for fibre varieties. However in order to achieve the optimum plant population in oil crops these may use 80–100 kg/ha. 1000 seeds weigh about 3–4 g for fibre varieties, 7 g for oil varieties.

Seed spacing: 1000–1300 seeds per square metre are sown, in rows 15 cm apart for oil crops, closer together for fibre crops.

Depth: 2–2.5 cm, in a fine seedbed.

Germination: seed can remain viable for 5–10 years if stored dry and cool.

GROWTH CONDITIONS

Day length: there are both long-day and day-neutral varieties of linseed.

Growth period: 120–150 days.

Temperature: linseed is a cool weather crop, and is sown in late autumn in regions with mild winters and hot summers. Seedlings may be killed by a heavy frost, which also reduces yields if it occurs at flowering. During and after flowering, about 32°C is the maximum. Seed of linseed should be stored at less than 10% moisture.

Rainfall: oil varieties can grow with only 300 mm, but are normally grown in 450–750 mm/year rainfall areas, or with irrigation. Fibre varieties need more water than oil ones.

Rotation: should not be grown on the same land more often than every five or six years. In a rotation, linseed can substitute for cereals, such as after legumes or pasture. If weeds can be controlled it is a good first "pioneer" crop on newly cleared land.

Pests: rarely a problem, but yields are sometimes reduced by grasshoppers, cutworms, armyworms, chinch bugs, stinkbugs and flaxworms.

– Flax Fleabeetle—this can completely decimate the crop, as the seedlings are just emerging. Control is with seed dressings, insecticide and/or later (warmer) planting for more rapid seedling emergence.

Diseases: three of the most troublesome, all caused by fungi, are:

– Flax Wilt—stems of young plants turn brown and dry up, normally in small areas all over the field, or on individual plants. Worse when linseed has been grown for a long time on the same land. Control: crop rotation, but as spores can remain viable in the soil for 25 years it is better to use resistant varieties.

– Flax Rust—bright orange or red spots appear on all aerial parts just before flowering which become black and shiny. Worse in wet conditions. Control is with seed dressings, crop rotation, or, best of all, resistant varieties, though these may not resist all races of the fungus.

– Pasmo—leaves have dark brown patches which then spread to the stem where the patches are brown or black. Occurs above about 20°C, spread by rain. Control: clean seed (treatment is not effective), destroy infected plants, crop rotation and resistant varieties.

– Botrytis, Mildew, Alternaria and **Sclerotinia** (see Horse Bean) can also be damaging.

There are also two virus diseases—**Aster Yellows** and **Curly Top** - both are transmitted by leafhoppers, which can cause some loss of yield.

YIELD

Linseed crops raised for oil are normally harvested when most of the seed capsules are mature. The stalks are very tough and fibrous and need sharp knives to cut them.

The average yield globally was estimated by FAO at 726 kg/ha in 2004, Tunisia recording the highest average of 2.14 MT/ha, and Latvia the lowest average of 259 kg/ha.

UTILISATION

- **Seed**—contains 30–45% of a drying oil and 20–25% protein. The seed is hard and must be either crushed or softened by soaking and boiling before feeding to animals—it is used as a concentrated energy feed for ruminants and pigs.
- **Linseed oil** is not palatable and is mainly used in the production of paints, varnishes, lino, oilcloth, printing inks, artists' oil paints, soaps and patent leather.
- **Linseed cake** - this is often ground up to make linseed cake meal, used as animal food it gives them a nice shiny coat, or "bloom". "Linseed meal" is the crushed, unextracted seed and contains about 35%,10% or 3% oil depending on the source of the "meal": ground, unextracted seed, ground linseed cake, or meal from the extraction process, respectively. Linseed cake is toxic to poultry, unless it makes up less than about 3% of their meal. Toxicity (linamarin) can be removed (detoxicated) by soaking the meal in water for 24 hours, or by adding pyridoxin, a B-vitamin, to the meal.
- **Straw**—this is comparable to wheat or oat straw in feeding value but is not very palatable.
- **Fibre**—stronger than cotton or wool, and with other special qualities which make it suitable for linen, thread, towels, clothing, fabrics, cigarettes, Bible pages, currency, etc.

LIMITATIONS

- Linseed plants compete very poorly with weeds, so it is often planted late, after the first weeds have been destroyed. It is, however, very tolerant of most herbicides.
- The seedbed must be fine and have adequate soil moisture in the upper horizon;
- Yields are rather low.
- Prussic acid. Immature linseed contains the glucoside linamarin which may be released under warm, acidic conditions. This can be toxic to animals, but is destroyed by heat—ten minutes boiling makes the feed safe.
- There should be long breaks—five or six years—between linseed crops in the rotation.

Niger Seed
Guizotia abyssinica

Inga Seed, Blackseed,
Guizotia Oléifere (French); Gingellikraut (German); Alashi (Oriya);
Hechellu (Kannada); Karale (Marathi); Neehoog/Neuk (Tigrinya); Noug (Amharic);
Payellu (Tamil); Ramtil (Hindi and Panjabi); Sarguza (Bengali); Sorguja (Assamese)

Niger seed originated in Ethiopia, where it is now mainly cultivated, on approximately 250,000 Ha. It is also grown in marginal areas in India, and to some extent in East Africa and the West Indies. It is the most important edible oil crop in Ethiopia, supplying about half of their oilseed production.

It is a member of the *Asteraceae* (alt *Compositeae*) family. It is a short-day plant, an annual, 1–3 m tall (up to 15 m), which is frost tolerant, drought resistant and adapted to a wide range of soils. There are three main types: dwarf, semi-dwarf and giant.

It is propagated by seed, which is about 3.5–5 mm long, contains 30–50% of a yellow semi-drying oil with a pleasant taste, and a protein content of about 20%. The oil from Ethiopian crops contains about 70% linoleic acid, while oil from crops grown in India contains about 50%.

For farmers, it is much kinder plant to grow than safflower, as the plants do not have spines, and harvesting could be mechanised. More research is needed on this potentially useful source of edible plant oil.

Detailed information is available online from the publications department of the Plant Genetic Resources Institute.

PLANTING
Propagation: Niger is self-sterile and needs bees for cross-pollination.
Soil: Niger seed grows well in poor soils, if they are neither very acidic (it is classified as "sensitive" to soil acidity) nor waterlogged. Ph range 5.5–7.5. In fertile soils the plants may lodge (fall over) and have a prolonged growth period. It is not normally fertilised though it does normally respond to both fertilisers and manure.
Seed rate: 5–8 kg/ha in rows 440–5°C m apart, 8–12 kg/ha when broadcast (when NPK fertiliser is often broadcast together with the seed, then harrowed into the soil).
Seed spacing: normally broadcast, sometimes in rows 35-50 cm apart.
Depth: covered with light harrows when broadcast, otherwise about 1 cm deep in a fine tilth.
Germination: the seed can be stored for a year or more without losing much viability.
Intercropping: commonly done, with finger millet (ragi), cereals, legumes and other annuals.
Rotation: works well with wheat and/or maize.

GROWTH CONDITIONS
Day length: intermediate response, varies with type, but most are short-day (do not flower or set seed until daylight hours average 13 hours or less).

Growth period: 100–150 days.
Temperature: very frost tolerant. Semi-dwarf types are adapted to temperate climates.
Rainfall: about 600–1800 mm a year for all types. Dwarf types are more drought resistant.
Altitude: the optimum is around 2000–2200 m, but it can be cultivated from about 1600–2600 m.
Pests and diseases: rarely a problem; locusts, grasshoppers and armyworms sometimes attack.

YIELD

When intercropped with finger millet in India Niger seed yields about 100–200 kg/ha. Pure stand yields have been recorded in both India and Ethiopia of about 300–400 kg/ha, 400–600 kg/ha in Kenya. 1.2 MT/ha is possible with good growing conditions.

Oil yields are 30–50% of dried seed yield.

UTILISATION

- **Seeds** of Niger seed are crushed, giving about 30% of a clear, edible semi-drying oil which is yellow in colour and tastes of nuts. Seeds can also be used fried or as chutneys and condiments, or fed to caged birds. In Ethiopia they are pressed with honey and made into cakes.
- **Oil** is used for cooking, for making soap and for lighting; some is used in making paints. It is used as a substitute for olive oil, can be mixed with linseed oil, and is commonly used as a (less expensive) adulterant for rape oil, sesame oil, etc.
- **Whole plant** can be used to attract bees, and also as a green manure (before flowering).
- **Presscake** is used as a high protein (30–35%) food for animals, especially cattle. This black oilcake is comparable in feeding value to undecorticated groundnut cake. Up to 30% can be added to laying poultry rations. It is sometimes used as manure and/or soil improver.

LIMITATIONS

- Yields are rather modest.
- There is a shortage of improved varieties and of large quantities of good quality seed.
- The growing crop needs a well prepared seedbed, with very few or no weeds.
- The seeds are very small, and it is more difficult for subsistence farmers to extract oil from Niger seed than from other oilseed crops.

Sesame
Sesamum indicum (Syn. *S.orientale*)

Simsim, Benne, Benné, Benniseed, Til, Gingili, Gingelly (from Hindi *jingali*)
Sésame (French); Sesam (German); Sésamo, Ajonjoli (Spanish); Sésamo, Gergelim
(Portuguese); Simsim (Arabic); Til (India); Gingelly (Sri Lanka); Sillit, Simsim
(Ethiopia); Utolo* (Umbundu, Angola); Kunzeleh (Pashtu)
*Also used for sunflower.

Sesame has been cultivated since the earliest days of agriculture in the hot, dry parts
of Africa, from where it was taken to India, China, Burma, the Mediterranean, India,
Mexico and the Far East. These days it is grown throughout the tropics and
subtropics. There are thousands of different varieties, some of which are very
specific to a small area or soil type.

It is an annual plant, 50–250 cm tall, with a taproot and a dense surface mat of
feeding roots which improves soil structure. Seed colour may be white, yellow, grey,
red, brown or black.

It is very resistant to both heat and drought. Perhaps it is because of its rather low
yields that in global terms it is only about the ninth most important of the vegetable
oil crops. The FAO estimate of global production in 2004 was 3.3 million MT.

The main producer countries are India, China, Sudan, Mexico, Venezuela,
Burma and Ethiopia. About half of all the sesame produced is grown in Asia.

PLANTING
Propagation: by seed. Mainly self-pollinated, but some limited cross-pollination, no
more than about 5%, may occur.
Soil: sesame can often grow well on certain poor soils, but prefers deep draining
soils like sandy loams. Very intolerant of waterlogging. Fertiliser is generally not
used, though improved varieties do respond to fertiliser; a typical application is 200–
300 kg/ha of 15:15:15. It grows well on the same soil types that produce good
sorghum crops, but it is more sensitive to salinity than sorghum.
Rotation: it is a good crop to grow before wheat, or after an irrigated crop.
Intercropping: in India it is often grown mixed with cotton, maize, sorghum, millet
or groundnuts. In Africa it is often intercropped with maize or sorghum.
Seed rate: 10–15 kg/ha. 1000 seeds weigh 3–5 g.
Seed spacing: not very critical, as sesame plants compensate well - ie they tend to
grow to the size of the land area and light available. It is normally not sown in rows,
but the seed is "broadcast". When it is sown mechanically, 40–100 cm between rows
and 10–30 cm between plants, depending on available soil water and the variety
chosen.
Depth: 2–4 cm

GROWTH CONDITIONS

Day length: in general sesame is very sensitive to photoperiod. There are both long- and short-day varieties, and even a few day-neutral varieties, such as "Venezuela 51".

Growth period: 70–160 days, mainly depending on the variety - the average is about 105 days. The pods and seed mature from the bottom of the plant upwards; harvest should be timed to start when most of the lower pods have turned from green to yellow. If harvest is delayed, the mature pods dehisce and some of the seed is lost on the ground.

Temperature: the crop needs hot weather, and can tolerate very high temperatures, about 27°C is optimum. The plant is frost sensitive. For germination soil temperature should be about 20°C.

Rainfall: moderately drought resistant; it is extremely sensitive to excess soil water, and so is normally only grown in low rainfall areas, around 350–1100 mm per year. Some early varieties in retentive soils can produce a crop with stored soil moisture alone. Heavy rainfall soon after sowing can lead to losses from soil capping and the seeds being swept away or buried too deep.

Grows poorly in high rainfall, when *Botrytis* tends to develop, causing the heads to drop off.

Altitude: 0–1500 m

Pests: six of the insects which commonly cause some damage to sesame plants are:

– **Til Leaf-Roller**—specific to sesame. Larvae feed on the leaves and bore into the stem and pods, killing plants at the seedling stage.

– **Tobacco Capsid**—brownish yellow insects that suck the plant sap.

– **Sesame Gall Fly**—larvae feed on young flower buds, causing them to form galls which fall off.

– **Aphids**—transmit various virus diseases.

– **Common Red Spider**—a small 8-legged mite. Larvae cover the undersides of leaves with a fine webbing and suck their sap; leaves become brittle and fall off.

– **Spheraylia sesami**—larvae eat the flowers.

Diseases: humid conditions favour the spread of most of these:

– **Leaf Curl**—a virus transmitted by the White Fly. The leaves curl downwards, and become thick, brittle and dark green. Weekly pesticide sprays may be needed to stop it.

– **Phyllody**—a virus transmitted by a jassid. Plants develop abnormal vegetative growth, and the floral parts are transformed into green, leafy structures. The plants become so heavy that they bend over. Resistant varieties are the best cure.

– **Leaf Spot**—appears at flowering time, producing spots on the leaves, which fall off early. Can be controlled with hot water treatment of seed. Destroy all infected plants.

– **Fusarium Wilt**—leaves turn yellow, and wilt, and the plant eventually dies. A brown discolouration is seen when a transection is cut through the plant, which spreads from the roots upwards. Controlled by using resistant varieties and clean seed.

– **Root and Stem Rot**—caused by fungi. A dark discolouration appears at the base of the stem, which may break off at ground level, and the roots may become rotten. Control with clean seed.

YIELD

Sesame yields are generally low, although it should be noted that sesame is very often grown under the most marginal conditions. Dryland crops rarely yield more than one MT/ha, and average irrigated crop yields are less than two tonne/ha. In many dry, infertile areas yields average 200–300 kg/ha.

FAO's estimate of global average yield for 2004 was 489 kg/ha, the highest average being recorded in Lebanon (3.13 MT/ha), and the lowest in Cameroon (138 kg/ha).

One major problem with sesame is that most varieties are "dehiscent" - their seed pods open when they are mature (they are said to "shatter") and the seed falls to the ground and is lost. However the pods are carried nearly upright on the stem so that if the plants themselves are also kept upright after harvest only a few seeds are lost. If the plants bend or fall over, or if they are not carried upright and handled carefully before threshing, large yield loss can occur.

Unfortunately the "non-dehiscent" varieties that are currently being developed often have lower yields and may be difficult to thresh. Clearly, the yield of oil is also low.

UTILISATION

- **Grain**—a valuable food, containing 20–25% protein, normally made into halvah, stews, soups, porridges or sweets, or sprinkled onto bread, cakes, etc.
- **Oil**—the grain contains 45–60% of a highly unsaturated oil, which is very stable and does not turn rancid or smell bad due to the presence of phenolic material. It is mainly used for cooking, and also for making perfumes and medicines. *Sesamin* is extracted from the oil and is added to the natural insecticide Pyrethrum to increase its effectiveness. In India the oil is used for lighting. The presscake (oilcake) is valued as animal food.
- **Young leaves**—sometimes included in soups and stews.
- **Whole plant**—in India the plant has several interesting medicinal uses.

LIMITATIONS

- Dehiscent, or *shattering*, seed pods (hence the expression "Open sesame!") mean that much of the seed is often left in the field. The non-dehiscent varieties tend to have lower yields and may be difficult to thresh.
- Sesame has a reputation for producing low yields, though this is not altogether surprising as it is often grown in the most marginal conditions.
- It is sensitive to excessive soil water and frost, and needs plenty of heat for oil production.
- The seedlings are weak, and susceptible to attack by many pests and diseases; they also need to be well weeded.
- A very fine seedbed must be prepared, which must have adequate moisture and a minimum soil temperature of about 20°C. A soil crust, or "cap", hinders seedling emergence.

Sunflower
Helianthus annuus

Tournesol, Soleil (French); Sonnenblume (German); Girasol (Spanish & Portuguese); Mant (Latin America); Abbad ash-shams (Arabic); Surajmukhi, Suryamukhi (Hindi); Elmer Mekhay (Pashtu); Farenji Suf (Ethiopia); Utolo (Angola—"Utolo" is also used to describe sesame)

Both the sunflower and its close relative the Jerusalem artichoke (*Helianthus tuberosum*) are native to temperate America, either Peru or Mexico, and were used by the Indians as food plants; their seeds have been found in an archaeological site dated 2000–3000 years old. These days sunflowers are grown widely, from the equator to as far North as 55°N.

The plant is an annual, a member of the *Asteraceae* (*Compositae*) family, and boasts a massive tap root that can penetrate down to 3 metres.

There are three main types: Giant (1.8–4.2 m tall), Semi-dwarf (1.35–1.8 m tall) and Dwarf (0.6–1.35 m tall). The giant types are generally late maturing and have low oil content. The dwarf types have the highest oil content, more than 50% with improved varieties.

The oil is classified as being "semi-drying", polyunsaturated with a high percentage of linoleic acid. Protein content of the grain varies from about 13 to 27%.

Both the leaves and flower heads are *heliotropic* ie they respond to the direction of the sun's movements.

The flowers are pollinated by insects and there should be plenty of bees and other insects at flowering time in order to set seed properly.

Many of the cultivated plants do not grow true from seed and will eventually revert back to the wild type plants after a few generations.

Sunflowers are one of the most important oil crops, and according to FAO were grown on 21 million hectares throughout the world in 2004. They are widely grown in Russia, and in Rumania, Bulgaria, China, France, USA, Canada and South America, particularly Argentina. FAO estimated a global production of 26.1 million MT in 2004.

PLANTING
Propagation: by seed. Only grains that are full and fat should be used as seed, which can remain viable for several years if it is carefully dried and stored.

Sunflowers can be grafted onto their close relative the perennial Jerusalem artichoke, *Helianthus tuberosum*.

Soil: sunflowers can grow on most soils and need less fertile conditions than maize. The plants are fairly salt tolerant, but will not survive in either very acid soil (they have "intermediate tolerance" to acidity) or waterlogged soil.

The crop responds well to phosphatic fertilisers, but high nitrogen levels should generally be avoided.

Seed rate: 4.5–15 kg/ha for grain, 35–40 kg/ha for silage. There are about 6000–20,000 seeds per kg.

Seed spacing: 60–100 cm between rows, 15–30 cm between plants. The normal plant population is about two or three plants per square metre.
Depth: 2–4 cm

GROWTH CONDITIONS
Day length: day-neutral.
Growth period: the average is 125–160 days. Some dwarf varieties can mature in 45 days.
Temperature: sunflowers are fairly resistant to both heat and cold, and are not damaged by mildly freezing temperatures that would kill maize or soybean. For germination the soil temperature need only be about 8–10°C, much cooler than that needed for maize or sorghum.
Rainfall: sunflowers are fairly drought resistant, and some varieties need only 400 mm if this is well distributed through the growing season.
Altitude: lowlands and medium altitudes are preferable.
Rotation: to avoid root rots sunflowers should not be grown on the same land more often than every four years, nor should they follow peas, beets or potatoes.
Pests: not normally a serious problem, though cutworms, wireworms, grasshoppers, aphids, weevils and other insects do feed on sunflower plants. Often the worst damage is caused by the Sunflower (Head) Moth, whose larvae attack the seed.
The plant is also damaged by the Sunflower Leaf Beetle which feeds on the leaves, and the Sunflower Maggot which tunnels into the stem.
Sunflowers are also very susceptible to damage by birds and rodents.
Diseases: the three most important are:
– **Rust**—dark brown spots, which turn black, on the underside of leaves, which turn yellow, dry up and fall off. Up to 50% yield loss can occur. Some varieties have good resistance; crop rotations and dusting with sulphur also limit this disease.
– **Grey Mould**—worst in warm, humid conditions. Infects both the leaves and the ripening seed head.
– **Stem Rot (Wilt)**—caused by *Sclerotonia sclerotiorum.*
Other diseases include Powdery and Downy Mildews, White Rust, Verticillium Wilt, Charcoal Rot and Black Rot. Virus diseases are also sometimes seen.

YIELD
The global average sunflower yield in 2004 was 1.22 MT/ha, according to FAO. In that year, the highest national average was recorded in Austria, at 2.69 MT/ha, and the lowest average in Zimbabwe, at 308 kg/ha.

Giant sunflower varieties yield more seed per hectare than dwarf varieties, but their seed normally contains a lower percentage of oil.

A typical yield of oil is around 225 kg/ha.

UTILISATION

- 90% of sunflower grain produced is crushed for **oil**—the nutritive value equals that of olive oil. It is used for margarine, cooking oil and in industry for soaps, lubricants, paints, varnishes and cosmetics. Modern varieties produce grains with 50% oil and more.
- In many countries they are grown mainly for **human food**, either raw or roasted.
- Whole grains are used as a high protein (13–27%) **poultry food**.
- The leaves by themselves make good **silage**, but silage that is made from the entire plants is less nutritive and palatable than silage made from maize.
- The **presscake,** or oilcake, is a useful animal food, containing 30–40% protein; it is also sometimes used as fertiliser.
- The **stalks** are used for pulp in the paper industry, and as a source of cellulose. When they are shredded and incorporated into the soil they improve soil fertility, mainly Nitrogen, Calcium and Potassium, as well as increasing the organic matter content.
- The **flowers** are used to produce a yellow dye.
- **Honey** is often produced as a by-product. Bees are the best pollinators of sunflowers, so beehives are often placed in sunflower fields.
- The **dried heads** can be used for animal food, containing about 9% protein and 3% oil.
- The crop is sometimes used as a **green manure.**

LIMITATIONS

There are relatively few problems growing sunflowers, which are hardy plants that can adapt to a wide range of growing conditions. However:

- **Uneven maturity** of the grain makes the timing of harvest critical, and not easy for the farmer to decide. While many of the seeds will be dry and starting to fall out, others may still be at 50% moisture.
- **Loss of germination and/or damage by fungi** can occur if the moisture content of stored seed exceeds about 8–9%, especially in warm conditions.
- **Damage by birds** can be serious; a number of different **diseases** can also reduce yields considerably.
- It may not always be possible to select, or obtain seed of, the **appropriate variety.**

2D. ROOT CROPS

A wide range of species are commonly referred to as "root crops", their common feature being their fleshy, underground storage organs, which may be either a true root, or a tuber or a corm—the difference between the various types of "roots" is discussed in **1Eh**, page 40.

Root crops mainly contribute energy to the diet, being high in starches and sugars but low in oil and protein. Some are eaten as vegetables, such as carrots, radish and parsnips, and are more useful as sources of vitamins, flavour and subtle nutrients than as sources of energy. Other root crops such as sweet and Irish potatoes, cassava and sugarbeet are more useful as sources of energy. The global production of all root crops, including those for animal food, is only slightly less than that of cereals, although in terms of dry nutrients the root crops are much less important due to their high moisture content.

Wild species of many plants produce edible roots and rhizomes that can be eaten in times of famine. Some species are known and eaten only in very limited regions—in the Peruvian Andes for example at least six species of root crops are eaten which are hardly known in the rest of the world. Similarly, in Africa "fra-fra potatoes" (*Coleus dysentericus*), aroids (*Araceae*) and sedges (*Cyperaceae*)—see below—are eaten only in certain relatively small regions. With some effort and imagination these and other root crops could be introduced and promoted in many parts of the world where there is a shortage of starch in the diet.

The three most important root crops grown for food are the **Irish potato**, in temperate climates, and the **cassava (manioc)** and **sweet potato** in tropical climates. The other root crops described in this section are the **taro** and the **yam**.

Listed below are some of the minor root crops:
Anu (Mashua, Cubio, Quecha, Apina-mama or **Ysano)** *Tropaeolum tuberosum* (Tropaeolaceae or nasturtium family). Looks very similar to Oca (see below) which grows in the same area (high Andes). Hardy, climbing plants up to 2 m long. Reddish flowers, 5-lobed leaves, similar to but smaller than the garden nasturtium. Dry tubers contain about 14–16% protein, 80% carbohydrate, 9 microgramme/100 g beta-carotene and about 480 mg Vitamin C/100 g.
Chufa (Tiger Nut) *Cyperus esculentus* (Cyperaceae or sedge family). Grown since early Egyptian times they are now mainly grown in West Africa and Spain. The so-called "nuts" are small underground stem tubers, eaten either raw or roasted, which contain about 24% fat, 30% starch, 4% protein and 16% sucrose. They are used also to make certain non-alcoholic drinks.
Jerusalem Artichoke *Helianthus tuberosus* (Asteraceae/Compositae or daisy family). Originated in North America, it was taken to Europe in the early 1600s. A perennial herb with irregularly shaped underground stem tubers 5–20 cm long, with white or red skin, eaten boiled, baked or in stews and soups. Contains fructose, which is acceptable to diabetics as a substitute for glucose. The plants can grow to 3 m high; in temperate climates it flowers only after a long warm summer.

Oca *Oxalis tuberosa* (Oxalidaceae or wood sorrel family). White, yellow or red smooth skinned cylindrical tubers about 10 cm long with grooves and bulges, with long deep eyes. Staple food in high altitude parts of the South American Andes, also cultivated in Mexico and New Zealand. The tubers are nutritionally equivalent to potatoes, and they are both cooked in similar ways. Some varieties contain calcium oxalate crystals, which can be removed by drying in the sun.

Ulluco *Ullucus tuberosus* (Basellaceae family). Endemic in the Andes, very frost resistant. 2–15 cm long tubers are yellow or pale magenta, red or purple and either small and round or long and curved. The fresh tubers contain about 14% starch and sugars, 1–2% protein, 23 mg/100 g Vitamin C. Prepared and eaten like potatoes, or dried and stored. The leaves are also eaten.

Cassava/Manioc
Manihot esculenta (Syn. *M. utilissima, M. aipi, M. dulcis, M. palmata*)

Brazilian Arrowroot, Cassata, Mandioc(a), Imanoka, Katela boodin, Maniba, Manioc, Muk shue, Shushu, Yuca, Tapioca. Manioc (French); Manioc (German); YMCA {Brava}, Mandioca, Guacomote (Spanish); Mandioca, Maniçoba, Aipim (Portuguese); Utombo, Ombuta, Okambuta (Angola); Muhogo (Kiswahili); Omuwogo (Luganda); Aypu, Boniato (Caribbean).

Cassava was probably first cultivated by the Maya in Yucatán, Mexico and is now the most important of all the tropical root crops in terms of global production. FAO estimated the global production in 2004 was 101 million MT.

The plant can grow in the poorest, most infertile soils, and many varieties produce tubers that can remain in the ground for up to three years. Its root tubers often serve as a famine reserve crop for subsistence farmers. Compared with other tropical staple crops, only yams produce more carbohydrate per hectare. Raw tubers contain about 35% starch but only about 1–3% protein.

The plant is a short-lived perennial shrub, 1–5metres tall; it is normally grown in the lowland tropics. Cassava is a member of the *Euphorbiaceae* (spurge) family, which also includes the Castor oil plant (*Ricinus communis*) and the Poinsetia.

Recent research has involved managing cassava as a perennial forage crop, repeatedly harvesting the leaves at 2–3 month intervals. The roots are not harvested but serve as a nutrient reserve to support the forage regrowth.

IITA (the International Institute for Tropical Agriculture) in Ibadan, Nigeria maintains a germplasm collection for Africa. IITA is one of 16 research centres that come under the auspices of the CGIAR—the Consultative Group on International Agricultural Research.

The largest germplasm collection is at the International Centre for Tropical Agriculture (CIAT), in Cali, Colombia, and the largest national collection is in Brazil, under the direction of the Brazilian Agricultural Research Network (EMBRAPA). All three institutions have breeding programmes.

There are two main types of cassava—sweet and bitter. Most varieties of both types contain the glycoside linamarin which breaks down under enzyme action to give the poison **Hydrocyanic glucoside** (HCN, hydrocyanic, or prussic, acid), a cyanide producing sugar derivative:

Sweet types—HCN (potentially 30–100 mg HCN/kg) is present only in the two outer layers of the tuber, which can be eaten raw after peeling. The tubers have a soft, white flesh. They normally mature in about six months, and can then be left stored in the ground for almost a year.

Bitter types—HCN (potentially 1350 mg HCN/kg) is present throughout the tuber, which must be cut up and boiled, the HCN squeezed out, and then reboiled in clean water; or it can be roasted or fermented. The tubers have a firm, yellow flesh which normally mature in twelve months or more and can be left in the ground for up to four years.

The HCN content also varies with the soil type, for example increasing in soil which is deficient in potash, and also with the climate. Changes in toxicity can also occur when varieties are introduced from one country to another. Old, stale tubers are more toxic than freshly harvested ones.

Both types of cassava are virtually immune to attack by the African Migratory Locust. The bitter types are also only rarely eaten by wild game such as baboons, pigs, rats and hippos, and they are sometimes the only crop that can be cultivated in areas with this problem.

PLANTING

Propagation: by stem cuttings, 20–30 cm long, often cut from the middle of the parent stem to avoid Mosaic Virus which is often present on the lower parts. About half the length is pushed into the soil, where it sprouts readily within about 7–14 days.

Soil: best on sandy or sandy loam soils, but cassava can grow almost anywhere if the soil is not waterlogged, saline or too shallow or stony. Some cassava varieties are adapted to dry areas with alkaline soil, others to acidic mud banks along rivers.

In heavy clays or very fertile soils, or with high Nitrogen levels, vegetative growth can be excessive and at the expense of tuber growth. Potassium is the most important element, although Nitrogen, and to some extent Phosphates, also normally produce a good response. However fertiliser is rarely used on this crop.

Plant spacing: about 12,000 per hectare, 1–1.5 m apart. Sometimes two or more stems are planted together, either on the flat or on ridges or mounds. They are sun-loving plants.

Intercropping: very common, with a wide range of annual food crops. Legumes such as peanuts are commonly used. Trials have shown that intercropping with trees such as the acid tolerant leguminous shrub *Flemingia macrophylla* increases overall yields per hectare, improves the soil fertility and provides fuel wood.

Rotation: since cassava can grow on exhausted soils which are unsuitable for many other food crops, it is often grown at the end of a long period of monocropping (monoculture), or as the last crop planted in shifting farming systems.

GROWTH CONDITIONS

Growth period: 6–12 months for short-season sweet types and improved varieties. Long-season types are often left in the ground for one or more years. Both types are normally harvested as and when required for eating.

Temperature: the optimum is about 25–30°C. Intolerant of frost and cold weather.

Rainfall: 1000–2000 mm per year is optimum, but cassava can grow with 500 mm or less. It can withstand long periods of drought, except at planting time.

Altitude: cassava is basically a lowland crop, though it is grown up to 1600 m near the equator and up to 1000 m in the humid tropics.

Pests: not usually a big problem, but sometimes one of the following can cause problems: Green Spider Mite and Mealy Bugs (spread from South America to Africa in the 1970s), Stem Borers, Weevil Borers, Red Spiders and Scale Insects. Wild game can also devastate sweet cassava crops.

Diseases: viruses are the most serious problem; two are described below:

– Mosaic Virus—young leaves become white or yellow (chlorotic) and are smaller and distorted. Yields can be reduced by 95%. Transmitted by the Whitefly and by planting infected cuttings. Control: use resistant varieties and clean stem cuttings for planting, destroy infected plants.

– Brown Streak Virus—mature leaves become chlorotic, but without distortion; the stems have brown streaks, become shrunken and may die; roots become discoloured. Transmitted in the same ways as Mosaic Virus. Common in the coastal regions of East Africa, and mainly associated with low temperatures. Controlled with crop rotation; some varieties have some resistance.

– Bacterial Blight (or Wilt)—caused by *Xanthomonas manihotis*, it is now more serious than Mosaic Virus in tropical Africa. Controlled with resistant varieties and improved drainage.

YIELD

The tubers of cassava can rapidly rot or spoil when they are above ground, so they are normally left in the field and only harvested when needed as food. However the tubers can be stored for much longer periods if they are peeled, sliced and dried in the sun.

Yields vary according to variety/growth period, soil type, cultivation method and severity of disease attack, particularly by virus. It has been observed that varieties with long thin leaf lobes often yield more than varieties with short wide leaf lobes.

In ideal conditions 80 MT/ha can be achieved, but the global average is closer to 10 MT/ha; the African average is about 8 MT/ha.

In 2004 FAO reported that the Cook Islands achieved the highest national average yield, 25 MT/ha, and that Burkina Faso had the lowest, 2.0 MT/ha.

UTILISATION

- **Tubers**, after detoxication, can be eaten in many different ways: ground into flour, fermented slightly and dried into a flour known as *garri* in West Africa, or made into a coarse meal known as *farinha* in South America, *gaplek* in Indonesia, etc.
- **Tapioca** is made from fine starch extracted from the tubers, while in industry **cassava starch** is used in adhesives and cosmetics, and as a source of acetone, sugars and starches.
- Rural and other communities around the world are also cheered by the large range of **beers and other alcoholic beverages** made from the tubers of this noble plant.
- **Whole plants** can be used, planted very close together in staggered lines to form living fences for protection of crops, property, etc from animals. If nothing else is available the stems can be used as low quality building or shade material.
- **Leaves** unfortunately may contain HCN, which is a pity as they also have a high content of protein and Vitamin A and C. They can be very useful for protecting the soil from erosion and also reducing weed growth—many modern varieties have a very high leaf area index and so are especially useful for this purpose.

LIMITATIONS

- Cassava tubers often contain HCN; they also have a high fibre content and a low protein content (0.7–3%).
- The plants are susceptible to virus diseases, and sometimes also to insect pests and predation by wild animals.
- Long growth period, though this is not always a problem as cassava is a good famine reserve food, which can be stored safely in the field and underground, or sun dried.
- The tubers can deteriorate rapidly, within one or two days, unless they are stored in good conditions.

Irish potato
Solanum tuberosum

European Potato, English Potato, Common Potato, White Potato, Solanum Potato - to distinguish it from the Sweet Potato (*Ipomoea batatas*). Pomme de Terre (French); Kartoffel, Erdapfel (German); Patatas, Papa (Spanish); Batatas, Batata Reina (Portuguese); Ekapa (Angola); Dinish (Tigray); Dinich (Ethiopia); Aaloo, Alu (Hindi); Kachalu (Pashtu and Dari).

The potato originated in South America, and its name comes from the Spanish *patata*, from *batatas*, an American Indian word for the sweet potato. Potatoes are now grown throughout the world, principally in Europe; large areas are grown in Russia, Germany, Poland, India and China. In temperate climates it is the most important root crop, while the sweet potato is king in warmer climates.

The FAO estimate for global production in 2004 was 328 million MT.

Many thousand separate varieties of potatoes are grown, and the description below is only a rather brief synopsis of the vast array of literature available on potatoes.

Although plants can propagate from true seed the planting material used by food producers is the swollen underground stem, or tuber (which for planting purposes is known as *seed* or, occasionally, *setts*). Thus each new potato plant formed exactly resembles its parent in its genetic makeup and, therefore, characteristics.

As a general rule, plants with white flowers have tubers with white skin, while plants with pink, red, blue or purple flowers have tubers with coloured skin, normally pinkish.

In the tropics potato plants do not normally flower, and are normally only grown there as a cool season crop or at high altitudes.

The tubers contain about 80% water, 2% protein, 18% carbohydrate (mostly starch), and a number of minerals; freshly dug tubers contain about 20 mg/100 g Vitamin C though this decreases during storage.

PLANTING

Propagation: by means of the whole tubers ("seed potatoes" or "setts") or parts of tubers, cut to include at least one bud or "eye". If they are cut, the tubers should be cut at right angles to the main axis so as to avoid apical dominance (the inhibition of lateral buds to form branches). Ideally, seed should be "chitted" before planting; place the seed in single layers, "rose end" up—the end with the most eyes. Protect from frost and/or direct sunlight. Before planting, remove all but 3–4 chits at the rose end.

Soil: should be well drained. Optimum pH is 5.5–6. They do not normally need lime, but do need a good supply of nutrients, manure being particularly beneficial. They are moderately susceptible to saline soils.

Fertiliser is commonly used, up to 110 kg/ha Nitrogen and 225 kg/ha Phosphate. Potash is also normally needed, but not always, and in smaller quantities than nitrogen or phosphate.

Dormancy: potato tubers are normally dormant for at least six weeks, and up to about 10 weeks, after the tuber is fully grown. Some time after this they may begin to develop sprouts, or *chits*, indicating that they are ready for planting (see "Propagation", above).

Seed rate: for maincrop about 1.5–3 MT/ha, for earlies about 3–4.5 MT/ha of tubers, which for planting purposes become known as *seed* or *setts*.

Spacing: 20–30 cm between plants; 70–120 cm between rows for maincrop, 40–60 cm for earlies.

Depth: 5–15 cm. Potato plants should be *earthed up* or *ridged up* as they develop, to improve the development of the tubers and to prevent the upper ones from turning green ie developing solanine when exposed to light, and also to reduce the build-up of blight disease. Solanine is an alkaloid which is toxic in high concentrations, causing vomiting and other symptoms.

Intercropping: potatoes are normally grown in pure stand, but are occasionally mixed together with maize or beans. A fast growing catch crop (**1Gc**, page 62) such as lettuce can sometimes be grown on the ridges before the potato haulm takes over.

GROWTH CONDITIONS

Growth period: 95–150 days, depending on the type/variety, normally classified as being either 1st early, 2nd early or maincrop.

The haulm (**1Ic**) should be burned off or removed a few days or weeks before harvest to allow the tubers skins to harden up, and so bruise less easily.

Day length: there are both short-day and long-day varieties of potatoes; the former types are mainly grown in the tropics.

Temperature: varies greatly depending on the variety, though the optimum soil temperature is about 20–25°C. The higher the temperature, the more the aerial parts develop at the expense of tuber development. At about 30°C tuber development virtually stops. The plants are frost sensitive, especially when young.

Rainfall: some varieties can produce a small yield with only 350 mm during the growing season, though the optimum is about 25 cm per week during the growing season. Drought can be disastrous for potatoes, especially if this occurs when the tubers should be *bulking up*.

Altitude: in the tropics, potatoes grow best above about 1800 m. For most varieties the maximum is about 3000 m, though some Andean varieties grow best at 2000–4000 m.

Diseases: volunteer plants, including the tubers, should be removed from the field as soon as possible, to minimise spread of diseases. The dreaded **Potato Blight** (sometimes called Late Blight) is caused by the *Phytophthora infestans* fungus and is the most devastating of all potato diseases. In many ways it was responsible for the Irish famine in the 1840s. Symptoms are almost always seen on the plants to some extent, but the disease really only spreads rapidly and causes loss of production in warm and humid conditions.

Symptoms of Potato Blight: leaves develop irregular brown necrotic (dead) patches.

In warm, humid weather the whole plant becomes affected, the tubers becoming rotten and horribly smelly. Tuber rot is more common in temperate than in tropical regions. The disease is mainly transmitted by the fungal spores being carried from plant to plant, suspended in air or water droplets; and also by infected tubers being planted.

Spores can survive in the field even without a potato crop because they also live and multiply on wild Solanum species and volunteer potato plants.

It can be *controlled* with fungicides—Copper Sulphate or "Bordeaux mixture" works well, sprayed every 2–3 weeks during the growing season as a preventative; resistant varieties are available, uninfected seed (tubers) only must be planted, and it may be possible to choose the planting date to avoid warm, wet weather during the later stages of the potato's growth period. The haulm should also be burned off and/or removed as soon as it becomes seriously infected.

– Bacterial Wilt—a very serious problem in that land which is infected with this bacteria can never be used again for growing potatoes. Symptoms: the plants wilt, despite a moist soil, and a white mass oozes from the base of the stem or tuber when it is cut. Bacterial Wilt does not occur in Europe, but virtually all varieties grown there are susceptible to it, so potato plant material should never be moved from country to country unless it has been reliably certified to be free from this disease. The only practical control is by using varieties with some resistance. In areas where there is a marked dry season, some control is obtained by a bare fallow (nothing planted). Fumigation works well, but is very expensive.

– Virus Diseases—Leaf Roll, Potato Virus X and Y are the most damaging, but not serious unless the tubers are kept for seed. Seed potato crops are therefore grown in cold, aphid-free areas.

– Scab—parts of the surface of the potato become brown and cracked, but this mainly causes only cosmetic "damage". Worst in alkaline/heavily limed soils with low organic matter content. Controlled by watering and by incorporating plenty of humus into the soil before planting.

Pests: potatoes are very attractive to **slugs**, which can be controlled on a small scale with various traps, or as a last resort, with Aluminium Sulphate or other insecticide pellets applied to the soil near the plants. This should always be covered, as protection against the elements and so that wild life cannot eat the pellets.

– Potato Aphids—see Virus Diseases above. They are not normally sprayed.

– Nematodes—two types: the type found in temperate regions (potato root eel-worm) forms cysts on the roots and can become a major problem if potatoes are grown too often on the same soil. The other type, found in the tropics, causes neither cysts nor any damage.

– Colorado Beetle—yellow, with four black stripes on its back. The larvae eat the leaves and can destroy the entire crop. Controlled with derris, pyrethrum or nicotine.

– Potato Tuber Moth—larvae attack the haulm, then enter the tubers; or they attack the tubers in storage. Worse when potatoes are grown out of season.

– Cutworms, Wireworms and **Epilachna Beetle** can also cause some damage.

YIELD

The global average yield of potatoes in 2004 was estimated by FAO at 17.6 MT/ha, but yields vary widely from about 5–10 MT/ha in low input systems in Africa and elsewhere to over 50 MT/ha in advanced farming systems.

According to FAO the highest national average yield in 2004 was achieved by Belgium, with an impressive 48.4 MT/ha. The lowest was in Angola and Benin, both reporting 3 MT/ha.

The UK average yield for 2004 was estimated by FAO at 43 MT/ha.

UTILISATION

- Virtually all potatoes are grown for more or less local consumption, normally boiled, fried or roasted. In parts of Africa mashed potatoes are eaten with maize and legumes.
- Potatoes are widely enjoyed in alcoholic form as vodka.
- Small and/or diseased tubers are often fed to animals.
- In industry, potatoes are used as a source of starch, and for spirits and industrial alcohol.

LIMITATIONS

- Potatoes are susceptible to many diseases, especially Blight.
- The growing plants need quite specific temperature conditions as well as a regular rainfall pattern or irrigation.
- Low protein content of about 2% (though the protein is richer in lysine than cereals).
- The crop is labour intensive to cultivate, especially in unmechanized systems.
- The tubers are bulky and expensive to transport, 80% of the weight is water. In store they are prone to become either rotten and/or damaged by insects, rodents, etc.
- All green or sprouting plant parts contain poisonous glycoalkaloids, known as solanine, which may cause acute nausea, vomiting, diarrhoea & prostration, usually of a mild degree.

Sweet potato
Ipomoea batatas

Yam (North America) Patate Douce, Patate de Malaga (French); Susskartoffel (German); Papa Dulce (Latin America); Batata de Malaga, Boniato, Camote (Spanish); Batata douce (Portuguese); Batatas (Arabic); Kamote (Philippines); Kumala (Fiji); Shakarkandi (Hindi); Mucur Dinish (Tigray); Tafach Dinich (Ethiopia); Usambé, Katata {Katito—small} (Angola); Kachalu-e Shirin (Dari)

Sweet potatoes have a great range of types, and show wide variety in characteristics, but basically there are two types, based on the consistency of the tubers when they are cooked. The one type has firm, dry, mealy flesh which is normally yellowish or purplish; the other type has a softer, moist, whiter flesh—often incorrectly called "Yams", especially in the USA. Hundreds of named, improved varieties of both types are available.

The plant is perennial but is usually cultivated as an annual. It is a member of the Convolvulaceae (morning glory) family; the flowers are monoecious (male and female flowers both on the same plant).

In the subtropics and temperate regions the tubers are mainly harvested all at the same time, then stored for winter use and subsequent planting. In the tropics stem cuttings are taken and planted during virtually 12 months of the year.

Some varieties produce trailing or twining stems 5 metres or longer, and the leaves are very variable between varieties; even on the same plant older leaves may look very different from younger ones.

Both the tuber and the aerial parts are eaten, and sweet potatoes are an important part of the diet in many tropical countries. Sweet potato tubers contain the pigment beta-carotene, which the human body can convert into Vitamin A. The tubers contain about 16% starch, 6% sugars and are a good source of Vitamin C (23 mg/100 g).

In 2004 more than 127 million MT were grown, according to FAO. The biggest producers are China, Asia, Africa, North and South America, Japan and New Zealand.

PLANTING

Soil: sandy loams are ideal for sweet potatoes, though there are varieties that will grow well in most soil types. Sweet potatoes are sensitive to waterlogging and alkalinity—pH6 is optimum ie they are tolerant of acidity, and are moderately susceptible to salinity.

Fertiliser is rarely used, and although the plants do not have a high demand for nutrients sweet potatoes normally respond well to high Phosphate and Potash fertilisers. Typical applications are, in kg/ha: N20–80, P80–150, K80–170. K is generally more important than N or P, and too much N may encourage stem growth at the expense of tuber growth. Organic manures are very beneficial.

Propagation: stem cuttings about 20–45 cm long are almost always used in the tropics, as they are less valuable than tubers and are also free from soil-borne diseases. The lower half is pushed into the soil at an angle—sometimes only the central part of the cutting is buried.

In some countries in temperate regions small tubers are first planted close together in a nursery; they are then sprouted, and become known as "slips" which are transplanted into the field after 4–6 weeks when about 30 cm long. With this system only the best tubers from the least diseased and insect damaged plants should be used.

Spacing: normally on ridges about 90 cm high with about 30 cm between plants. Sometimes on mounds or hills about 60 cm high about one metre apart, with several plants per hill. The optimum plant population is about 25–30,000 plants (vines) per hectare.

GROWTH CONDITIONS

Day length: short-day; most varieties will only flower when the day length is less than 11 or 12 hours per day, though flowering is not of any great relevance to farmers.

Growth period: 3–8 months, depending mainly on variety and rainfall. Plants are mature and ready for harvest when the leaves turn yellow, but the tubers are best left "stored" in the ground until needed. In arid areas smaller yields are produced, in about 3 months.

Crop care: the vines should be lifted and turned back on to the ridges from time to time. Although this practice will allow more weeds to grow in the unshaded inter-row areas, it does prevent the vines from forming roots, so the plant can produce fewer—but more uniform and larger—tubers.

Temperature: 25°C is optimum, but most varieties will still grow, slowly, at 10°C. The maximum is about 32°C. Sweet potatoes need a 4–6 month frost-free growing period; a slight frost may kill young plants of some varieties. The foliage of older plants may be damaged, even by a light frost, but the plants normally recover.

Rainfall: the optimum is from about 750 to 1300 mm per year, minimum is about 500 mm. If the plants are well established they can tolerate long dry periods. They will usually also tolerate very high rainfall, especially if planted at the end of the rainy season; a mature crop covers the soil surface with vines and leaves, reducing both soil erosion and weed growth.

Altitude: from sea level to 2500 m, on the equator.

Pests: Sweet Potato Weevil is the most serious pest of sweet potatoes. The larvae feed mainly on the tubers but also on the leaves and vines, especially with late planted or dry season crops. Controlled with crop rotations and destroying or feeding all infected plant material; plant only clean cuttings or slips.

Other, normally less damaging, insects include the Sweet Potato Leaf Beetle, wireworms, grasshoppers, fleabeetles and termites. Leaf-eating caterpillars can be controlled with insecticides.

Diseases: not usually a big problem in the tropics. There are two main types:

a) *field diseases* of leaves, vines and tubers; Black Rot, Soft and Dry Rots, Leaf Spot, Scurf (Rust), Stem Rot (Wilt) and Soil Rot (Pox) are the most common. Mainly easily kept under control, if necessary, with fungicides.

b) *storage diseases*, especially of damaged tubers. Soft Rot is the most common, control is by using resistant varieties, crop rotations and planting clean cuttings/slips. Dry Rot, Java Black Rot and Charcoal Rot can give problems. As mentioned, the tubers are best left stored in the ground.

(Some diseases appear at both stages).

YIELD

The tubers are ready for harvest when the leaves turn yellow and begin to drop. Although the potential yield of sweet potatoes is well over 50 MT/ha they commonly yield no more than 15 MT/ha or less, mainly because they are very often grown on poor land and are not carefully cultivated.

The global average estimated by FAO in 2004 was 14.8 MT/ha, from the highest average in Israel of 35 MT/ha to a low in Mauritania of 1.0 MT/ha.

UTILISATION

- Sweet potato **tubers** are an important part of the diet in many countries in the tropics and subtropics, mainly for subsistence purposes. They are normally either boiled or baked. Tubers are also often fed to animals. For storage they can be sliced and dried in the sun. Products manufactured from the tubers include starch flour, syrup, glucose and alcohol.
- The **young leaves** and tender parts of the stem are sometimes eaten as a vegetable.
- The **haulm** (vines and leaves) can be either fed directly to animals or ensilaged.

LIMITATIONS

- Sweet potatoes have a high labour requirement for land preparation and harvesting.
- They are also sensitive to waterlogging, salinity and alkalinity.
- The plants are "heavy feeders" and need plenty of calcium, boron and magnesium as well as the major elements to produce good crops.
- The tubers do not store well, unless dried, and lose much moisture within a few weeks.

Taro
Colocasia spp.

Taro (French); Kolokasie (German); Malanga Islena (Spanish); Kechalo Pakistani (Pashtu); Assipi (Angola)

The **taro, dasheen, eddoe, cocoyam** ("old" and "new"), **malanga, tania, elephant ear,** etc. are all members of the large botanical arum-lily family the *Araceae*, or "Aroids" or "Arum", with about 100 genera and 1500 species found throughout the world. Farmers have selected these useful plants since prehistoric times, each type being adapted to grow well in certain parts of Asia, Africa, Oceania or Latin America.

The Aroids are an important food crop in many parts of the humid tropics, such as the Pacific, West Indies and West Africa. Global production is estimated at about 6 million MT per annum.

They are grown mainly for their edible *corms* or *cormels*—a "corm" is a short, solid, swollen underground stem which lasts for about one year; the next years corm grows at the top of, or close to, the old corm. A "cormel" is a corm which arises vegetatively from a parent corm.

There is great confusion and disagreement among botanists and agriculturalists about the classification of the Aroids. The common names used for them often depend on the country—in West Africa for example the name "Cocoyam" is used for both *Colocasia* and *Xanthosoma* species. In the Pacific, most of the species on the table below are known as "Taro". This table includes only the most important species of Aroids:

Species (spp.)	Common Names	Comment

Asian Origin
 1. *Alocasia macrorrhiza*
 2. *Alocasia indica*
 	
 3. *Colocasia esculenta*
    var. *esculenta*
 4. *Colocasia esculenta*
    var. *antiquorum*
    (=var. *globulifera*)
 5. *Cyrtosperma*
    chamissonis
	Giant Taro,Ta'amu, Ape
	 Sometimes
	 considered the same
	 species as 1)
	 Dasheen, Taro, Eddoe,
	 Cocoyam
	 Eddoe, Dasheen (Asia
	 and Pacific),
	 Elephant
	 Ear, Akhi (Hindi)
	 Giant Swamp Taro
		1. 3-4m tall; Stem
		 above ground is
		 eaten
		 2. 2m tall. Important in
		 India
		 Species 3. and 4. are
		 described in the
		 text below
		 5. 3-4m tall; Tubers up
		 to 60 kg after ten
		 years

South American Origin 1. *Xanthosoma* *atrovirens* 2. *X. sagittifolium* 3. *X. violaceum* 4. *X. brasilense*	All four of these species are known as Tannia, Tanier, Yautia or (new) Cocoyam	1. Yellow tubers; found mainly in Puerto Rico 2. Makes good "fufu" 3. Large plants, tubers of little food value 4. Only the leaves are eaten

Both species of *Alocasia* are very hardy and high yielding.

Colocasia and Xanthosoma are similar crops; they are complementary, not competitive, because they are adapted to different growing conditions. Xanthosoma are more drought tolerant than Colocasia, and they grow better in the shade. Both types are highly resistant to attack by pests and diseases.

The description below refers to the two species:
Colocasia esculenta var. *esculenta*—normally called the **Dasheen**;
Colocasia esculenta var. *antiquorum*—normally called the **Eddoe**.

PLANTING
Propagation:
Dasheen—either by the leaf-bearing tops of mature corms, especially in the Pacific, or by small side tubers or suckers;
Eddoe—small cormels (60–150 g) are planted whole.
Soil: the best are forest soils that have been cleared and burned for the first time. Soils should be wet, heavy and fertile. Eddoes need less fertile soils than dasheens. Some varieties are tolerant of flooding and salinity. Both normally respond well to potash as well as nitrogen and phosphate, normally applied in two or three split dressings.
Spacing: roughly 60–90 cm if grown in pure stands.
Intercropping: this is the normal method, mixed with okra, pigeon peas, maize, etc.

GROWTH CONDITIONS
Growth period: 5–6 months for eddoes, 8–10 months for dasheens.
Temperature: they need hot conditions; eddoes withstand low temperatures better than dasheens, which need about 6 or 7 frost free months. The optimum daily temperature is 21–27°C.
Rainfall: dasheens need about 2500 mm per year, eddoes need less.
Pests: Taro Leafhopper (*Tarophagus. proserpina*), can be biocontrolled with Mirid bugs; Gabi Moth larvae can cause major defoliation; Root-knot nematodes are a serious pest in some countries.
Diseases: Taro Leaf Blight is usually the most serious, especially in the Pacific. Symptoms are round, water-soaked dead spots on the leaves, then the plant collapses and dies. Some control with fungicides such as Dithane. Root rots and virus diseases may also be a problem.

YIELD
Dasheen—small side suckers are left in the ground after harvest and allowed to regrow, or *ratoon*. The ratoon crop is often grown only for the leaves. Average yields of corms are 15–20 MT/ha for the first harvest and 10–15 MT/ha for the first ratoon crop.
40 MT/ha and more is possible.
Eddoe—this crop is not ratooned. The corms and cormels can be left "stored" in the ground until needed during the dry season. The main, central corm is relatively small but has many side cormels.

UTILISATION

- **Corms**, or "tubers", are not only rich in Vitamins B and C but are also easily digested and are therefore suitable for feeding to infants, old people and those with gastric disorders. Raw corms contain about 25% starch and up to 13 mg/100 g Vitamin C. They can be a good substitute for Irish potatoes, sliced and fried into chips. The corms can be roasted, boiled or baked, or made into meal or flour. In Polynesia the famous Poi dish is made from crushed, fermented taro.
- **Young leaves** are often eaten as a vegetable, a bit like spinach, though many varieties have a very bitter taste. An excellent source of Vitamin C (52 mg/100 g in raw leaves), the young leaves also provide protein, calcium, phosphorus, iron, potassium and vitamins A and B.
- **Young shoots** are sometimes blanched in steam and eaten like asparagus.

LIMITATIONS

- Dasheen corms can normally only be stored for a few weeks—at about 10°C they last up to six months. Eddoes can be stored for several months, but only if dried and well stored.
- Leaf blight and virus diseases can be very damaging.
- Taro plants need high rainfall and heavy, fertile soil unless heavy fertiliser applications and/or irrigation are used.
- The plants take a long time to produce a crop, especially dasheens.
- The corms and leaves of some varieties have bad tasting calcium oxalate crystals which must be removed during cooking, and make them inedible when raw.
- Low protein content.
- Very little scientific or agronomic effort has been given to Taro, and modern science has played virtually no part in the development of this potentially very productive food crop. The reason may perhaps be partly due to the confusion over the naming and description of the different species, sub-species and varieties.

Yam
Dioscorea spp.

D. alata Greater Yam, *D. bulbifera* Air Potato, *D. cayenensis* Yellow GuineaYam, *D. dumetorum* African Bitter Yam, *D. elephantipes* Elephant's Foot or Hottentot Bread, *D. esculenta* Lesser Yam, *D. hispida* Asiatic Bitter Yam, *D. nummularia & pentaphylla* Yam, *D. opposita* (or *D.batatas*) Chinese Yam (or Cinnamon Vine), *D. persimalis* Bush Yam, *D. rotundata* White Guinea Yam, *D. sansibarensis* Wild Yam, *D. trifida* Cush-cush Yam; Igname (French); Yamswurzel (German); Name, Yame (Spanish);
Aja, Yampi, Mapuey, Inhame (Portuguese); Khamba Alu (Hindi)

The Sweet Potato (*Ipomoea batatas*) is sometimes known as "yam" (from the African *nyami*), especially in Americanised English and when they have orange flesh. To add to the confusion the edible tubers of some other species are also sometimes known as yams. The description below refers to plants of the genus

Dioscorea. Yams should not be confused with the Yam Bean *Pachyrrhizus erosus*, a leguminous climbing plant grown mainly in India and South-East Asia; the tubers are either thinly sliced and eaten raw in salads or cooked or pickled; the young pods are edible, but the seed is poisonous.

The global production of yams was estimated by FAO at 40 million MT in 2004, about 75% from West Africa, especially Nigeria. Yams are also widely grown in southeast Asia, the Pacific and the Caribbean, although production of yams is declining overall as they are becoming replaced by cassava and sweet potatoes, both of which are often able to produce more food with less labour input. Since they are a climbing crop they need vertical support, but then don't we all. Ideally yams should be grown up stakes.

Nutritionally the yams are more useful than cassava, particularly in protein. They contain about 28% starch and about 5 mg/100 g Vitamin C; carotene is present in varieties with yellow flesh. Cases of kwashiorkor have been seen to increase when yam is replaced with cassava in the diet. They were used on long sailing journeys, including slave ships between West Africa and the New World; they contain enough Vitamin C to help reduce scurvy.

There are about 600 species of *Dioscorea*, of which three are particularly interesting:

Dioscorea alata—Greater Yam, Water Yam, Winged Yam, Ten-month Yam, Asiatic Yam or White Yam. A native of southeast Asia, it is now grown throughout the tropics. The highest yielding species of them all. Requires high rainfall, about 1600 mm a year is ideal. In West Africa the White Guinea Yam (*D. rotundata*) is more popular as it makes better fufu. Stores well, for five or six months at least. Stems twine to the right.

Dioscorea rotundata—White Guinea Yam, Eight-month Yam. A native of Africa, where it is the most important yam, mainly grown in West Africa. Not grown in Asia. Very similar to the Yellow Guinea Yam (*D. cayenensis*), though it has better drought resistance and the tubers have a longer dormancy period, and so can be stored longer. Stem up to 10 m long, twines to the right.

Dioscorea cayenensis—Yellow Guinea Yam, Twelve-month Yam, Cut-and-come-again. A native of West Africa where it grows well in the forest zone with a short dry season. Not grown in Asia or the Pacific. Can be harvested throughout the year. Stores poorly. Stems twine to the right.

There are at least eight other important sub-species, listed below:-

Dioscorea esculenta—Lesser Yam, Chinese Yam, Pana Yam. Mainly grown in Asia and the Pacific. Produces up to 20 small tubers per plant. Does not store well. Stems twist to the right.

Dioscorea trifida—Cush-cush Yam, Aja, Yampi, Indian Yam, Mapuey. The only food yam native to South America. Now grown in the Caribbean and South America where it is highly prized for its flavour, though it yields less than the introduced *Dioscorea alata*. Produces a group of small tubers 15-20 cm long. Stems twine to the left.

Dioscorea dumetorum—American Bitter Yam, Cluster Yam, Forest Yam. Occurs wild throughout Africa between 15°N and S; mainly cultivated in West Africa. Many of the wild forms contain a poison which must be removed by soaking and boiling, though they are often used as famine food. Tubers are either single or in clusters. Stems twine to the left.

Dioscorea bulbifera—Potato Yam, Aerial Yam, Bulbil-bearing Yam. Widely grown in the tropics of Africa and Asia. Subterranean tubers are small, hard and bitter (or may be absent), while the aerial tubers are good to eat, though they may need to be detoxified. Stems twine to the left.

Dioscorea hispida—Asiatic Bitter Yam. Similar to the African *Dioscorea dumetorum*. Large tubers, near to the soil surface, which are very toxic and are used for hunting or criminal purposes; the toxin can be removed, and the tubers are used as a famine food in the tropical East. Stems twine to the left.

Dioscorea nummularia—grown in Indonesia and Oceania. Similar to *Dioscorea cayenensis*, but tubers are formed deep in the ground, and are allowed to grow for 2–3 years before they are dug up. Stems twine to the right.

Dioscorea opposita (or *D. batatas*)—Chinese Yam, Cinnamon Yam or Vine. Used medicinally in China. The most resistant to cold of all the yams. Grown in China, Korea, Taiwan and Japan.

Dioscorea pentaphylla—occurs both wild and cultivated in the warm, moist parts of Asia. Non-toxic tubers. Often grown as hedges around fields.

PLANTING

Propagation: normally done vegetatively, using 100 g–2 kg parts of a large tuber, or whole small tubers. Rich topsoil and compost should be put in the planting hole. About 2 MT/ha of tubers are needed. Many wild yams reproduce freely by seeds, but many cultivated varieties rarely or never produce viable, fertile seed.

Germination: yams are monocotyledons, though some botanists believe that some varieties have two cotyledons, one of which remains within the seed.

Soil: should be fairly fertile, loose, deep and well drained. Responds well to Nitrogen but very often does not respond to Phosphorus. Normally, yams give an economic response to NPK fertilisers at about 600 kg/ha, and manure. The plants also benefit from mulching.

Plant spacing: 10–15,000 setts/ha. For sub-species with larger tubers, either in a 1×1m square grid or with 50 cm along the ridges and 1.5 m between ridges, giving about 10,000 setts/ha. Planted somewhat closer together when the tubers are small.

Stakes: yams should be allowed to grow up stakes at least 2 m high and with branches, which can increase yields by 60% or so.

Rotation: often the first crop planted during shifting cultivation, or after a fallow. Yams yield well after a leguminous green manure crop.

Intercropping: almost always done, with one or more crops such as maize, okra or cucurbits.

GROWTH CONDITIONS

Growth period: *D. alata* and *D. rotundata* produce the first edible tubers in 8–10 months. *D. cayenensis* needs about 2 more months, and can then be harvested continuously.

Day length: long days favour vine (stem) development, short days favour tuber growth.

Temperature: yams do not tolerate frost and grow poorly below 25°C. *D. opposita* is the most tolerant species to cold. Should not be stored below about 10°C.

Rainfall: optimum is 1500 mm per year or more, though some species can survive with 1000 mm. There should ideally be at least 2–4 months of dry season.

Pests: in West Africa Yam Beetles are the worst; adults feed on the tubers, making round 1–2 cm lesions and may also damage the growing points of newly planted setts. Other pests include the Yam Weevil and Scale Insects. The Yam Nematode, which is also a storage pest, can be controlled with rotation and by only planting clean tubers.

Diseases: rarely a serious problem. Leaf Spot is the most common, causing brown or black spots on the leaves and stems. Anthracnose and Witches Broom occur in West Africa, and Rust occurs everywhere. Mosaic virus diseases have been reported in West Africa and the Caribbean, controlled by destroying all infected plants. Storage rots can cause major losses, though the severity depends on the species, variety, degree of injury of the tubers and the number of yam nematodes that are present.

YIELD

There is enormous variation in yields of yam, according to the growing conditions, crop management, species and variety.

The global average is about 10 MT/ha (9.1 MT/ha in 2004, according to FAO), but yields of about 20–30 MT/ha are common. National average yields in 2004 ranged from a high of 22.3 MT/ha in the Solomon Islands to 2.7 MT/ha in Ruanda.

The most commonly grown yam species produce 1, 2 or 3 tubers, each weighing from 5 to 15 kg. Some plants produce up to 50 kg of tubers.

UTILISATION

- Yams are grown almost exclusively for the human consumption of the tubers. Small tubers can be peeled, then boiled or roasted; larger tubers are also peeled, cut into smaller pieces, then either fried in oil, roasted or boiled. They are sometimes added to stews. The tubers can be stored for longer periods if dried up and ground into flour. In West Africa, they are normally eaten as "fufu", peeled, cut up and boiled tubers pounded in a wooden mortar.
- Some wild species such as *D. floribunda* and *D. villosa* contain usable proportions of diosgenin, a precursor for the synthesis of fertility and anti-inflammatory drugs.

LIMITATIONS

- Yams are very labour intensive, for both cultivation and food preparation.
- The tubers are often rather expensive to buy.
- The plants require high rainfall, or irrigation, and fertile and well drained soil.
- The plants are also sensitive to frost.
- Fungi and/or nematodes can often cause large storage losses.
- Some species contain toxic alkaloids.
- Modest energy value of about 100 calories per 100mg edible portion.

2E. VEGETABLES

The word "vegetable" is an imprecise term—many crops that are often described as "vegetables" are described elsewhere in this booklet as, for example, "Root Crops" (Irish potato) or "Legumes" (peas, beans, etc). For the purposes of this publication, vegetables are defined as being plants eaten whole or in part, either cooked or raw, and which do not usually constitute the main part of the meal or diet.

Vegetables generally provide very little in the way of energy, but are useful in making high energy food such as cereals and root crops more interesting and palatable to eat. They are also very often valuable sources of fibre (both soluble and insoluble), vitamins, especially A and C, and certain trace elements.

The most important plant families that provide vegetables are listed below:

Alliaceae (Liliaceae). The onion family; leeks, garlic and chives.

Asteraceae (Compositae). The daisy sunflower family; both the Jerusalem and Globe artichoke, chicory, endive, lettuce, salsify (oyster plant) and sunflower.

Chenopodiaceae. The beet family; spinach, leaf beet, sugar beet, beetroot, Good King Henry, kaniwa, Lambs Quarters, orach, quinoa and Swiss chard and orach.

Cruciferae (sometimes called *Brassicaceae*). The cabbage (or mustard) family. The vegetables in this family are mainly in the Brassica Genus, such as cauliflower, cabbage, broccoli, Brussels sprouts, cauliflower, collard/kale, rape, turnip and kohlrabi. These are sometimes called "cole crops", a term reflecting the fact that these crops are primarily stem crops. Cruciferae which are not also Brassicas include garden cress, watercress and radish.

Cucurbitaceae. The gourd family; marrows, courgettes, cucumbers, melon and watermelon, pumpkin, squash and chayote.

Labiatae. This is mainly a family of herbs used for flavouring, such as mint, sage, thyme and oregano, and so are not true vegetables. The few vegetables present in this family have edible tuberous roots, found in some species of Woundwort (*Stachys affinis*), the Chinese artichoke, and in the African genus *Plectranthus*.

Leguminosae. The legume, or pulse, family; mainly the beans and peas, both of which have varieties grown as field crops as well as vegetables—French beans, mange tout peas and snowpeas have edible pods, and a few species such as *Apios tuberosa* also produce edible root tubers.

Liliaceae (Alliaceae). The onion family; leeks, garlic and chives.

Poaceae(Graminae). The grass family produces most of the world's food, in the form of wheat, maize, rice, sorghum, millet and so on. However only a few of them are eaten as vegetables; these include sweetcorn, lemon grass and wild rice (*Zizania aquatica*).

Solanaceae. The nightshade family; Irish potatoes, tomatoes, peppers and aubergine.

Umbelliferae(Apiaceae). The carrot family;carrot, caraway, celeriac, celery, coriander, fennel, anise, dill, coriander, caraway, cumin, ginseng, hemlock, parsley & parsnip.

Three other less important families, *Araceae, Convulvulaceae* and *Dioscoriaceae* produce the Taro, Sweet Potato and Yams respectively. Plants from at least another 15 plant families are also grown to be eaten as vegetables. The common and botanical names of many of these are listed in Section **3A**, pages 288–309.

Cabbage
Brassica oleracea var. *capitata*

Common Cabbage, Head Cabbage
Chou (cabus) (French); Kohl (German); Col (Spanish); Couve (Portugal), Repolho, Repolhuda (Brazil); Acovi (Angola); Patgobhy (Hindi); (Shna)Karam (Pashtu and Dari).

Cabbage is just one of the races (or botanical varieties) of the species *Brassica oleracea*, the Wild Cabbage. Brassicas are the most important Genus of the cabbage (or mustard) family the *Cruciferae*. There are several other races of *Brassica oleracea*:

- var. *acephala*—Kale, Collard, Borecole
- var. *botrytis*—Cauliflower
- var. *alboglabra*—Chinese Broccoli
- var. *gemmifera*—Brussels sprouts
- var. *gongylodes*—Kohlrabi
- var. *italica*—3 types of broccoli: Sprouting Broccoli, Calabrese and Romanesco. Also perennial ("Nine Star") broccoli.

Chinese cabbage is normally classified as *Brassica chinensis* var. *pekinensis*

Wild cabbage plants are still found growing in southern England, western France and northwest Spain. Selections from this plant have been made by food producers for over 2000 years, and kale is mentioned in early Greek writings. Modern cabbage varieties are now grown throughout the world, including cooler parts of the lowland tropics.

The plant is a biennial, but the crop is grown as an annual. The thick, overlapping leaves are the part of the plant that is eaten. The "heads" formed by these leaves come in a wide range of shapes, sizes and colours—essentially, heads are either round or pointed, green or red, smooth or wrinkled—rather like people.

Cabbages can be produced twelve months of the year in many places around the world by planting successively a few of the eight types available (months refer to the UK):

Spring cabbage comes in two forms, either spring greens or spring hearting, both transplanted out in September in northern temperate regions, ready to eat in April and May. Some varieties, such as Durham Early are dual-purpose, providing first greens then small hearts.

Summer cabbage always forms a heart, either red or green, pointed or round. Transplanted out in spring they are eaten from mid to late summer.

Autumn cabbage is planted out in mid-summer, grows rapidly and is eaten in the autumn.

Winter white cabbage plants are large and grown to be stored; planted out in June, maturing in November and December, can be stored for months.

Savoy cabbage is the hardiest of all and some varieties withstand coldest winters. Planted out in early July, harvested October to March according to variety. Wrinkled leaves, good for coleslaw.

Savoy hybrids are crosses between savoy and white cabbages. Very hardy.

January King leaves are normally slightly crinkly and purplish, and are very tasty. Planted out in early June, mature by November/December.

Red cabbage is normally eaten either raw or pickled; if cooked it takes much longer than other cabbages, up to 2 hours boiling. Harvested in October/November, for winter storage if needed.

Cabbage leaves are a good source of vitamins A, C, B1, B2, B3 and D; in hard times they can also be useful in providing bulk to the diet. Cabbage is also high in potassium and calcium but loses most of these nutrients when the leaves are cooked. However **pickling** does not destroy Vitamin C—Captain Cook insisted that his crew stayed healthy due to a daily intake of sauerkraut.

Collecting seed from Brassica plants is not worthwhile as they all readily cross-pollinate and the hybridised offspring are often useless, so you need to buy fresh seed every year or use the previous years purchased seed if it was well stored and still germinates vigorously. A large range of varieties of all of the cabbage types listed above are available from commercial seed suppliers, each one adapted to grow in particular conditions. Hybrid seed of Brassicas has so far had only limited success.

Unless otherwise stated, the account below refers to the Common (or "Head") Cabbage, *Brassica oleracea* var. *capitata*.

PLANTING

Propagation: by seed (not home-grown—see above), normally planted in boxes or nurseries then transplanted when about 15 cm tall, at about 1–2 months. At the time they are being planted out many people dip the roots in a bucket containing thin mud and a little lime.

Germination: viability of the seed rapidly diminishes in hot and/or humid conditions. Germination is normally quite fast, in 3–6 days.

Soil: cabbages need rich, fertile and well drained soils. Sandy loams are ideal. They like plenty of manure, but lime may be needed in more acidic soils (or plant cabbages after a legume). Clubroot disease is more prevalent in acidic soils. Both yield and quality are increased with compound fertilisers provided that the Nitrogen level is not too high, but cabbages should not need any. The soil should be firmly compacted around the transplanted plants.

Seed rate: 3.5–5 kg/ha. There are about 300–375,000 seeds per kg.

Plant spacing: either in a 35–50 cm square grid or 50–70 cm between rows and 40–60 cm between plants. Fast growing crops such as lettuce or French beans can be interplanted. They are sun plants and so prefer an open, unshaded position.

Depth: 0.5–2 cm

GROWTH CONDITIONS

Growth period: highly variable, depending on type and variety, and growing conditions.

Temperature: varieties adapted to temperate regions can withstand −7°C frost. In hot, dry regions they are grown in the cooler season and/or at high altitudes, as they grow poorly in heat.

Rainfall: medium-high rainfall regions are ideal. Drought affects the plants badly, which should not suffer any check to their growth. Irrigate if possible whenever dry.

Altitude: in the tropics cabbages are grown in the highlands.

Pests: cabbages are extremely attractive to insects; often the worst damage is caused by the **Cabbage root fly**, which causes plants to wilt, sometimes also to develop bluish leaves with yellow edges, and occasional death. The larvae burrow down from the soil surface, then into the roots and stem, which they attack at ground level. The best control is by placing a "collar" of material such as roofing felt, carpet—preferably greased or paraffined—around individual plants. A square of about 15 cm is enough for the collars. Some control also comes from applying high N fertiliser, supporting the plants by banking up soil around their stems, hoeing or moving the soil regularly so that birds can find and eat the larvae—and destroying all infected plants.

Large and Small Cabbage White butterflies commonly attack all Brassicas. Caterpillars of the Large White are yellow and black and highly conspicuous as they also feed on the upper side of leaves. The Small White's however are small and green and inconspicuous. Both can either be picked off and dealt with, or sprayed with a salt solution, which also deters slugs.

Cabbage Moth larvae eat the central leaves, control with derris or pyrethrum.

Cabbage White fly look similar to the greenhouse whitefly. They overwinter on Brassicas; eggs look like small brown scales and are laid on the lower leaves, which should be removed.

The flies can be discouraged by planting lovage and/or fennel near to the cabbage plants, which attract the parasitic wasps *Aphelinus*.

Leatherjackets are grey-brown, legless larvae of Daddy Long-legs which sometimes eat seedling roots, causing the plants to wilt and die. Dig or plough soil several times to allow birds to eat the larvae.

Cutworms are very small worms which eat small plants at the ground level. Control with a collar—as for root fly control—or pile wood ash around the stems.

Diseases: if the dreaded **Clubroot** disease is present in your soil, you have a problem, as the fungal spores can lie dormant for at least seven years. The roots become large and malformed, and look similar to cabbage gall weevil damage, identified by cutting open the root galls and looking for the larvae. Plants may wilt and even die. Control or minimise spread by rotation, no Cruciferae of any kind, including shepherd's purse or charlock, for at least four years. Apply lime to acidic soils, burn all infected roots and never put Brassica roots on the compost heap. One ingenious method is to spray soil that is not planted with Cruciferae, but may be planted later on, with water in which Brassica plants have been boiled. This stimulates the fungal spores to break dormancy, but they will soon die because there are no suitable Cruciferae to live on. Clubroot is mainly a problem of the temperate regions. In the tropics, Blackleg, Black Rot, Downy Mildew and Cabbage Mosaic Virus can devastate crops of cabbage equally badly.

YIELD

Cabbages can produce enormous yields of 100 MT/ha and more, although about 90% of this consists of water. The global average is around 25 MT/ha.

An average crop is about 0.5–0.7 kg per 30 cm of row. One head can weigh 3 kg or more.

UTILISATION

- Eaten by humans, either fresh or stored, for twelve months of the year in temperate climates. They can be eaten raw or cooked, or pickled to produce Sauerkraut.
- Coleslaw, a salad made with grated cabbage, originated in Holland.
- Cabbage soup is "enjoyed", or otherwise, throughout Europe and elsewhere.
- Cabbage can be successfully stored, particularly the late maturing Danish types.
- Eaten by animals, especially cattle—the Drumhead types are favoured.

LIMITATIONS

- As food, cabbages have low nutritional and calorific value.
- Cabbage plants are heavy feeders, and need strong, fertile soil.
- The plants are also susceptible to attack from both disease and insects.
- Some varieties are very slow to mature.
- Cabbages do not travel well, and only certain types can be stored for long periods.

Carrot

Daucus carota

Carotte (French); Karotte, Mohre (German); Zanahoria (Spanish);
Cenoura (Portuguese);
Ocenola (Angola); Gajar (Hindi); Gazar, Gazaray (Pashtu); Zardak (Dari).

The modern cultivated carrot is derived from selections made by farmers for hundreds of years from the wild carrot, which normally has a white root. Nowadays you can find carrots not only in their normal carroty-orange colour but also red, yellow, white and even crimson.

The roots have some nutritional value in that they contain not only high levels of sugar (about 7%) but also carotene, the orange pigment, which is converted into Vitamin A when eaten, beneficial to eyesight.

Yellow carrots were first recorded in Turkey in the tenth century, and the Romans also grew them. Today carrots are grown virtually everywhere in the world where other vegetables are grown. Although they are basically a cold season crop, carrots can be grown in the subtropics and tropics, using very early, small varieties such as *Early Horn* or *Early Gem*. At higher altitudes vigorous quick growing *Nantes* or *Chantenay* types can be grown.

The plant is a biennial, and if left in the ground will normally flower in the second summer. It grows from about 30 cm to about 1m tall, and has a swollen taproot, either short and stumpy or long and tapering according to the variety. Plant breeders focus on the two main plant types, either the slower growing higher yielding long root types or the faster growing lower yielding types with short, stubby roots.

Sometimes the plants behave like an annual and *bolt* in their first year, ie they form a seed head, which is fed from the food reserves in the root which becomes thin and shrivelled, useless as food in fact. The best way out of this dilemma is to grow varieties that are adapted to the area.

Carrots are members of the *Umbelliferae* (or *Apiaceae*) family, which includes a number of poisonous species such as poison hemlock, water hemlock and fools parsley, as well as edible plants such as arracacha, carrot, caraway, celeriac, celery, coriander, fennel, anise, dill, coriander, caraway, cumin, ginseng, parsley and parsnip.

Some hybrid varieties are available, a few of which claim to have some tolerance to the dreaded Carrot Fly. This pest is normally the single biggest problem with growing carrots, and is discussed on the following page under "Pests".

PLANTING
Propagation: by seed, which should be sown quite thickly as the germination rate is often low. The seed is very small, so to make plant spacing more regular the seed is often mixed with dry earth and/or lime when it is planted. Emergence can take ten days or more.

Soil: carrots are not very demanding and will grow in most soils if they are not too acidic or heavy. Sandy loams are the best, although some of these can be deficient in some minor or trace elements, especially the deeper soils. Carrots are tolerant of some soil acidity; ideally, the pH should be about 6–7, but prefer sandy, chalky well limed soils.

Manure and/or compost should be applied during the previous autumn so that it is well rotted into the soil before the carrots are developing, to avoid "forking" of the roots. The soil should be left light and fluffy, and not too compacted.

Seed rate: 1–5 kg/ha, equivalent to 4–5 seeds per 2.5 cm. There are about 890–910,000 seeds/kg.

Spacing: 30–40 cm between rows. Plants in the row are first thinned to about 4 cm between plants, and later to about 8 cm. The thinnings can (and should!) be eaten—organic carrots taste divine. After thinning, the soil should be drawn up around the remaining plants to discourage carrot flies from re-entering.

Depth: from about 1 cm in deep and retentive soil to 2.5 cm in dry and light soil.

Intercropping: Carrots can be interplanted with clumps or rows of Alliums such as onions, shallots, perennial onions and garlic to hoodwink the nostrils of the wily fly. Both the Carrot Fly and the Onion Fly are kept away if the Allium "guards" are thick enough and well positioned.

Rotation: ideally, five years should elapse between carrot crops to avoid disease buildup, especially where disease/s have been significant.

GROWTH CONDITIONS

Day length: long-day.

Growth period: the first thinnings can be eaten in about 35 days. The crop can then be thinned again, and continuously harvested when needed for a period of several months.

Temperature: carrots are a cool season crop, although there are varieties available which are suitable for growing in highland areas of the tropics and subtropics, normally planted in the autumn and winter.

Rainfall: the plants should be kept moist but not wet. If irrigated, this should be plentiful so that the water goes deep down, encouraging the roots also to go down deep.

Pests: the commonest and worst is the **Carrot Root Fly**; its maggots are about 8mm long and eat into the outside of the carrots, leaving ugly marks which may also later develop into rot. Leaves of infested plants tend to turn red. The adult fly can smell carrots from several kilometres away, especially when the plants are bruised, so both thinning and weeding should ideally be done in the evening and on wet days, preferably in a light rain, or after a thorough watering. Methods to reduce the fly problem include placing a barrier of netting 60–80 cm high all around the carrot plants, eliminating all nearby weeds of the *Umbelliferae* family as these are alternate hosts, putting soot or wood ash on the soil around the plant stems, spraying a mixture of 30 cc of paraffin in 4.5 litres of water onto plants, or mixing paraffin with sand and applying it on the soil around plants. Probably the most satisfactory way of dealing with this almost inevitable problem is to try to deter the flies with interplanting sage, onions and other Allium plants so the flies cannot detect the distinctive carrot smell.Best of all, try earthing up.

Plant breeders have had some success in breeding carrot fly resistant varieties, based on the partially resistant variety "Sytan" crossed with the Libyan species *Daucus capillifolius*. Other partially resistant varieties include "Fly Away", "Resistafly" and "Maestro".

Diseases: the fungus called **Carrot Disease** causes brown spots to form on the roots, then tiny red spores on the soil surface. Minimised by burning all infected roots, applying a mixture of 2:1 sulphur:lime to infected soil and rotating five years between carrot (or *Umbelliferae/Apiaceae*) crops.

– Storage Diseases—carrots should be stored dry, and so they should not be washed before storage—all earth and plant tops should be removed, and the roots stored in a cool, well ventilated place—2–5°C is ideal.

YIELD

Yields of carrots are very variable, according to the variety, growth period and conditions, soil, rainfall, etc. In poorer conditions 3 or 4 MT per hectare is common; modern cultivation techniques can produce yields 10–12 times higher.

Maincrop carrots for winter storing can be left in the ground until well into the winter. Carrots can be stored, but for this to succeed the roots must not touch each other and must be dry (unwashed) - sand or peat is ideal, in any container such as a barrel or tea-chest, which must be ventilated; damaged roots should not be stored.

UTILISATION
- **Human food**—carrots are a good source of Vitamin A, converted from carotene; the roots are used as a vegetable, and in soups, stews, curries, etc. They can also pickled, canned and dehydrated. The juice can be extracted, which can be added to orange and other juices.
- **Animal food**—roots and tops are sometimes fed to animals.
- **Seed**—contains an essential oil, used for flavouring and perfumes.

LIMITATIONS
- Carrots have low nutritive value, apart from Vitamin A and about 7% sugars.
- The plants are susceptible to damage by the Carrot Root Fly.
- In warm or humid conditions carrots have a short storage period.

Cucurbits
Cucurbita spp.

Chayote, Courgette (Zucchini), Cucumber, Cushaw, Gherkin, Gourd, Loofah, Marrow, Melon, Pumpkin, Squash (winter and summer), Watermelon; Potiron, Citrouille, Courge {Courge à la moelle = marrow} (French); Ayote {=*C. mixta*}, Gartenkurbis {=marrow} (German); Calabaza (Spanish); Abóbora {=pumpkin}, Abobrinha {=courgette}, Cabaça {=gourd} (Portuguese); Sitaphal (Hindi); Kadu (Pashtu and Dari).

The gourd family *Cucurbitaceae* has a great range of species and sub-species, some of which have evolved to survive in extreme climatic conditions, such as the Tsava melons of the Kalahari which survive years of drought as seed.

Cucurbits are relatively robust and easy to grow, and from their origins in northern Mexico and southwest America have spread throughout the warmer regions of the globe. They were one of the first crops cultivated by man, and *Cucurbita pepo* (vegetable marrow, gourd, pumpkin or summer squash) is thought to have been cultivated since 7000 BC.

Identification of all the many genera, species and sub-species is often difficult and confusing, especially as different people often use different names to identify the same crop. The plants are mainly the vigorous climbing or trailing types, and there are both perennial and annual species. They are often named according to the season in which they mature, or according to their shape, providing for example the crookneck squash, (Turks) turban gourd and banana squash. To add yet further to the confusion there are a number of local or popular names, such as Hubbards, scallops and acorns.

Some Cucurbit species can provide *anthelmintics* (aka *vermifuges*), which cause the evacuation of parasitic intestinal worms, in the form of a substance called *cucurbitine*, contained in seeds of some cucurbits, *C. maxima* in particular.

There are about 25 species in the genus Cucurbita; some are described below:

Summer Squash—also known as Summer Crookneck, Autumn Pumpkin, Vegetable Marrow or Gourd - the immature fruits of *C. pepo* are eaten, used as a vegetable.

Winter Squash—mature fruits of *C. pepo, C. maxima, C. mixta* and *C. moschata*. Used as a vegetable, for baking, in pies, for jam and for animal food. Can be stored for six months or more. Fine-grained flesh, and mild flavour, suitable for baking.

Pumpkin—normally have soft rinds and hard stalks, while marrows have hard rinds and soft stalks. Like Winter Squash, Pumpkins are the mature fruits of *C. pepo, C. maxima, C. mixta* and *C. moschata*, and are used in pies and soups and as animal forage. The coarse flesh and strong flavour mean that it is not normally eaten as a vegetable.

English Marrow—fruits of *C. pepo*, eaten both immature (Courgettes/Zucchini) or mature (for jam, and for storage and cooking in winter).

Malabar or Fig-leaf Gourd *C. ficifolia* is a perennial species in frost free areas; it has some local importance in Mexico and central and southern America. The plants tolerate cool climates, and are cultivated at high altitudes in the tropics.

Cushaw ("Ayote" in German and Spanish) *C. argyrosperma* (syn *C. mixta*)—mature fruits with striped green and grey warty rinds, used for baking and for animal forage.

Some other pumpkin-like species within the *Cucurbitaceae* family include:

Buffalo Gourd *Cucurbita foetidissima*—described in **2G**, "Under exploited Crops".

Bottle Gourd *Lagenaria siceraria* (Syn. *L. vulgaris*)—the Calabash or White-flowered Gourd. Normally not eaten,but used for containers, musical instruments, fishing-net floats, rafts and spoons.

Loofah *Luffa* spp.—the Luffa, Vegetable Sponge, Sponge Gourd or Dishcloth Gourd; the immature fruits are sometimes eaten (*L.acutangula* 'Chinese Okra'), in India and the East, but Loofahs are mainly used as highly efficient cleaning pads for human skin, pots and pans, etc and as filters in industry *(L. aegyptiaca)*.

Bitter Gourd *Momordica charantia*—the Bitter Cucumber, Balsam Apple/Pear; Carilla Gourd etc is eaten as a vegetable and is also used in traditional medicine (p 294).

Chayote (Choyote) *Sechium edule*—the Christophine, Choko or Shu-shu; eaten by the Aztecs, and nowadays grown throughout the tropics and sub-tropics. A vigorous, perennial climbing plant with pear-shaped fruit 10–20 cm long, containing one large flat seed 3–5 cm long. The large, tuberous roots (20% carbohydrate) and the fruits are eaten, and sometimes also the young leaves & shoots.

Snake Gourd or **Serpent Gourd** *Trichosanthes cucumerina* (Syn. *T. anguina*). The immature fruit is eaten, which is more nutritious than most Cucurbits. The ripe fruit becomes red and is fibrous and bitter. Cucumber-like pods, up to 2 m long—a weight is often tied to the end to encourage them to grow straighter.

The following description applies to the four most commonly grown species of Cucurbitaceae, *Cucurbita pepo, C. maxima, C. mixta* and *C. moschata*, unless otherwise stated. These are widely found throughout Europe, the USA and the drier parts of the tropics, especially in Africa.

PLANTING
Propagation: normally by seed, though Cucurbits can be grown from cuttings taken from rooted nodes.

Soil: Cucurbits need strong rich soil with plenty of humus, manure being particularly beneficial. The plants do not tolerate waterlogging and prefer slightly alkaline conditions, so lime can usefully be applied when the soil pH is about 6 or lower. They are, however, sometimes classified as having "intermediate tolerance" to acidity.

Seed rate: 3.5–5 kg/ha. Several seeds are often placed per station, then thinned to the best plant.

Spacing: highly variable. Many types can be trained to grow upwards, on poles, fences, bushes, etc. This way the plants use up vertical space, leaving space underneath for other crops which benefit from the shade, and increase the amount of food produced per unit area. This climbing habit can be very useful in small gardens, window boxes, urban food growing allotments, etc. Climbing plants and their uses are discussed in **1Gd**.

GROWTH CONDITIONS
Day length: day-neutral (*C. ficifolia*, the Fig-leaved or Malabar Gourd, Chilicayotl is short-day).

Growth period: Cucurbits are mainly fast growing and the immature fruit is usually ready to harvest in 45–60 days; production often continues for a further two or three months. Pods mature in about 3 or 4 months.

Temperature: sub-tropical in origin, though many varieties have been selected for temperate regions. Sensitive to frost—*C. pepo* types are the hardiest.

Rainfall: although Cucurbits need regular and plentiful water during their short growing season they suffer in prolonged wet, humid conditions. Low or medium rainfall regions, no more than about 1000 mm per year, are therefore preferred.

Pests: Root-knot nematodes can be harmful but are controlled with crop rotations. Vine Borer hollows out marrow stems, causing the leaves to wilt and die. Cutworms often attack young seedlings; protect with collars of cardboard around each plant. The Cucurbit Fly is a major pest in Africa, and both the yellow/black adults and larvae of the Cucumber Beetle attack cucumbers.

Diseases: the most serious are Fusarium Wilt, Anthracnose, Downy Mildew and virus diseases.

YIELD
Immature fruits/pods of many Cucurbit crops yield about 7–12 MT/ha, mature fruits/gourds, etc. yield about 10–30 MT/ha.

UTILISATION
- The vast range of uses for the remarkable *Cucurbitaceae* family is also discussed in the introduction above. Human beings have grown and used this crop for thousands of years; the Bottle Gourd, for example, was one of the first crops to be cultivated, and can be a valuable crop to grow in certain circumstances such as in refugee camps where there is a shortage of containers, to hold both solids and liquids.

- Fruits of the Cucurbits can be used for food either fresh, cooked, pickled, candied, dried or in sauces and curries. Their tough skins can be used for containers, water and drinking vessels, musical instruments, penis sheaths, ornaments, floats for fishing and rafts, etc.
- Some species have seeds which are rich in oil and protein, which can be roasted and salted to make a nutritious snack, or fried to make *pepitos*. *Cucurbita pepo* seed kernels contain about 45% of an unsaturated fat, 25% protein and useful amounts of minerals and the Vitamin B complex.

LIMITATIONS
- Very few—there is usually one or more species of Cucurbit which is adapted and can grow in almost all of the extremes of climate, although they do need warm growing conditions and most species are killed by frost.
- A further reason that Cucurbits are not more widely grown is that it can be difficult to identify the most appropriate species or variety for a particular location, or, even when this information is known, it is not always possible to find a supply of enough good, reliable seed.

Okra
Abelmoschus esculentus (Syn. *Hibiscus esculentus*)

Bindi, Gombo, Gumbo, Lady's Finger, OkroKetmie {Comestible}, Gombo (French); Rosenpappel, Gombo (German); Quesillo, Quimbombó (Spanish); Quiabo{chifre-de-veado}, Caber (Portuguese); Bamies (Greece); Ochingombo (Angola); Bamia (Sudan); Bhindi (Hindi); Binday, Layloo (Pashtu); Bamia (Dari)

Okra originated either in tropical Africa (Ethiopia and West Africa) or in India and is now grown widely around the world in the hot lowland tropics and subtropics, mainly for local consumption. The FAO estimate of the total global production in 2004 was 5 million MT, grown mainly in India, Brazil, Thailand, Turkey and Spain.

It is closely related to the fibre crop Kenaf (*Hibiscus cannabinus*), to the Rose of Sharon (*H. syriacus* or *Althaea syriaca*) and to the Roselle or Jamaican Sorrel (*H. sabdariffa* var. *sabdariffa* and var. *altissima*). Together with Cotton, these are all members of the cotton or mallow family *Malvaceae*.

The plant is an erect annual herb 1–4 m tall, with large, single, bright yellow and red flowers. These develop into 10–30 cm long, slim, finger-like fruits, or "pods", which are eaten as a vegetable when they are dark green and immature. The fruits are usually five-sided in cross-section, normally green but sometimes red ("Burgundy" and "Red Okra") or white ("White Velvet") and have a sticky mucilaginous texture inside.

The nutritional value of okra is not sensational, about 86% water and only approximately 2-3% protein, 0.2% fat, 3–10% carbohydrate, 1% fibre and 0.8% ash. It contains reasonable amounts of carotene and about 21 mg/100 g Vitamin C. Nevertheless okra has magical culinary properties, giving body to sauces, such as enjoyed in the Creole dish called gumbo. The word "*gumbo*" is taken from the Bantu word for okra.

Plant breeding takes place mainly in India and in North America, where about thirty named, improved varieties are available. F1 hybrids such as Annie Oakley and Cajun Delight have been developed which normally outperform open pollinated varieties.

PLANTING
Propagation: by seed, which is round, about 5 mm in diameter, and dark green to dark brown in colour. The seed should be soaked in warm water for 24 hours before planting, sown directly into its growing place as it is easily damaged if seedlings are transplanted.
Germination: this can be very slow, especially in cool soil (minimum 16°C), so soak the seed - see "Propagation" above. Germination rate normally falls rapidly, to 0% in 1 or 2 years even when stored cool and dry. There are about 15–18,000 seeds per kg.
Soil: okra will grow in almost any soil, but prefers well-drained loams that are not too moist. If sweetcorn grows well, then okra will also normally grow well. The optimum pH is about 6–7 ("intermediate tolerance" to acidity). Okra likes plenty of potash so it can be a good crop to plant after a fire. Too much manure produces many leaves but little fruit.
Seed rate: normally 2 or 3 seeds are planted on ridges, sometimes in furrows or on the flat, and sometimes then thinned to the best plant. An average rate is about 4–8 kg/ha. Seed is often soaked in water for a few hours before planting. Soil temperature and moisture should be high.
Spacing: final spacing should be 60–90 cm between rows, 20–40 cm between plants.
Depth: 2–5 cm

GROWTH CONDITIONS
Growth period: 60–120 days to first harvest, which continues for another 2 or 3 months.
Temperature: many varieties of okra need four months of hot weather, and okra only grows slowly when the temperature falls much below about 25°C. The plant is killed by frost.
Rainfall: okra does not thrive with very heavy rainfall, and needs irrigation in arid regions.
Altitude: okra is normally associated with the lowland tropics and subtropics.
Pests: Root-knot nematodes can cause big yield loss but can be controlled with crop rotation, and some varieties have some resistance.
Diseases: Virus (Leaf Curl, Mosaic and Yellow Vein), Anthracnose, Powdery Mildew, Dry Rot, Stem Rot, Leaf Spot, Leaf Blight and Black Leaf Mould. Botrytis may arise in cold and damp conditions.

Intercropping: okra is normally planted together with other food crops such as cowpeas, maize, millet, sorghum, groundnuts and root crops.

YIELD
Okra pods are best eaten when young and fresh, when they are about 5–8 cm long for most varieties. They should be picked regularly, every other day ideally, to stimulate the plants to produce more.

Yields of fresh pods from intercropped okra are about 0.5–1 MT/ha, and from monocultures about 4–7 MT/ha. The FAO estimate of the average global yield in 2004 was 6.4 MT/ha; the highest national average recorded was Cyprus at 22 MT/ha and the lowest was Cameroon at 1.8 MT/ha.

The yield of seed grown in a monoculture is about 500 kg/ha.

UTILISATION
- Okra are mainly grown for the tender, young, **green fruit (pod).** Traditional varieties have pods 10–25 cm (4–10 in.) long, dwarf varieties are somewhat smaller. The pods can be either boiled, deep-frozen, canned, pickled or sliced and fried - eaten as a vegetable. Okra has the valuable attribute of strengthening stews and soups, which are thickened by the okra's mucilage, in dishes such as *Callaloo* in Trinidad and *Gumbo* in many African countries, southern USA and elsewhere. The pods can also be sliced and dried, while the young shoots and leaves can be eaten or used as a herb.
- The **ripe seed** contains about 20% edible oil; sometimes used as a coffee substitute.
- The **foliage** can be used as a low grade animal food.
- The **stem fibre** can be used locally but is of poor quality.
- The fruit (pod) is used to make a **plasma replacement** or **blood-volume expander**.
- The leaves and immature fruit are used in the East to make **poultices for pain relief**.
- The mucilage from the roots and stem is used for **clarifying sugarcane juice**, in **gur manufacture** in India, and for **sizing paper** in China.

LIMITATIONS
- Okra has low nutritive value.
- The crop needs a relatively long, hot growing season.
- Germination rate of okra seed falls rapidly (1-2 years).
- The pods have a high mucilage content and can be sticky and messy to harvest and cook.

Onion
Allium cepa

Oignon (French); Zwiebel (German); Cebolla (Spanish); Cebola (Portuguese); Osapola (Angola); Piyaz (Hindi, Pashtu and Dari); Shinkurt (Amharic), Shinkurtee (Oromifa)

Onions have been grown and enjoyed by humans since the very earliest days of food production and were a popular staple food in ancient Egypt—the people who built the pyramids were fuelled by the humble onion The ancients regarded its shape as being a symbol of the universe, and the name "onion" is thought to be derived from the Latin "*unus*", meaning "one".

Garlic and onions are surrounded with folklore regarding their magical properties; among other attributes, they are reputed to increase fertility in men and milk in nursing mothers, and to cure headaches when mixed with milk and vinegar.

Onions are also highly beneficial in nutritional terms, providing a rich source of Vitamins B_1, C and E and certain trace elements. Their carbohydrate content ranges from 5% to about 11%; 100 g of edible portion provides about 36 kcal energy value.

Plants of the *Alliaceae* (Allium) family contain alkaloids known as *Alliins*. The best source is garlic. Alliins are now known to have a positive effect on heart diseases and also to prevent growth of malignant (carcinogenic) cells, and it can be argued that the main role of the Onion family is anti-carcinogenic and cardioprotective rather than nutritional.

Shallots and onions are often regarded as being the same species, *Allium cepa*. Other authorities argue that the shallot is a separate species, *A. ascalonicum*. The two are both closely related to garlic (*A. sativum*), chives (*A. schoenoprasum*) and leeks (*A. porrum*). Most species in this family are biennial, but are mainly treated as if they were annuals. Other members of the Alliaceae family include ornamentals such as hyacinth, tulip and lily of the valley. Unfortunately for food producers, the *Allium* family sometimes behave like annuals, by setting seed—"bolting"—in their first year (**1Ea**). This tends to happen either if the variety is not adapted to the area or if the growing conditions change for the worse, eg drought. The seed heads are, on the other hand, highly decorative, and provide plentiful nectar for honey bees and other insects. Other onion types such as the Welsh and Tree Onions (see below) are perennial.

Unlike most of the other vegetables onions are monocotyledons; they grow from the base of their leaves, pushing their leaves up and out from this base. Dicotyledons, by contrast, grow outwards from the edges of the leaves.

Classification,or taxonomy, of the myriad onion varieties is not straightforward and botanists continue to dispute the issue. The genus *Allium* was for many years included in the *Liliaceae* family, though these days most botanists broadly agree that onions should be placed in the *Alliaceae* family, and that there are three groups, or botanical varieties, of *Allium cepa:*

– var. *cepa*—the **Common Onions**. Commercially the most important group of onions. This type of onion has been selected for thousands of years, so botanically they are now highly variable. Includes the **Spring** or **Bunching Onions**.

– var. *aggregatum*—three types, which form lateral bulbs or shoots for vegetative reproduction:

Potato (Multiplier) Onion—only flowers rarely, but has lateral bulbs which produce separate bulbs and tops in the second year;

Ever-ready Onion, Welsh Onion (The name comes from the German *welsche* "foreign"),

Japanese Bunching Onion, Scallions (a name also used for Shallots) or **Ciboule**.

– var. *proliferum*—the **Tree Onion** or **Walking Onion** which have only small and undeveloped ground bulbs, but instead produce "bulbils" on the flower heads. These bend down to the ground, take root and so propagate in this way in a circular form around the parent plant. No seed is normally produced. Tree onions are not important commercially.

The description below refers to the Common Onion *Allium cepa* var. *cepa*. Hybrid seed of this species is widely used, while other Allium species mainly rely on non-hybrid seed.

PLANTING

Propagation: either by seed or by dry bulbs, or "sets". When by seed, seedlings are often transplanted at about 10–12 weeks (sometimes pruned, tops and roots, to 15–25 cm long). Small sets have a tendency to bolt less readily than larger sets.

Germination: viability falls rapidly in hot, humid conditions—may be 0% in less than a year. F1 hybrid seed adapted to a wide range of conditions is readily available.

Soil: soil should have plenty of nutrients in the upper 10 cm or so profile and should retain moisture well. Onions adapt to most soils provided moisture is not limiting. "Intermediate tolerance" to soil acidity. The soil should ideally be well manured or composted some months before the seed or sets are planted. In heavier soils, coarse sand can be placed under each set.

Seed rate: 3.5–4.5 kg/ha when transplanted, 4–6 kg/ha when direct seeded. Approximately 250,000 seeds/kg.

Spacing: 25–40 cm between rows for bulbs, 10 cm for salad onions; 7.5–10 cm between plants.

Depth: 1–2 cm.

Other crops: they make good companion crops to other vegetables such as carrots, as their smell can keep away harmful insects such as the Carrot Root Fly.

GROWTH CONDITIONS

Day length: sensitive to photoperiod. Most varieties are long-day, some are short-day and some day-neutral. The long-day varieties that are adapted to temperate regions will not form bulbs in the tropics, where short-day varieties are needed. The photoperiod does not affect bolting, which occurs as a function of the onion variety and the temperature. If "new" (previously untested) varieties of onions are introduced into an area it is wise if there is time to try them out for a year or so in a small way, to compare them with the local types.

Growth period: 9–150 days depending on whether seed or sets are planted, the season/temperature when planted, rainfall, day length and the variety chosen. Plants tend to mature faster when they are stressed, such as when they are planted too close together.

Temperature: varieties vary greatly in their temperature requirement, but low temperatures encourage bolting, if the bulbs are large enough. Cool, wet weather is best for early growth, with warm, dry weather for harvest.

Rainfall: requirement depends on variety. Often irrigated. If the soil is liable to become waterlogged, onions should be planted on raised beds.

Pests: maggots of the **Onion fly** can stop the plants developing altogether. Seed-sown crops are worst affected, setts rarely suffer badly. Eggs are laid in the bulbs of young onions, the yellow maggots eat the bulb and the leaves go yellow and die. Hot weather and dry soil favours spread of this pest. To prevent the flies from laying eggs, either sprinkle soot or wood ash along the row, or spray water mixed with paraffin (1 fluid ounce/1 gallon water). As soon as this pest is identified, infected plants should be uprooted and burned, to avoid leaving maggots in the soil.

– **Thrips**—the leaves develop silvery blotches.

– **Eelworms**—very tiny worms, they make the tops of the plants wilt. Control by burning/destroying all infested plants and strict rotation of 6 years without allowing either other Allium species or Chickweed, an alternative host, to grow in or near the infected area.

Diseases: Downy Mildew causes leaves to develop a greyish or violet coloured streaks on the leaves, especially in humid conditions. Can be controlled with Bordeaux powder.

– **Purple Blotch**—leaves develop white sunken spots, which grow larger and become purple with a yellow halo; **Pink Rot & White Rot**—roots are attacked, then the leaves go yellow and die. Soil-borne; **Yellow Dwarf Virus**—worldwide. Yellow streaks on leaves. Spread by contact, or by sets or aphids, but not by seed.

– **Storage Diseases:**

Black Mould (*Aspergillus*) and **Neck Rot** (*Botrytis*) are both common. Neck Rot attacks onions in store. A grey mould forms on the skins, and later on the centres turn brown. Prevent by thoroughly drying the onions before storing and store in a cool airy place.

YIELD

In the tropics 7–10 MT/ha of freshly harvested onions is about the average, while sophisticated farming systems in ideal growing conditions can produce more than 40 MT/ha.

At the time of going to press, the world's largest onion (7.03 kg, 15 lb 5 oz and 33 inches in diameter) was grown in Fife, UK in 1997. The variety name is Kelsae Sweet Giant.

UTILISATION

- Onions are eaten all over the world, both cooked and raw. The dried, or "cured" bulbs can be stored for several months. Alkaloids known as *alliins* are said to have healing properties - see Introduction, above.
- Used in traditional medicine in many countries; to treat intestinal worms, high blood pressure, gastroenteritis, and as antibacterial, antifungal and antitumour agents.
- A home-made insecticide can be made by soaking the unused leaves and parts of Allium species in water for a few days.
- Allium species planted closely to other more vulnerable crops such as carrots discourages insect attack on those crops.

LIMITATIONS

- Onion plants are very sensitive to photoperiod and temperature, with the result that many varieties are only adapted to small, specific regions. This is important to know when selecting varieties for introduction to a different country or region, especially in the tropics. As a result of these specific requirements, onions are liable either to bolt or to not form proper bulbs if they are grown in unsuitable conditions.
- Onions cross-pollinate very readily, so if possible food producers should obtain seed from a reliable supplier and not grow their own seed, which may produce poor and variable results.
- The growing crop must be kept free of weeds, particularly in the early stages.

Peppers
Capsicum spp.

Sweet Pepper, Green Pepper, Garden Pepper, Wrinkled Pepper, Capsicum, Cayenne, Aji, Paprika; Poivron (French); Pepperoni, Paprikasdroten (German); Pimienta (Spanish); Pimenta do Reino—for hot peppers, Pimentão—for sweet peppers (Portuguese); Tchindungu {fruit}, Tchindungueiro {plant} (Angola); Lal Mirch (Hindi); Shatta (Arabic); Tarsa Murch—for hot peppers, Shireen Murch—for sweet peppers (Pashtu and Dari/Persian); Berberreh (Ethiopia)

There are so many species and sub-species of peppers that they have been classified in several ways by different botanists at different times. These days two main types or species are generally recognised, though the classification of the genus is very confused:

Capsicum annuum—the Sweet Peppers, Bell Peppers, Paprika, Pimento, Jalapenos, Chiltepin and Chillies. Mainly annuals, 30–150 cm tall. Fruits borne singly, 1–30 cm long, green or purple and ripening to yellow, orange, red, brown or purple. There are many botanical varieties, such as Cherry Pepper or Capsicum, Cone P., Cluster P., Guajillo P., Long P., Mulato P., Paprika, Papri Mild P., Sweet Pepper, Tree Chilli (P. de Arbol), etc, which can intercross to produce intermediate types.
Capsicum frutescens—the Chillies or Bird Chillies, Hot and Cayenne Peppers, Tabasco Peppers. Mainly shrubby, short-lived perennials (2–3 years). Fruits are small, 2–3 cm, usually red and borne in groups. These are mainly the "hot" or

"pungent" peppers. The "hotness" depends on the presence of *capsaicin*, which varies with both the variety and the temperature, the hotter the weather the hotter the pepper. Tabasco sauce and Cayenne pepper are made from the fruits of *Capsicum frutescens.*

Other species include the **Habanero ("Jamaican", "Scotch Bonnet") Pepper** *C. chininse* with square/heart shaped fruit 5 cm long, some very hot, *Capsicum pubescens*, which includes the South American rocoto peppers and *Capsicum baccatum*, which includes the South American aji peppers.

Capsicum peppers are members of the nightshade family *Solanaceae* and should not be confused with their unrelated namesake the White or Black Pepper, the product of *Piper nigrum*. This type of pepper was grown in the West Indies and sold at inflated prices in Europe, principally as the hot ingredient of curry powder. Following the introduction of Capsicum peppers, their dried and ground up leaves were used as a less expensive alternative.

☞Both types of Capsicum Peppers are good sources of vitamins, including B1 and B2, as shown below:

Species / type	Vitamin C / Ascorbic acid (mg per 100 g)	Vitamin A - Carotene (IU per 100 g)
• *C. annuum* (83% moisture)	50–280	100–1200
• *C. frutescens* (10% moisture)	2–50	200–20,000

F1 hybrid seed of Sweet Peppers is readily available, and can be produced without great cost. Capsicums are mainly self-pollinated, though about 16% cross-pollination also occurs.

PLANTING
Propagation: by seed, normally sown in nurseries, under glass unless the weather is very warm, and transplanted at 4–6 weeks when 8–15 cm tall. Farmers can easily produce their own seed, though there is often some cross-pollination. The seed can also be taken out from any capsicum fruit you like the look of which you buy in a shop, preferably from locally grown plants.
Germination: seed can remain viable for 2–3 years. 6–10 day germination time.
Soil: peppers adapt to a wide range, although ideally the soil should be well drained, light, loamy and rich in lime. Moderately susceptible to salinity. Good response to organic manures.
Seed rate: 140–170 seeds per gram. About 0.5–1.5 kg of seed produces enough plants for 1 ha.
Spacing: 60–90 cm square grid, either on the flat, or on ridges if water-logging is likely.
Depth: 0.5–1 cm.

GROWTH CONDITIONS

Growth period: 60–90 days to the first harvest, then harvested at 1–2 week intervals for 3–5 months. In the tropics and subtropics Capsicum peppers can survive as perennials for a few years. The fruit should be cut off, not broken, leaving about 2.5 cm of stem.

Temperature: slightly hardier than tomatoes, but still killed by frost. 19°C minimum at flowering.

Rainfall: 600–1000 mm/a. Heavy rainfall or water-logging causes poor fruit-set, and rotten fruit. Keep the fruits dry if possible, and any irrigation system should only apply water at the soil level, on to the roots and not on to the fruit or plant.

Altitude: from sea-level to about 1800 m.

Pests: Thrips transmit virus diseases, and Cutworms attack seedlings.

Diseases: the two virus diseases, Mosaic and Leaf Curl, are the most serious. Loss of yield can also be caused by Fruit Rot, Powdery Mildew, Bacterial Wilt and Anthracnose. To avoid Anthracnose, peppers should be grown far away from beans; if peppers do become diseased they become rotten and should be destroyed.

YIELD

The fruits of pepper can be either eaten fresh, or sun-dried for 3–15 days.

Typical yields of dried fruit, from plants grown in a pure stand, are:

275–825 kg/ha rainfed, 1680–2750 kg/ha irrigated. {45 kg of fresh fruit = approx. 10–12 kg of dried fruit.}

UTILISATION

- Peppers are normally used as some kind of spice, but in different ways depending on how hot, or "pungent", they are. Although they have low nutritional value, apart from Vitamins A and C, they are valuable in making staple foods such as sorghum more palatable to eat. Tepin Pepper *C.annuum v. aviculare*—tiny round fruit, is the hot one! (though some say the Red and Orange Habanero Peppers are hotter…).
- *Capsaicin* improves digestion, helps heal stomach ulcers and is an antiseptic, used on some sticking plasters.
- Extracts from certain Capsicums known as oleoresins are used to flavour ginger-beer and other drinks. These are also used in some aerosols to deter muggers.
- Extracts from *Capsicum frutescens* (the Chillies or Bird Chillies) are used in medicines, both internal and external.
- Cayenne pepper is used in some poultry food.

LIMITATIONS

- Peppers are killed by frost and are susceptible to Botrytis and other moulds.
- Too much rain causes poor fruit set and rotten fruit.
- The tap root is easily broken or damaged during transplanting.
- Water-logged soil, even briefly, causes leaves to fall, low fruit set and the fruits to rot.

Tomato

Lycopersicon esculentum (Syn. *Solanum lycopersicum*)

Golden Apple, Love Apple, Peruvian Apple, Tomatl;
Pomme d'amour{*Pomme des Mours* = eggplant}, Tomate (French); Tomate
(German and Spanish); Pomodoro {"Golden Apple"}(Italy); Tomate {Grande}
(Portugal/Brazil); Náhuatl (Mexico, from Aztec *tomatl*); Tammatter (Hindi);
Banjan-e Rumi (Pashtu and Dari); Ematya (Angola); Teemateem (Ethiopia)

Tomatoes were already widely cultivated in what is now Peru, Ecuador and Mexico
when the Spanish arrived there. When the Spanish introduced these new fruits to
Europe they were greeted with suspicion because they are members of the
Solanaceae (nightshade) family, which includes Deadly Nightshade, Tobacco,
Henbane, Thorn-apple and Belladonna, as well as potatoes, egg plants, peppers,
tamarillos {tree tomatoes}(*Cyphomandra betacea*) and tomatillos (*Physalis
philadelphica* and *P. ixocarpa*).

Since then tomatoes have become one of the main vegetable crops in many parts
of the world, and global production is estimated at around 60 million MT per year.

A large number of varieties are available, adapted to grow in a wide range of
climates and conditions, though they are not well suited to the wet tropics. Many of
the older varieties have an indeterminate growth habit (the *cordon* types eg
"Gardeners Delight" and "Shirley"), while plant breeders have been mainly
producing determinate varieties (the *bush* types eg "Roma", "Marmande" and
semibush types eg "Glacier"), particularly useful for mechanised harvesting.
Tomatoes are almost always self-pollinated; in tropical regions, and sometimes
elsewhere, the stigma protrudes beyond the anthers, allowing some cross-pollination
to take place. Hybrid seed is commonly used.

The plants are either half-hardy annuals or short-lived perennials in warm
climates. In temperate and other regions they are grown as annuals. They have stems
70–200 cm long, or even longer with some trailing (indeterminate) varieties, which
normally benefit from support, growing up sticks or string, against a wall, etc.

The roots and leaves are poisonous, containing the neurotoxin *solanine* which
can cause vomiting and nausea if eaten to excess. Solanine is also found in potatoes,
and deadly nightshade; it has both insecticidal and fungicidal properties and is one
of the plant's natural defense mechanisms.

Stems can readily establish adventitious roots when they are in contact with
moist earth, which can be done deliberately by the farmer/gardener if there is a large
gap to be filled amongst the tomato crop, a procedure often known as *layering*.

PLANTING

Propagation: by seed, though cuttings and graftings also work well. Seed is normally
planted in nurseries, indoors in temperate climates, and transplanted some 4–6 weeks
later when the plants are 15–20 cm tall. Seed can easily be obtained from home grown,
ripe tomatoes.

Germination: tomato seed can remain viable for 3–4 years if it is stored cool and in airtight containers. In real life these conditions are rarely found and seed viability can fall dramatically, especially in warm and humid conditions. Germination normally takes about 7–10 days. The effect of temperature on germination of tomato seed is discussed in Section **1Fa**, "Germination", page 46.

Soil: should be fertile, and both compost and manure are highly beneficial. Lighter soils usually signify shorter growing and fruiting periods than heavier, more retentive soils. Optimum pH is about 6–7 (classified as having "intermediate tolerance" to acidity), and tomatoes need potash as young plants and a balanced fertiliser with more Nitrogen as they grow.

Seed rate: About 250–500 g/ha., 250–350 seeds per gram. Rarely planted in large areas by subsistence farmers. Often two or three seeds are planted together, then thinned to the best plant.

Plant spacing: very variable, but average spacings are 90 cm between rows and 30 cm between plants. The plants can grow very large if they have adequate space, nutrients, light, etc.

Depth: 0.5–2 cm. The transplants should be buried well down deep to encourage development of adventitious roots from the hypocotyl area.

GROWTH CONDITIONS

The taller, indeterminate types need support, and should be firmly supported on poles, string, fences, etc. The side-shoots are normally pinched out, to encourage the plant to set fruit and to make a tidy plant. The tip is also often pinched out after the plants have developed a certain number of trusses, 4–10 or more depending on the quality of the growing conditions.

When the lower truss begins to ripen the lower leaves can be removed, especially if they are showing signs of fungal disease, using a sharp knife or secateurs to minimise wounding. This simple act not only increases the air-flow and light supply to the lower canopy, but also makes the ripening tomatoes more easily visible to growers when they check for weeds, pests, diseases, ripeness and readiness to harvest.

Day length: tomatoes are day-neutral and in many parts of the world they are cultivated 12 months of the year.

Growth period: three crops a year can be grown, in the tropics. In temperate climates about 130 days are normally taken to produce the first ripe fruits. Indeterminate types can then be continuously harvested over a very long period. When winter approaches, the green fruits can be taken indoors and ripened in a warm place.

Temperature: most varieties need 3 or 4 months of warm, frost-free weather. Some new varieties can set fruit at low temperatures (down to about 4°C), others at high temperatures (up to about 32°C). Frost sensitive.

Rainfall: badly affected by drought periods. Best in medium or low humidity.

Pests: Cutworms can bite off the young plants at ground level. Protect them with collars of material (cardboard, stiff paper, etc) or wood ash, as with cabbages. Root-knot nematodes are also common, especially in nurseries. Greenflies, Whiteflies and Red Spiders can be damaging, especially in greenhouses. Regular sprayings with a soap solution will keep their numbers down, and both Whitefly and Red Spiders can be controlled by introducing their natural predators. Tomatoes are often protected from pests by interplanting them in various configurations with protective ("companion") plants such as onions, garlic and marigolds.

Diseases: Blight can be just as big a problem with tomatoes as with potatoes. Controlled with two-weekly sprays of Bordeaux mixture (copper sulphate).

Bacterial Wilt makes plants wilt very rapidly, without yellowing, and die; worse in the wet tropics and in waterlogged soil.

Fusarium Wilt. The plants wilt, the leaves become yellow. The symptoms of **Sclerotium Wilt** are that plants wilt rapidly, and feathery white growths appear on the stem, mainly at ground level.

Verticilium Wilt ("sleepy disease") is caused by a fungus which attacks the roots and base of the stem, especially in cold wet soil. Earthing up larger plants with good soil may save them, as new roots may be formed above the infection; otherwise the plants droop and die.

Septoria Leaf Spot causes brown water-soaked spots to appear on the leaves, which may then fall off.

Grey Leaf Spot may cause complete defoliation.

Virus Diseases include Tobacco Mosaic, Spotted Wilt, Cucumber Mosaic, Curly Top and Tomato Yellow Top—all spread by aphids and handling, and may originate in nearby plants of other Solanaceae plants, or cucumbers. Some virus disease may even originate in tobacco, so cigarette smokers may transmit this disease to the tomato plants while working with them.

Blossom End Rot occurs on the young fruit when vigorously growing plants are stressed, by either too much or too little water, or when the roots are damaged. The tip of the fruit dies and turns dark brown; the whole fruit may become infected.

Greenback causes fruit to remain green and unripe on the base around the stalk; unripe area appears hard. This disease develops most rapidly in strong sunlight, and is associated with potash deficiency. Susceptibility to Greenback varies according to the variety.

YIELD
The global average yield of tomatoes grown on a field scale is about 10 MT/ha. More than 25 MT/ha is produced in more sophisticated systems.

UTILISATION
- The **ripe fruit** of tomatoes is eaten raw or cooked. It is about 90% water, but is rich in Vitamins A, C and others as well as certain minerals, particularly potassium (250 mg/100 g). Carbohydrate content (fructose and glucose) is about 3% and there is very little fat or protein.
- The **green fruit** is used in pickles and preserves.
- Tomatoes are **processed** into soup, juice, sauce, ketchup, purée, paste and powder. There is also a very large tomato canning industry.

- The **seed** contains 24% of a semi-drying oil—discussed in **2C** "Oilseeds" page 189—which is extracted industrially from the pulp and residues of the canning industry. The oil is used as a salad oil, and in soap and margarine; the **presscake** is used for animal food and fertiliser.

LIMITATIONS
- Tomatoes are **low in protein and calories**.
- The plants are **susceptible to diseases**, especially in warm, humid conditions.
- The fruit is **bulky and squashy**, in other words it is not easy to transport.
- The **storage time is short**, unless the fruit is preserved. The "Flavr Savr" type was genetically engineered in the USA to delay the ripening process, so they can be picked closer to full ripeness than most other tomatoes. Genetic engineering (GM) has also lead to varieties with reduced amounts of the cell wall softening enzyme, resulting in a longer shelf life.

2F. FRUITS

Botanically speaking, a fruit is defined as a mature, fertilised ovary (and its associated structures) so that, for example, tomatoes, cucumbers and peppers are, from a botanist's point of view, "fruits" and not "vegetables". Other examples of "fruits" include **drupes** (stony seeded fruits) such as coconuts, cherries, peaches and mangoes, **legume fruits** such as peanuts and beans, and **multiple fruits** such as raspberries and mulberries.

Most fruit producing plants and trees are insect-pollinated, which is the reason they have such attractive flowers. The seed of fruits is normally dispersed by birds and animals, which is the reason that they have attractive, edible fruits.

Seedless Fruit
Some varieties of some fruits naturally produce fruit without their flowers being pollinated and fertilised. This produces seedless fruit and is known as *parthenocarpy*; it occurs with, for example, banana, cucumber, grape, grapefruit, orange, persimmon and pineapple. This natural phenomenon has been mimicked by scientists in a process called *induced parthenocarpy* which manipulates various growth hormones, applied either in paste form, or spray or by injection to produce seedless fruit.

Seedless fruit from non-parthenocarpic varieties and species, and also out of season fruit production, is also induced artificially using pollen which is either dead or has been altered, or is from another species.

Genetic engineering is also used these days, with watermelon for example. The seed for seedless watermelon is produced by crossing a normal diploid watermelon with one that has been genetically modified into a tetraploid state. The triploid plants that result from this produce seedless fruit when pollinated by normal watermelon pollen.

In common with vegetables, most fruits have only limited food value, though they do normally contain useful amounts of vitamins and minerals. For many subsistence farmers (and other food producers...) fruit can also serve as a useful cash crop.

Most of the edible fruits grown in **temperate climates** are produced by seven plant families:

- *Rosaceae*—apples, pears, quinces, cherries, peaches, nectarines, apricots, plums, damsons, raspberries, blackberries and strawberries.
- *Rutaceae*—oranges, tangerines, kumquats, lemons, limes and grapefruit.
- *Grossulariaceae*—blackcurrants, red currants, white currants and gooseberries.
- *Moraceae*—figs, white and black mulberries, hops and breadfruit.
- *Ericaceae*—bilberries, blueberries, cranberries and huckleberries.
- *Oleaceae*—olives.
- *Vitaceae*—grapes.

This section on Fruit describes seven of the **tropical and sub-tropical fruits** which are the most likely to be useful for subsistence or small scale food producers in warmer climates:

avocado *(Lauraceae)*, **banana** *(Musaceae)*, **citrus** *(Rutaceae)*, **guava** *(Myrtaceae)*, **mango** *(Anacardiaceae)*, **papaya** *(Caricaceae)* and **watermelon** *(Cucurbitaceae)*.

Other subtropical fruits include: coconuts *(Palmae)*, dates *(Palmae* or *Arecaceae)*, figs *(Moraceae)*, olives *(Oleaceae)*, melons *(Cucurbitaceae)*, langsat *(Meliaceae)* and pineapples *(Bromeliaceae)*.

Avocado
Persea americana (Syn. *P. gratissima*)

Alligator Pear, Avocado Pear, Butter Pear, Midshipman's Butter, Summer Pear, Vegetable Butter Avocat, Avocatier, Poire d'avocat (French); Ahuacatl, Aguacate, Cura, Cupandra, Palta—Mexico (Spanish); Abacate—fruit, Abacateiro—tree (Portuguese); Abocado (Ethiopia)

Avocados are the most nutritious of all the fruits, and have been called "the butter of the poor". They are rich in Vitamins A, B complex, C, D and E, and some varieties contain more than 3000 calories per kilogram. The mono-unsaturated oil (3–30%) is very digestible.

They originated in Central and South America, where they are still an important food crop, and these days are also grown in most tropical and subtropical countries.

It is said that the name "avocado" comes from the Spanish *aguacate*, from the Náhuatl/Aztec word for testicle, *ahuacatl*.

Avocados have recently become an important food crop. Global production in 2004 was estimated by FAO at 3 million MT, grown on a large scale in USA, South Africa, Israel, Australia and Southeast Asia.

There are three ecological races, or types, of avocado:

Mexican—the leaves smell of anise. Small, slender, pear-shaped fruits (90–240 g). The skin is light green and almost smooth. The most hardy of the three races, it will survive down to minus 6°C. Not important commercially except when it is hybridised with other types. Some botanists consider this type to be a separate species, *Persea drymifolia*.

Guatemalan—medium-large (240–2200 g), compact, spherical fruits, on long long stalks, with medium oil content of about 8–15%. Slightly less cold tolerant then Mexican types, down to about minus 4.5°C. These are the so-called alligator types. Sometimes classified as *P.nubigena* var. *guatemalensis*.

West Indian—large fruits (1–2 kg), on short stalks, with light coloured leaves. The least hardy type, killed below about minus 2°C, it is adapted mainly to the hot, low tropics. Sometimes classified as *P.americana* var. *drymifolia*.

Hybrids between these three types are widely grown—for example, the well-known varieties *Sharwil* and *Fuerte* are **Mexican X Guatemalan hybrids**. There are hundreds of varieties available, although many of them are only adapted to small, specific growing conditions. Some varieties, such as *Collinson*, depend on other varieties for pollination; other varieties are not actually dependent on others but they do benefit from the additional pollen provided by another variety.

The plant is an evergreen tree, up to 20 m tall, with a shallow root system. The growth of the tree is not regular, but occurs in "flushes" alternating with periods of very slow or zero growth.

PLANTING

Propagation: although avocados do grow very easily from seed, commercial growers use vegetative propagation in order to produce more uniform, predictable & earlier maturing plants. Guatemalan types and their hybrids can be successfully grafted onto rootstocks of all types, but Mexican type rootstocks are incompatible with West Indian type grafting material (scion).

Germination: before planting seeds the skin should be removed, in order to produce better and faster results. Seed normally remains viable for 2–3 weeks after removal from the fruit, or longer still if it is stored in a cool place, in dry peat.

Soil: avocados will grow in a wide range of soils but they are extremely sensitive to waterlogging, and for this reason are often planted on mounds or other high spots. They are classified as having only "intermediate tolerance" to salinity; optimum pH is 6–7.

Fertiliser needs are similar to those for citrus, giving a good response to manure and Nitrogen; a typical application would be 2 kg/tree/annum, at 3-month intervals, of 12:6:10.

Spacing: 6–12 metres, depending on the variety. Some avocados are upright in their growth habit, others have a more spreading type of growth.

Intercropping: sometimes with pigeon peas or other annuals, grown under younger avocado trees. Cover crops or grass are often grown under the trees, to protect the soil and also the trees' shallow root systems.

GROWTH CONDITIONS

Growth period: with seedling trees, the first fruit can appear in 5–6 years. Vegetatively propagated trees produce fruit earlier. They tend to be *biennial bearing* ie they normally only produce a good crop of fruit every other year.

Temperature: requirements vary according to the type/race — see introductory paragraph, above.

Rainfall: minimum is about 750 mm per year, ideally not evenly distributed but with alternate wet and dry seasons.

Pruning: little is necessary, other than to remove suckers from the rootstock and any dead wood. Upright varieties often have the central shoot pruned so as to produce a more spreading growth habit. The branches of spreading varieties are sometimes thinned or shortened. The tops of tall trees are sometimes removed to make harvesting easier.

Wind: windbreaks are often necessary, as the wood is soft and breaks easily.

Pollination: most varieties are cross-pollinated, mainly by bees. Some varieties either require or benefit from another variety growing nearby to act as a pollinator, especially in hot regions. Other varieties are self-fertile.

Pests: some damage can be caused by scale insects, mites and mealy bugs. The sugar-cane root weevil can also give problems, in Puerto Rico for example.

Diseases: Root Rot (*Phytophthora cinnamomi*) is the most serious, sometimes killing the tree; often associated with wet soil. Other diseases damage the fruit and leaves, such as Cercospora Spot, Anthracnose (Black Spot) and Scab.

Trace elements: deficiency of Zinc and Copper can cause symptoms (**1Cd**, page 23).

YIELD

Typical weights of fruit:
>   Mexican types: 90 g–240 g;
>   Guatemalan types: 240 g–2.2 kg;
>   West Indian types: 1–2 kg.

The average yield of avocado is 100-500 fruits per tree per annum. In North America and elsewhere, good orchards can produce more than 13 MT/ha per annum.

UTILISATION

- Avocados are normally eaten as a salad fruit; they are most nutritious when eaten raw, such as in guacamole, ice cream and milk shakes.
- The pulp can be preserved by freezing.
- Avocado oil is used in cosmetics and toiletries.
- The cocktail avocado is grown in Chile, South Africa and Israel. The bullet shaped fruit is 5-6 cm long and contains no seed.

LIMITATIONS

- Avocado trees have a tendency to biennial bearing habit.
- With seedling trees, 5 or 6 years may elapse before the first fruit is produced.
- Seedling trees are not always satisfactory, while vegetative reproduction techniques require specialist knowledge.
- Some varieties need other varieties as pollinators.

- The trees do not tolerate waterlogged soil or excessive wind.
- The root system is shallow and so is easily damaged, when intercropped for example, and the trees need regular rainfall or irrigation.
- The fruits must be already almost mature before they are picked; they then bruise and rot easily in transport, and have a short storage life.

Banana
Musa spp.

Platano (Spanish); Fig (West Indies); Kela (Pashtu & Dari/Persian); Muz (Ethiopia)

More bananas are grown every year than all other fruits except grapes. In 2004, FAO estimated the global world production to be 71 million metric tons (MT), of which only about 12 million MT reach international trade. Nevertheless, bananas are big business; in terms of the world trade in agricultural produce they are fifth, after cereals, sugar, coffee and cocoa. About two-thirds of world exports are controlled by three corporations, Chiquita, Dole and Del Monte which in effect between them control the supply and pricing of that sector of exported bananas.

About half of the bananas grown are eaten as a raw fruit, and the other half, the plantains, are eaten as a cooked vegetable. Plantains are a truly multi-purpose crop, as described in the penultimate paragraph of this piece, "Utilisation".

When grown in large plantations, insect and disease damage is controlled by applying a cocktail of insecticides, fungicides, herbicides and disinfectants. In poorer countries, in central America for example, excessive amounts are often applied which results in serious health problems for many of the workers in the plantations, as well as accumulation in the soil of contaminants left in the plant residues, reduction of biodiversity and even soil sterility. Organic methods of production will inevitably increase dramatically as consumer awareness also increases.

Outside of the tropics two varieties only predominate: **Gros Michel**, which is susceptible to Panama disease (wilt) and leaf spot (Sigatoga) but tolerant of nematode worm; and **Cavendish** which is resistant to wilt but susceptible to leaf spot and worms. Varieties with good resistance to all three have not yet been developed.

Nutritionally, one banana provides more than an adult's daily requirements of potassium (about 380 mg) and also a large amount of energy (90 calories per 100 g). Ripe fruit (75% water) contains up to 22% carbohydrate, mainly as sugar, and is a good source of Vitamins A and C.

Botanical Classification, or taxonomy, of bananas is complicated, and various versions exist. The following interpretation is one of the most widely accepted of this complex plant family:

Bananas are members of the *Musaceae* plant family, which has two Genera, *Ensete* and *Musa*:

Ensete—there are six or seven different species, grown in tropical southeast Asia and Africa. In south Ethiopia *Ensete ventricosa* is a staple food crop, grown mainly at 1500–3000 m above sea level. Known as Ostafrik Anische or Wildbanane in German. The pseudostems and corms are cut up, and the pulp either eaten fresh or fermented in silos and then usually made into bread. Sacks and ropes are also made from the fibre extracted from the pseudostems. The plant is harvested about 4–5 years after propagation from buds, before the inflorescence (flower head) is formed, which would use up the starch.

Musa—about 40 species, divided into five groups:

– *Eumusa*—the largest group, with 13–15 species, including the Common Banana *Musa sapientum*; also the most important and widely distributed group, including the sub-group the Plantains, described below.

– *Callimusa*—small plants, grown in Indochina, Malaya and Borneo.

– *Rhodochlamys*—grown from India to Indonesia, sometimes as ornamentals.

– *Australimusa*—grown from Queensland to the Philippines. Includes the species *Musa textiles*, Manila hemp, or abaca, used for making marine rope, fishing gear, in the mining industry, in making strong sacks and paper (used for example in movable paper walls in Japanese houses), etc. Also the Feci banana, eaten in Polynesia.

– *Incertae sedis*—'of uncertain position/seat'—a taxonomic group where its broader relationships are unknown or undefined. Includes the wild indigenous species *Musa ingens* ('the world's tallest herb').

The Eumusa group provide virtually all of the edible bananas that are cultivated, and is the group which is described below.

Plantains

An important sub-group of the Eumusas (sometimes classified as a separate species, *Musa paradisiaca*, as opposed to the "Common" banana, *Musa sapientum*), plantains provide an important food source in south India, Africa and tropical America. They are not important anywhere East of India except for some "Horn Plantain" types grown eastwards to the Phillippines and the Pacific.

Their fruits are acidic, starchy and unpalatable until cooked. Nutritionally they are similar to the sweet potato.

The plants are resistant to the notorious Panama Disease, and also Leaf Spot, but are very susceptible to attack by the Banana Weevil (or Stem Borer). For details of the utilisation of plantains, see the penultimate paragraph of this sub-section on bananas, "Utilisation".

Botany of Bananas

Plants of the Musa species are large, perennial, tree-like herbs 2–9 m tall, with tightly rolled overlapping leaf sheaths forming a pseudostem of about 30 cm diameter, above the basal corm. The root system is very shallow, with most of the adventitious roots formed in the top 15 cm of the soil profile, and spreading laterally for 4–5 m.

About 2 months after planting a sucker, which was itself about 7 months old at planting, it develops a growing point at the base in the heart of the pseudostem which is transformed into an inflorescence. This is carried up from close to ground level by the true stem. In this flower head (inflorescence) the first rows of flowers produced are the larger female ones and the later flowers produced are the smaller male ones (there may be intermediate types in between, either hermaphrodite or neuter). The male flowers of most commonly grown varieties serve no useful function, as pollination rarely occurs. In wild species pollination is done mainly by bats, and also by birds and large insects.

About 2–4 months after the "shooting" of the bunch it is ready for harvest. The fruit bunch is sometimes called a "stem"; each cluster of fruits at a node is a "bunch", and the individual fruits are the "fingers".

PLANTING
Propagation: vegetative, except in breeding work. Five types of planting material are used in different parts of the world:
– *Peepers*—very young suckers, with scale leaves only.
– *Sword suckers*—small, pale, pointed leaves, planted out when about 75 cm tall.
– *Maiden suckers*—taller, with broad leaves. Lifted from the parent plant at 5–8 months old, the central meristem is destroyed, the roots are trimmed, and the sucker cut back to 10–15 cm above the corm.
– *Corms*—these are dug up, the aerial parts removed, and pieces of 2 kg or bigger are planted 25–30 cm deep.
– *Water suckers*—originating near the soil surface, with large, wide, green leaves.
The first crop after planting is known as the "plant crop"; subsequent crops are known as "ratoons".
Planting hole: should be at least 60 cm deep and wide, and filled with topsoil mixed with manure or compost and about 250 g of single superphosphate or equivalent.
Spacing: in pure stands:
– Fertile soils—3.6 × 4.5 m between plants (480–750 plants per hectare);
– Less fertile soils and in dry areas—3.6 × 2.7 m between plants (890 plants per hectare);
– Dwarf varieties—2.4 × 2.4 m between plants (1680 plants per hectare).
Soil: a wide range are suitable, but should be pH 5–8 (pH 5.5–6.5 is optimum, bananas are classified as "tolerant" to acidity), free draining but still retentive, and fertile. Fertiliser should be applied even in the most fertile soils. As a rough guide to fertiliser needs, the critical levels of phosphate and potash are: P_2O_5 20 ppm, K_2O 300 ppm. About 250 g of ammonium sulphate scattered around the plant, and 30–60 cm away from the stem, is often beneficial, applied once or twice during the rains. A good crop of bananas usually indicates good soil.
Interplanting: very common, with other food crops such as maize, beans, sweet potatoes, pumpkins, yams, coconuts, sugar-cane, etc. Bananas are also used as a nurse crop for cocoa. They are not suitable for interplanting with coffee as the bananas use up too much of the available soil nutrients.

When working the soil, care should be taken to avoid any damage to the banana plants shallow root system.

GROWTH CONDITIONS
Day length: no known response.
Growth period: for the plant crop, 9–18 months from planting to harvesting, depending on variety, growth conditions, etc. From the shooting of the inflorescence to harvest, 2.5–4 months. The period between plant crops varies from 4–6 years (commercial fields) to 10 years (central America) to 50–100 years (traditional farming). In theory, banana fields can be in permanent production due to the continuous production of suckers, but in practice are rarely kept for more than 60 years. Sometimes the plants are replanted every two years.
Ripening of the fruit is often done artificially after the fruit has reached the country of destination by exposure to ethylene gas.
Temperature: 27–29°C is optimum. Growth is reduced below 21°C. Frost sensitive.
Rainfall: to grow well, needs a minimum of 25 mm/week. When irrigated, 150–200 mm/month are applied. Optimum annual rainfall is about 2000–2500 mm, but some varieties are grown with much less or much more water than this.
Wind: can be a major cause of crop loss, by tearing the leaves and, more importantly, by breaking and uprooting plants. Windbreaks are therefore often necessary.
Weeds: important to remove all perennial weeds, and also to provide a mulch. Bananas are very susceptible to hormonal weedkillers, though Dalapon and some others can be used.

Mineral deficiencies: symptoms are commonly seen when insufficient fertiliser is used (see Section **1Cd**, page 23):
– **Calcium (Ca)**—leaf margins become yellow, then brown—similar to Potassium deficiency;
– **Potassium (K)**—leaf margins become yellow, then general leaf necrosis (unprogrammed death of cells or tissue); small bunches with thin fingers. Can be shown up by nematode attack;
– **Zinc (Zn)**—bunchy tops;
– **Phosphorus (P)**—slow growth, then leaf margins become scorched and the base of the corm roots; leaves are short; few hands and fingers per bunch;
– **Nitrogen (N)**—slow, stunted growth; leaves become yellowish;
– **Iron (Fe)**—young leaves chlorotic between the veins. If severe, general chlorosis.
– **Magnesium (Mg)**—young leaves chlorotic between the veins; leaf margins wavy. If severe, leaves become dark brown and hang down, and leaf sheaths split.

Altitude: sea level up to about 1800 m is optimum. They can grow up to about 2400 m but development is then normally slow and weak due to low temperatures.

Pruning: a controversial subject, but in general pruning to 3–6 stems produces earlier, larger bunches and plants with better resistance to Borer (Weevil) attack.

Pests: more than 200 insects like to eat bananas and banana plants. The most important are as follows:

– Banana Weevil (Stem Borer or Beetle)—the most serious, all over the world, especially in neglected plantations. Adult weevils live for up to two years in the stems, the larvae burrow into the corms, leaving tunnels visible if the plant is cut open; the plants are weak, yields are low. Control: plant clean suckers and treat them with insecticides, cut off old stems at ground level and cover with soil, chop up the old stems and spread them on the soil as a mulch, and encourage healthy plant growth with fertiliser, mulching, weeding, etc. Traps can be made with old stems split and placed around the clumps—the insects can be collected from them by hand.

– Thrips—fruit skins become discoloured or cracked. Very difficult to control.

– Nematodes—a worsening problem in some areas where the tolerant Gros Michel clones are replaced with the susceptible Cavendish clones. The roots are destroyed, leading to wilting and sometimes plants falling over.

Control: plant only clean material, and crop rotations, such as with sweet potatoes. DBCP is sometimes used, 2–4 times a year, but this chemical is very dangerous and has caused widespread sterilisation in plantation workers from Central America and the Caribbean to the Philippines and West Africa.

– Other pests—Fruit Scarring Beetle (North America), Scab Moth (Pacific) and aphids, which carry and spread virus diseases.

Diseases: four of the most common diseases are described below:

– Panama Disease—one of the most catastrophic diseases in the world, it is found in most countries where bananas are grown, especially in acid soils. The fungal spores can survive in the soil for 20 years or more. Symptoms: the outer, lower leaves become yellow, then hang down and collapse; the inside of stems and rhizomes become purple.

Control: ideally with resistant varieties, also plant clean material into clean soil, or flood fallow to 1–2 metre depth for six months, or increase soil pH by liming, or improve drainage.

– Leaf Spot (Black Sigatoga)—a serious problem in the tropics, causing yield loss and premature fruit ripening. Symptoms: faint yellow spots, first on the leaves, the spots become necrotic at the centre and join together in lines parallel to the lateral veins; leaf margins die. Control: plant only resistant varieties, or spray Bordeaux mixture or mineral oil sprays every 2–4 weeks at 11 litre/ha.

– Cigar End Rot—fruit tips look like the ash on the end of cigars. Worse with cool nights. Control: remove the end of the inflorescence, beyond the developing fruits.

– Bunchy Top Virus—occurs in the Pacific, Far East, Australia, Egypt and the Congo. Symptoms: green streaks on the secondary veins on the underside of leaves and on midribs and petioles (leaf stalks); lower leaves become brittle and stunted; petioles are short, creating a "bunchy" effect. The fruits are spoiled, often unsaleable. Controlled by planting clean material and controlling aphids.

Other diseases include Black Leaf Streak, Bacterial Wilt (Moko disease), other virus diseases and a range of fruit rots and spots.

YIELD

With poor management, the plant crop usually gives the best yield, with lower and lower subsequent yields.

With good management the reverse is true; yields from the early ratoons are are usually higher than for the plant crop. Also, as less time is taken to produce a ratoon crop, the yields per unit of person-hours labour are greater.

An average annual yield is about 1000–1200 bunches/ha, or 16,000 to 27,600 kg/ha, assuming an average bunch weight[*]of 16–23 kg. The FAO estimate for the world-wide average in 2004 was 16 MT/ha; the top national average was recorded in Guatemala (52.5 MT/ha) and the lowest was in the Cayman Islands (1.3 MT/ha).

With good crop husbandry, including irrigation if necessary, 75 MT/ha is possible.

[*]In commercial plantations, the aim is to produce about nine hands per bunch, weighing between about 20 kg and 65 kg.

UTILISATION

- **Raw, eaten as fruit**—about 50% of bananas are consumed this way. They are especially useful for children and for people of all ages with gastric complaints.
- **Plantains**—these account for the other 50%. They are normally peeled, cooked and then mashed—in Uganda this is known as *matoke*. They can also be boiled in their skins and then peeled, or peeled and roasted in hot ashes, or fried with maize, beans, potatoes, etc. To store, they can be peeled, dried and ground into flour (sometimes called *ugali*).
- **Brewing**—very popular in certain parts of Africa. Flour made from sorghum or finger millet is often added to the brew, which is rich in Vitamin B and other useful nutrients, not to mention the alcohol itself.
- **Dried products**—many different types are made: *Banana chips*, for which the beer varieties are mainly used—the mature, unripe fruits are sliced and dried; in marginal areas such as parts of East Africa they are used as a famine reserve food. *Sweetmeats,* the so-called banana figs, are made from dried slices of ripe fruits. *Flour* is made from dried unripe fruits, and *powder* is made from dried ripe fruits.
- **Male buds**—in some places such as southeast Asia the ends of the inflorescence are eaten as a boiled vegetable. Occasionally available in specialised shops in the West.
- **Animal fodder**—various plant parts can be fed to animals, an important aspect for some groups such as the Wachagga in Tanzania.
- **Green leaves**—used for plates, umbrellas, wrapping material, etc.
- **Dried leaves**—used for tying material, thatching, screens, plant pots, head protection, etc.
- **Fibre**—extracted from the pseudostem, it is used for fabrics, rope, tea bags, etc.
- **Shade crop**—commonly used for cocoa and coffee plants.
- **Ink**—a brown indelible ink can be extracted from the sap.

LIMITATIONS

- Bananas are susceptible to damage by both insects and diseases, so that a cocktail of chemicals may be applied, often to the detriment of the plantation workers.
- The current oligarchical position of the big three banana companies can be a threat to the entire global market, and allows, all too frequently, unacceptable working conditions regarding pay and health care to be imposed on their employees.
- Sophisticated, expensive equipment is needed to store and transport the fruit in large quantities.
- The plants can be badly affected by wind and/or cold weather.
- A long time elapses from planting to the first harvest, and the plants then need a high level of management and labour to maintain yields.
- Growing requirements of the plants are quite stringent: good soil and/or high fertiliser input, high light intensity and temperature, and a large and regular supply of water.

Citrus

Citrus limon—lemon ("*citron*" in French)
C. aurantifolia—lime ("*limette*", "*citron vert*" in French)
C. sinensis—sweet orange
C. aurantium—sour (Seville) orange
C. paradisi—grapefruit
C. reticulata (Syn. *C. nobilis*)—tangerine, mandarin, satsuma (and Rangpur lime)
C. maxima (Syn. *C grandis, C. decumana* or *C. maxima*)—pomelo (shaddock)
C. medica—citron ("*cédrat*" in French)
Poncirus trifoliata—trifoliate orange

Hybrids: four examples:
C. reticulata X *C. paradisi*—tangelo, *C. reticulata* X *C. sinensis*—tangor
C. reticulata X *Fortunella* spp.—calamondin *(Citrofortunella microcarpa)*
C. sinensis X *Poncirus trifoliata*—citrange

Citrus species have been cultivated by man since the earliest days, selecting the plants with the juiciest fruits and the best flavours. Evidence has been found in China of the cultivation of citrus fruit in about 2200 BC. The various species hybridise readily, and there are an enormous number of complex hybrids. Mutations (bud sports or bud mutants) may also arise in different parts of the same tree, and within each species there are a large number of varieties.

They are grown in the tropics, subtropics and warm temperate regions; limes, grapefruits and pomelos (shaddocks) are more suited to the wetter areas, oranges and mandarins preferring drier, cooler conditions. The **kumquat** *(Fortunella* spp.) is not strictly a citrus fruit but is a member of the same family, the Rutaceae. The orange or golden yellow fruit is ovoid or round, 1–4.5 cm in diameter and contains about 9% sugars and 39 mg/100 g Vitamin C.

In certain American English speaking areas the word "pomelo" (a fruit with a large yellow, grapefruit like fruit) is used to describe the grapefruit. The true pomelo, or shaddock (*C. maxima*) has pale green or yellow fruit, when ripe, larger than grapefruit.

Citrus are one of the most difficult fruit trees to cultivate, and they normally need careful attention if they are to give consistently good yields of fruit.

PLANTING

Propagation: seed is generally not suitable as it contains out-pollinated genes and so it rarely breeds true. Instead "*marcottage*" (for limes, and sometimes mandarins and pomelos), cuttings or graftage is done. Marcottage uses the same principle as "*layering*"—soil is tied onto a branch using polythene or sacking, sometimes the bark is also ringed at this site, which encourages roots to grow.

Grafting onto rootstocks is more common, the rootstock itself being grown from seed. There are many types of rootstock, listed below, the first three being the most commonly used: **Sour Orange**, mainly for lemons and oranges, grown in medium and heavy soils; the root system is deep and vigorous. **Rough Lemon**, a lemon × citron hybrid, has a shallower root system for use in lighter, sandy soils; used in the Caribbean and tropical America; generally produces poor quality fruit. **Sweet Orange**, for rich, well drained soils; susceptible to foot rot; generally produces lower yields than rough lemon with mandarins and oranges but gives good fruit quality. **Rangpur Lime**, easy to transplant; tolerates wet conditions; resistant to gummosis and tolerant to tristeza virus; used in South America and the Far East. **Mandarins**, widely used in Asia; some resistance to tristeza, quite resistant to foot-rot and gummosis; mainly in humid tropical regions. **Trifoliate Orange**, the best for cold conditions; tolerates some waterlogging; resistant to gummosis and tolerant to tristeza; used in Japan for Satsuma oranges. **Citrange**, used in cold conditions where resistance to tristeza is necessary. **Grapefruit**, not much used, best when used for other grapefruit and pomelo varieties; not very compatible with oranges. **Limes**, not much used, except in Israel and Egypt; tend to develop and mature quickly producing early fruit but they are short-lived and many are sensitive to foot-rot. **Citropsis**, a native citrus in the Congo region of Africa, has potential as a rootstock due to its high resistance to foot-rot and gummosis.

Soil: citrus species will grow on most soil types provided they are neither saline nor waterlogged. Optimum pH is 5–7. Magnesium limestone is beneficial if the pH falls below about 5. Manure and nitrogenous fertilisers are very beneficial. With infertile soil, limes will sometimes survive where other species fail. All species respond well to fertilisers and manure—Nitrogen is the most important nutrient; excess Potash (K) can be harmful.

In poor soils, secondary nutrients such as Sulphur, Calcium and Magnesium and some trace elements may need to be applied as well as the major nutrients, N, P and K.

Spacing:

Type		Distance (m)	Trees/ha
• Orange	– small varieties	6	280
	large varieties	10	100
• Grapefruit	– small varieties & grafted	9	120
	large varieties	12	70
• Mandarins	– marcotts	4.5 × 6	280–500
	small varieties	10	100
• Pomelos	– large varieties	12	70
• Limes	– grafted	3–4.5	500–1000

GROWTH CONDITIONS

Day length: day-neutral.

Growth period: the trees normally produce their first fruit in 3–6 years.

Temperature: growth is much reduced below about 13°C and virtually stops below 10°C. Maximum tolerated is about 40°C. In order of their hardiness (tolerance to cold), with the least hardy species (Citron) first: citron, lime, lemon, grapefruit, sweet orange, sour orange, mandarin, kumquat, trifoliate orange.

Rainfall: 900 mm/a. minimum without irrigation, which is often used, especially for young trees. Not suited to the humid tropics—high humidity increases damage by pests and diseases. Mandarins are the species most adapted to high rainfall and humidity.

Pruning: as a minimum, this should be done in the nursery and on 2–3 year-old trees, to form an open canopy. Suckers and dead wood should be removed.

Wind: citrus species do not tolerate high winds, so windbreaks should be provided if necessary.

Altitude: mainly from 0–600 m. On the equator, they do not grow well above about 1800 m.

Pests: damage can be caused by: scale insects, aphids, mealy bugs, mites, thrips, fruit flies, moth borers and the false coddling moth. It has been shown that indiscriminate insecticide use can create more problems than it solves, discussed in **1J**, page 78.

Diseases: Foot-Rot (Root-Rot) (*Phytophthora* spp.)—worse in high humidity and in poorly drained soils. Sour orange and mandarin rootstocks show some resistance. **Gummosis**, the exuding of gum, is also caused by *Phytophthora*. **Tristeza Virus ("Quick Decline")** is the most serious and widespread of the virus diseases. Can be partially controlled with the careful use of insecticides in nurseries, Sweet Orange, Rough Lemon and some Mandarin rootstocks, and burning of diseased or suspected branches or entire trees. **Citrus Scab** is caused by the fungus *Elsinoe fawceti*, controlled with copper fungicides (sprayed before and after flowering) and resistant types. Worse in cool conditions. Sweet Orange and lime are very resistant; sour orange, lemon and some grapefruit species are very susceptible. **Melanose** is widespread. Brown, raised pustules on young twigs, leaves and fruits, especially on mature trees of all Citrus species. Controlled with copper sprays, remove all dead/diseased wood.

Trace Element Deficiencies: see also Section **1Cd**, page 23:
– **Boron:** lumpy fruit, leaves with water-soaked spots. **Copper**: die-back with gum pockets or blisters on young growing branches, leaves and fruit. Then leaves and fruit drop off, and twigs become distorted. Treat with sprays of 5 kg copper sulphate in 10 kg lime per hectare. **Iron:** chlorosis of leaves, veins remain green. **Magnesium:** very common. Chlorosis of whole leaf. *Control*: Epsom salts (magnesium sulphate). **Zinc:** known as "Mottle Leaf", it is fairly common. Leaves become mottled and reduced in size. Control: spray 2.25 kg zinc sulphate and 1.13 kg slaked lime in 455 litres of water – 4.5 litres per tree. **Molybdenum:** "Yellow Spot", controlled by adjusting the soil pH if possible to pH 5.5–6.5.

YIELD
Highly variable, according to species, variety, soil, climate, etc.
   Average yields, in kg/ha/year, are, approximately: limes 25–80, oranges 150–1800, and grapefruit 150–1000.

UTILISATION
- Citrus plants are mainly grown for the juicy fruits. They contain almost 90% water, and variable amounts of sucrose, glucose and fructose.
- The skins are a source of pectin, used in jam making and for the ascorbic acid/Vitamin C (90-150 mg per 100 g edible portion).
- Lemons and sour oranges provide citral and limonene, used in perfume manufacture.

LIMITATIONS
- Citrus plants need to be carefully and skilfully grafted, and also pruned in the early stages.
- Intolerant of poorly drained or saline soils, and low temperatures.
- Susceptible to attack by many pests and diseases.
- Trace element deficiencies are common, and these are not always easy to accurately diagnose and/or treat.
- 3–6 years before any fruit is produced.

Guava
Psidium guajava

Goyave (French); Guayaba (Spanish); Goiaba {fruit}, Goiabeira {tree} (Portuguese); Amrood, Jamphal (Hindi); Zaituna (Amharic, Ethiopia); Banjiro (Japanese)

The guava tree is indigenous to tropical America where it still grows wild. It is now found throughout the tropics and subtropics. It spreads easily, often by birds distributing the seeds, and can become a weed in some places such as Fiji and Queensland. The name "Guava" comes from the Spanish *guayaba*, from a South American Indian word.

The largest producer of guava is India, and they are also grown widely in the USA, Mexico, Pakistan, Colombia and Egypt.

The fruit contains variable amounts of Vitamin C, ranging from 23–492 mg/100 g, according to the variety and the growing conditions. Fruits also contain about 5% total sugars (glucose and fructose) and useful amounts of iron, calcium, phosphorus and Vitamin A and B.

The following description applies to the **Common Guava**, *Psidium guajava*. There are a few other species of *Psidium* which have edible fruit, the most important of which is the **Strawberry Guava** (or **Cattley**), *P. cattleianum* (syn. *P. littorale*), which has two types, one with bright yellow fruit and the other with reddish purple fruit; often grown as a garden plant, but not important commercially.

There are dozens of other *Psidium* species, including the **Guisaro, Cás**, and the **Brazilian Guava**, each with many local names and synonyms. This topic is well documented online at: www.plantnames.unimelb.edu.au.

The **Pineapple Guava** (the **Feijoa** or **Guavasteen**), *Feijoa sellowiana*, is a native of cooler parts of southern South America, sometimes cultivated for its fruit. Aka Goiaba do Campo & Goiaberra Serrana (Portuguese), or Guayaba Chilena (Spanish).

The Common Guava *Psidium guajava* plant is a shrub or small tree, 3–10 m tall, with many branches produced close to the ground, and a shallow root system. The guavas are members of the myrtle family *Myrtaceae*, which grow almost exclusively in hot and rather dry regions.

About 35% of the flowers are cross-pollinated by insects, the other 65% are self-pollinated. The fruits are very variable in size (2–8 cm) and flavour, and all have a characteristic musky smell, which normally disappears when they are cooked.

A wide range of improved, named varieties are available in a number of countries.

PLANTING

Propagation: can be done by seed, though this produces variable and unpredictable results and so vegetative reproduction is preferable. Veneer grafting, using young vigorously growing plants as rootstocks, is the preferred method.

Germination: seeds remain viable for about a year; germination takes 2–3 weeks.

Soil: guava adapts to a wide range, including poor, acidic soils and even tolerates some waterlogging. It responds well to fertiliser, especially Nitrogen—an 8:8:8 compound is often used.

Spacing: often grown as single, backyard plants. Ideally, about 5–7 metres apart, in squares.

GROWTH CONDITIONS

Growth period: the first fruits appear in about two years, the production increasing for about six to eight more years. Trees can continue producing fruit for 30 years or more.

Temperature: guava trees are susceptible to frost. They prefer warm weather and can tolerate extreme heat.

Rainfall: one or more varieties of guava can be found to grow in most rainfall zones, though ideally they should receive about 1000–3000 mm/a. They need plenty of water to grow well.

Pruning: suckers and shoots from the tree base should be removed regularly.

Altitude: from 0–1500 m; commercial plantations are normally below about 1000 m.

Pests: fruit flies (Oriental and Mediterranean), Guava flies (especially in Trinidad and the West Indies), mealy bugs, thrips and scale insects can all cause major damage. Root-knot nematodes can cause some damage and can be partially controlled with heavy fertilisation, irrigation and spraying nutrients on the leaves.

Diseases: Wilt is a serious problem in some parts of India, Bacterial disease is the major problem in the Caribbean. The alga *Cephaleuros virescens* causes fruit and leaves to rot, especially in wet areas; controlled to some extent with fungicides.

YIELD

Mature, grafted or layered trees of local, unimproved varieties each produce about 400-800 fruits per year, weighing a total of 60–120 kg.

The same type of trees of improved types yield about 1000–2000 fruits per hectare per year, weighing 150–300 kg. Seedling trees normally produce half or less that of grafted or layered trees.

UTILISATION

- Guavas are mainly grown for local consumption. The seed is normally removed from the fruit for making jam, jelly, paste, preserves, juice and nectar. In the West Indies and elsewhere guava cheese is made by evaporating the fruit pulp with sugar.
- Commercially, guavas are used mainly for making jellies. The fruit, juice and nectar are canned.
- The leaves are sometimes used in traditional medicine to treat diarrhoea; also for dyeing material and for tanning.

LIMITATIONS

- Very few, though under certain circumstances, usually in tropical conditions, the growth of guava trees may become so uncontrolled that they become a weed, in Fiji for example.
- Guava trees are sensitive to low temperatures and frost.
- Propagation by seed gives unpredictable, varied trees with low production. Vegetative propagation on the other hand requires specialist knowledge of grafting or budding.

Mango
Mangifera indica

Arbre de Mango, Manguier (French); Manga (Spanish); Manga {fruit}, Mangueira {tree} (Portuguese); Am, Amra (Hindi); Am (Pashtu & Dari/Persian)

Mangoes are the most popular fruit in many hot countries such as India, where they have been grown for at least 4000 years. They are a commonly used feature in the Hindu religion.

The tree is a large, evergreen perennial, 10–30 m tall, which can live for more than 100 years. It has a long tap root which can penetrate down six metres, and a dense surface mass of feeding roots.

Both vegetative and reproductive (flowering) growth happens in "bursts" or "flushes", one part of the tree may be producing young flush growth (green, red, yellow or purple leaves) while the rest of the tree is mature (dark green leaves).

Mangoes are members of the cashew family *Anacardiaceae*. The so-called **Wild Mango** (Oba or Dika) *Irvingia gabonensis* is unrelated botanically, despite its yellow, edible fruit which somewhat resembles mango. The oil-rich seed is used in the treatment of obesity and as a soup thickener, called Ogbono (Agbono) in Nigeria.

The trees normally flower in January–March in the Northern hemisphere and in June-August in the Southern Hemisphere. Each inflorescence may have up to 6000 flowers, mainly male, some hermaphrodite. The latter are insect pollinated but only about 0.1% set fruit. Some varieties are polyembryonic ie each seed produces, in addition to the sexual seedling, one to five nucellar seedlings which are genetically identical to the parent plant. Monoembryonic varieties produce only one embryo per seed, which, because it has arisen from a sexual process, will not grow "true to type" (ie it will not exactly resemble either parent). With these varieties therefore uniformity can only be achieved by vegetative propagation.

The fruit is produced 2–5 months after fertilisation. The trees are often biennial bearing (they only produce a good crop every other year, sometimes only every 3–4 years, or they may fruit only once a year instead of the normal twice a year, or on only a part of the tree at a time.

Mangoes are grown in most tropical countries, particularly in Asia, as well as in Florida, Queensland, Egypt and Natal. There are many thousands of named varieties available, though some are very specific to local conditions, and types are often chosen more for their colour and flavour than their yield. The production of hybrids is difficult but possible.

PLANTING
Propagation: can be by seed, transplanted after 4–12 months but this normally produces poor quality fruit, so the best methods are vegetative, either grafting, in arching (approach grafting) or shield budding. More efficient methods such as veneer grafting and chip budding are also used in commercial plantations.

Germination: seed remains viable for only about 4 weeks after removal from the ripe fruit, but ideally the seed is sown soon after removal. For this reason, mangoes were not widely and rapidly spread around the world. The husk should be removed to reveal any weevil larvae.

Soil: optimum pH is 5.5–7.5 (classified as "tolerant" to soil acidity), but not adapted to a much wider range. Should be deep and free-draining, and not too fertile because in very fertile soil they tend to grow very large but with few fruits. Fertiliser is beneficial for the first 3–4 years, and Nitrogen can be applied when the trees are bearing a heavy crop, to encourage a good subsequent crop.

Spacing: 6–12 m apart. Large planting holes should be dug, incorporating plenty of organic manure and/or compost.

Intercropping: very common, especially under young trees, with legumes, vegetables, pineapples and other annual crops.

GROWTH CONDITIONS

Growth period: the first mango fruit appears after 4–5 years in grafted or budded trees, and in 7–8 years when grown from seed. Fruit matures 2–4 months after flower fertilisation.

Temperature: optimum is about 24–27°C. Sensitive to frost.

Rainfall: 500–2500 mm per year. A marked period of dry weather is needed for pollination and fruiting, so mangoes are generally not well suited to the humid tropics, though recently some varieties have been developed for these conditions.

Altitude: 0–1300 m in the tropics, though some varieties can grow at 1800 m. Best conditions are normally below 600 m.

Pests: not normally a big problem; the **Mango Weevil** (*Sternocochetus mangiferae*) is the worst; larvae enter the fruit, leaving no mark on the skin, and attack the seed; young fruit falls off and older fruit rots. The Mango Hopper (Jassid), various fruit flies, thrips, scale insects, mites and mealy bugs also live on mangoes. Fruit flies are most easily controlled by destroying fallen fruit.

Diseases: also not normally a big problem; **Anthracnose** can cause significant damage, causing discoloured fruit, leaf spot, blossom blight and fruit rot. Partial control with copper fungicides, but hard to control. **Powdery Mildew** (*Oidium mangiferae*) can also infect flowers and young fruit, and can be controlled with fungicides. Both diseases spread fastest in warm, wet weather.

YIELD

Ten year old mango trees can produce 400–600 fruits in their "on" year, increasing until about their 20th year, when they can produce around 2000 fruits, and declining after they are about 40 years old.

Many varieties produce fruit erratically, one good year often followed by 2–3 poor years.

UTILISATION

- **Ripe fruit**—eaten raw and also to produce juice, squash, jams, jellies and preserves. They can also be canned. Mangoes are a good source of Vitamin A, though the amount depends on the variety and the colour of fruit—apricot coloured fruit has the highest level. The fruit also provides useful amounts of Vitamins C (37 mg/100 g) and D, as well as about 14% total sugars, mainly in the form of sucrose.
- **Unripe fruit**—rich in starch, used in pickles and chutneys. Also sliced, dried and seasoned with turmeric to produce *amchur*.
- **Seeds**—can be ground into flour and used as human food during famine—70% carbohydrate, 10% fat and 6% protein.
- **Leaves**—if little or nothing else is available the leaves can be fed to cattle, though they may die if given this diet for too long.
- **Timber**—for making boats and dugouts, and also in the construction of buildings.

LIMITATIONS

- Mango trees may only produce a good crop of fruit only every two, three or four years;
- The trees are also susceptible to insect damage, and they are sensitive to frost;
- Minimum period of four years from planting to first harvest—even longer for trees grown from seed, and the fruit is difficult to transport without loss of quality.

Papaya (Pawpaw)
Carica papaya

Papaw. Papaye (French); Mamão{fruit and plant}, Mamoeiro, Papaeira, Pinoguaçu {plant} (Portuguese); Papaya (Amharic, Ethiopia)

Papaya originated in tropical America, probably around present day Mexico and Costa Rica, and is now widely grown throughout the tropics and subtropics, and to some extent in warm temperate regions. The trees are mainly grown on a very small scale, in back yards and so on, but in places such as Australia, South Africa, Mexico, Brazil, Indonesia and Hawaii they are grown on a field scale. The annual production worldwide is estimated to be about 2 million MT.

Papaya is a member of the *Caricaceae* family and should not be confused with the other Pawpaw or Papaw, *Asimina triloba*, the Custard Apple, a member of the *Annonaceae* family, which produces small, fleshy 8–18 cm long fruit that looks rather like stubby bananas.

The Mountain Papaya (*C. pubescens* syn. *C. candamarcensis*) is a native of the Andes and is cultivated at high altitudes in the tropics. The small fruits are cooked before eating and can be made into jam.

Papayas are short-lived perennials; they usually live for only 5 or 6 years, and no more than 25. The trees grow rapidly, to a height of about 2–12 m; the wood is light and soft. The roots are also soft and easily damaged.

There are two types of papaya tree: **dioecious**, the most common type, with male and female reproductive organs (flowers) in separate flowers on separate plants, or **monoecious**, less common, with male and female flowers separate but on the same plant. Male plants can be recognised as soon as they flower because the flowers are borne on long stalks. If a dioecious variety is grown, some male trees should be grown near the females (about 1 male per 15–20 females) to ensure that pollination occurs.

Papaya fruit is a good source of Vitamin A and also contains about 60 mg/100 g Vitamin C. Total sugars are about 9%, equal parts of glucose, fructose and sucrose. The fruit is also the source of *papain*, a proteolytic enzyme extracted from the latex of the unripe young fruit, which is mainly produced in East Africa and Sri Lanka—see "Utilisation" and "Yield" below.

PLANTING
Propagation: normally by seed, either in nurseries or directly into the field. Farmers can use their own seed, selected from the best trees.
Germination: takes 2–3 weeks. Air-dried seed remains viable for 2–3 years.
Soil: papaya plants will not tolerate waterlogging, though the soil should always be moist. Ideal soils are well drained, sandy and fertile with pH 6–6.5 (classified as having "intermediate tolerance" to soil acidity).
Roots are soft and easily damaged, so weeding around the trees should be shallow and careful.
Seed rate: 14,000–15,000 seedlings per hectare, with about 5–7 seedlings per hole (see next).
Spacing: 2.1–2.7 metre square grid. 5–7 seedlings or seeds are planted per station and grown until the sex of the plant becomes apparent, in about six months. The best female plant is left, and also one male plant per 25–100 female plants.
Depth: 1 cm.
Rotation: when a papaya crop is finished, normally after 2–5 years, an unrelated crop should then be grown to minimise the build-up of nematodes and/or root rots.

GROWTH CONDITIONS
Growth period: first fruit in 9–14 months, then can live for up to 25 years, though commercially they are normally only grown for 3–5 years. Papain is normally extracted for only 2 or 3 years.
Temperature: the plant is killed by frost, and low temperatures lead to tasteless fruit.
Rainfall: the soil should be constantly moist, so irrigation is often necessary during dry spells. Poor drainage tends to lead to root rots.
Altitude: below 1500 m is best, though they can grow at over 2000 m. Papain is produced at lower altitudes, normally below 900 m for commercial tapping; latex flow decreases with altitude.
Pests: rarely a problem. Birds sometimes damage fruit. Mites can cause some damage.

Diseases: one of the main reasons for the short life of papaya trees: **Mosaic virus** is the most serious; leaves turn yellow and distorted, plants become distorted and die. Spread by aphids, which also spread other papaya diseases. **Pythium rots** are worse in wet or waterlogged soil. **Anthracnose**—the ripe fruit develops spots. Also: mildew, foot rot, damping off and leaf spots.

YIELD

Fruit: 10–15 MT/ha per year is possible with careful farming. Maximum is about 35 MT/ha per year, with each tree producing 30–150 fruits per year. Some fruits weigh up to 10kg or more.

Papain: 45–170 kg/ha per year—50% in the first year, 30% in the second and 20% in the third. Trees are then either replanted or grown only for the fruit. The average yield in Tanzania is 45–55 kg/ha per year.

UTILISATION

- **Fresh fruit**—produced throughout the year, eaten when ripe and also used to make soft drinks, jam, ice-cream flavouring and crystallised fruit. They can be canned in syrup.
- **Papain**—a protein digesting enzyme which is extracted from the dried latex taken by tapping young, unripe fruit. Papain greatly resembles the animal protein *pepsin* in its digestive action. It is used as a meat tenderiser, in manufacturing chewing-gum and cosmetics, as a drug for digestive disorders and for detecting intestinal cancer, in the hide tanning industry, for degumming natural silk, for clearing beer and to make wool resistant to shrinking. That is all.
- **Unripe fruit**—can be cooked as a substitute for marrow and for apple sauce.
- **Young leaves**—can be eaten as a spinach, and used to tenderise meat—the young fruit can also be used for this purpose.
- **Flowers**—occasionally eaten, as in Java.
- **Seed**—used in some countries as a vermicide (worm killing substance), counter-irritant and abortifacient.
- **Pectin**—extracted from the by-products of canning, it is used in making jams and jellies.

LIMITATIONS

- Papaya fruit is thin-skinned, easily bruised and so does not travel well.
- The plant is killed by frost.
- The plant is also susceptible to attack from nematodes, virus and other diseases.
- The shallow root system is easily damaged by cultivation and weeding activities, and means that trees are easily blown down by wind.
- Monoecious varieties bear fruit on only about half of the trees planted, and most of the male trees should be removed and replaced with female trees.

Watermelon
*Citrullus lanatus (*Syn. *Citrullus vulgaris, Colocynthis citrullus)*

Melon d'eau, Pasteque (French); Sandia (Spanish); Melancia (Portuguese); Wassermelone (German); Cocomero (Italian); Kalitangi (Umbundu, Angola); Tarbooj (Hindi); Hindwana (Pashtu); Tarbooz (Dari/Persian); Habhab (Ethiopia)

Watermelons can provide a valuable alternative to drinking water in desert areas and in other situations where water is in short supply. The edible portion, which constitutes about 60% of the whole fruit, contains about 94% water and very little protein or fat. They do however provide about 7% total sugars, some Vitamin A and about 5–8 mg/100 g Vitamin C, but are generally more useful as a cash crop than as a source of nutrition.

They originated in the sandy, dry areas of the Kalahari in Africa and are now grown on every continent except Antarctica, normally for local consumption. They are one of the most ancient cultivated crops; Egyptian art shows them being grown more than 4000 years ago.

There is great variation within the species, from small, hard, bitter and inedible fruits to the well known large, juicy, sweet fruits. A large number of named varieties, including hybrids (some of which are seedless), are available, with improved disease resistance and tough rinds for long distance transport.

The plant is an annual, a member of the *Cucurbitaceae (gourd)* family, with trailing stems up to 5 m long. The plants are both self-pollinated and cross-pollinated, mainly by honey bees. They are monoecious—the male and female flowers are on the same plant, but they are separate. Seedless fruit production is discussed on page 243.

Watermelons are related to the **Chinese Watermelon** or **Wax Gourd**, *Benincasa hispida*, which produce fruits shaped like melon or cucumber up to 40 cm long.

Farmers and other growers can save their own seed, selected from the most well adapted plants and fruits. The seed should remain viable for two years if kept dry and cool. To collect seed, the fruits should be left to mature on the plant, then left about two weeks after harvesting before removing the seeds. These should then be washed in clean water and dried in the sun.

PLANTING
Propagation: by seed, normally planted *in situ* into their final station. If transplanted, great care is needed to avoid disturbing the rootball. Hand pollination assists fruit setting.
Soil: watermelons prefer fertile sandy loam, near neutral pH (classified as "tolerant" to soil acidity) but will also grow in a wide range of well-drained soils. Sandy riverbanks are ideal.
Fertilisers: in common with the other Cucurbits, watermelons benefit from manure and compost. Inorganic fertiliser, particularly those that are high in Phosphorus, are also often applied. Nitrogen top dressings should be applied when the plants start making runners and/or when they start flowering or after fruit-set.

Seed rate: 1–3 kg/ha depending on variety and planting distances. 6–8 seeds often planted together on a mound, then thinned to the best 1 or 2 plants. Epigeal germination.

Seed size: very variable, eg *Sugar Baby* averages 41g per 1000 seeds, *Blackstone* averages 125 g. Seed dressing such as Captan or Thiram is often used to combat mildews.

Spacing: the mounds are about 2–3.5 m apart. Sometimes planted in rows, thinned to 1–1.5 m between plants.

Depth: 2–5 cm.

Rotation: should not be grown on the same land more often than every 4–5 years, nor after another Cucurbit crop, to avoid buildup of soil-borne insects(and fungal diseases to some extent).

GROWTH CONDITIONS

Day length: watermelon plants are day-neutral.

Growth period: 80–150 days. The fruit is ripe when it gives a dull thud when tapped with the hand and when the tendrils have withered.

Temperature: killed by even a light frost. For good fruit development watermelons need a long hot sunny period with a mean temperature of more than 20°C. For proper seed germination the soil temperature should be more than 20°C, and germination will not happen below about 16°C.

Rainfall: fairly drought resistant. If irrigated, this can be done infrequently as the deep, extensive root system ensures growth during long dry periods. The plants tolerate fairly humid conditions, though the fruit quality can suffer and diseases may also develop faster.

Pests: Cutworms and other soil inhabiting insects can be controlled with poisoned baits, such as Furadan. Fruit flies can be controlled with insecticide. Other pests include root-knot nematodes, Cucumber Beetle and aphids, controlled with a combination of rotation with insecticides.

Diseases: Mildews, both Downy and Powdery, may become serious. Also Anthracnose, Fusarium Wilt & Watermelon Mosaic Virus, all controlled with a combination of resistant varieties and sensible crop rotations. Genetic engineering in the USA has produced some virus resistance.

YIELD

A good crop of watermelons can produce about 1200 fruits/ha, each fruit on average weighing about 7–8 kg, varying from 1–2 kg to more than 20 kg. Each vine produces 1–15 fruits. 75 MT/ha is possible under ideal growing conditions.

The global average yield is about 15 MT/ha. Plant breeders aim to produce varieties with a larger number of smaller fruits, with tough skins, mainly to facilitate transport.

The yield of seed is about 200–250 kg/ha.

UTILISATION

- Watermelons are normally grown for the refreshing **ripe fruit**, eaten fresh. About 94% is water, but each 100 g contains Vitamin C (8 mg), Vitamin A (570IU), Vitamin B6 (0.045 mg) and some thiamin and riboflavin.
- The **seeds** are much more nutritious, containing 20–45% of an edible semi-drying oil and 30–40% protein; they are also rich in the enzyme urease. They can be dried and then chewed, as in southern China, or ground up and baked into bread, or roasted. The oil extracted from the seed is used in cooking and for lighting—the seed cake ("presscake") which remains is used for animal food.
- The **rind (skin)** is sometimes preserved as a pickle.

LIMITATIONS

- Low nutritional value of the fruit, mainly useful as a source of clean water and/or cash.
- The fruit is easily damaged & does not travel well, though some does enter international trade.
- The fruit has relatively short storage time ("shelf life").
- The plant is killed by frost and needs high temperature for germination and growth.
- The plant is also susceptible to Fusarium Wilt and Anthracnose, though there are many varieties available with good resistance to these, and other, diseases.

2G. UNDER EXPLOITED FOOD CROPS

At some time or another mankind has used at least 3000 plant species for food, of which only about 150 have been cultivated commercially. About 60 of these 150 species are described in detail in this document, in Section **2A–2G**, pages 107–287.

However, these days only about 20 plant species feed most of the people in the world, and it may be very risky to depend on such a small number of crops. Monocultures can be vulnerable to catastrophic failures arising from changes in climate or the uncontrolled development of pests or diseases.

One of the reasons that these days we are only cultivating a small proportion of edible crops may well be because these crops have been studied more intensively by plant breeders, agronomists and research workers than other "local" or "traditional" crops.

In colonial times, these workers were often trained and financed by European governments; market demands by European consumers dictated which crops were imported, and so which crops were encouraged in the colonies—examples include pineapple, banana, rubber and oil palm. In this way many traditionally grown crops, adapted over many generations by local farmers to the particular growing conditions of their area, were neglected or even deliberately suppressed.

This pattern has tended to continue even after independence, and mankind is only recently discovering the enormous possibilities of manipulating the ability of plants to adapt and succeed in different environments.

The ability of these under exploited crops to help to feed people in many of the marginal parts of the world deserves wider recognition from people involved in food production for survival. The current deficit in food, particularly protein, which exists in many of the poorer countries of the world will inevitably continue. One of the positive ways to assist farmers to produce more food is to introduce, test and then distribute the successful (if any) varieties of adapted crops of some of these less well-known species.

The best plant introductions are those that have at least one attribute, such as high protein or calorific content, which is in short supply in that area. Naturally these "new" plants also have to be acceptable to, and grown by, a proportion of "leader farmers" in the area before they become adopted to any significant extent. Many farmers are reluctant to experiment with new crops, or techniques; one way to help break down this reluctance is to arrange for demonstration areas to be planted in the areas involved. If a picture is worth a thousand words, then a demonstration area should be worth a million.

One problem, often the most serious one, concerns the preparation and cooking of any new and strange, or forgotten, crop. Very often the older women in the community will remember how to cook these almost forgotten food crops and they may be able to provide advice or give "in the home" demonstrations on the preparation and cooking of the "new" food.

Seven of these under exploited crops are described in the following section: **Amaranth, Bambara Groundnut, Buffalo Gourd, Leucaena, Lupin, Tepary Bean** and **Winged Bean (Four-angled Bean)**.

There are dozens of other plant species, many adapted to specific and often hostile growing conditions such as drought or high temperatures, including:

Adzuki Bean *Phaseolus angularis*
African Yam Bean *Sphenostylis stenocarpa*
(American) Wild Rice *Zizania aquatica*
Australian (Moreton Bay) Chestnut *Castanospermum australe*
Chinese Water Chestnut (Matai) *Eleocharis dulcis*
Kersting's (Hausa) Groundnut *Kerstingiella geocarpa*
Lotus *Nelumbo nucifera*
Mat (Moth) Bean *Phaseolus aconitifolius*
Pillepesara *Phaseolus trilobus*
Quinoa *Chenopodium quinoa*
Rice Bean *Vigna umbellata*
Water Chestnut *Trapa* spp.

Amaranth
Amaranthus spp.

Amaranthus Spinach, Chinese Spinach, Joseph's Coat, Inca Wheat, Love-lies-bleeding, Prince's Feather, Pigweed, Tassel Flower, Tumbleweed
Amarante {à Queue de Renard} (French); Amarant, Gartenfuchsschwanz (German);
Trigo del Inca (Spanish); Amaranto (Portuguese);
Jimboa, Otchimboa, Ofumboa (Angola).

Plants of the Amaranthaceae family are fast growing annuals that produce either high protein grain in large seed heads, and/or large protein-rich leaves which are eaten like a spinach. Grain Amaranths have been grown for thousands of years in South America, where the grain is still eaten, while the vegetable (leaf) Amaranths are of Asiatic or Indian origin.

The word *amaranthus* has its roots in Greek, meaning "unwithering"—the plant was regarded as a symbol of immortality. Ancient Greeks also considered Love-lies-bleeding (*A. hypochondriacus*) to be sacred due to its healing properties and used the plants to decorate tombs and sacred images.

Before the Spanish Conquest grain Amaranths were widely grown in the highlands of tropical and subtropical America; in fact they were the major grain crop in many parts of the Andes. The Spanish church suppressed the cultivation of Amaranths in an effort to eradicate them, partly because they were used in pagan Aztec religious ceremonies, and also to encourage the cultivation of maize.

These days the various species of the *Amaranthaceae* family are more commonly cultivated for their leaves by Asian hill tribes, and to some extent in the plains of India, the Caribbean and parts of Africa such as Angola and Zambia.

There are more than 800 species of *Amaranthus*, although their classification varies according to the source—Latin names do not always correspond to the commonly used name in English, which also vary according to country.

At least three species are grown for their grain and have good potential as sources of protein:

Amaranthus caudatus—(*A. edulis*, with club-shaped inflorescence branches, is one race of this species)—grown in the Andean regions of Argentina, Peru and Bolivia; known as Kiwicha. One form of this species *A. hypochondriacus* has red flower spikes, grown as a garden ornamental known as "Love-lies-bleeding". Very high lysine content of 6.2%, similar to soybean meal.

Amaranthus cruentus—occasionally grown in Guatemala and other Central American countries.

Amaranthus leucocarpus (Syn. *A. hypochondriacus* or *A. frumentaceus*)—the most widespread and important species. Grown mainly in Mexico and Guatemala. The grain contains 15% protein and 63% starch.

The smooth pigweed **Amaranthus hybridus** has leaves that are used as a vegetable similar to spinach—see penultimate paragraph "Utilisation" for details.

They are found all over the world, and many of the various species of Amaranths, including some wild species, are mainly regarded as weeds—hence the "Pigweeds"—Prostrate Pigweed (Mat Amaranth), Spiny Amaranth and Rough Pigweed or Redroot, and the well known spaghetti western weed Tumbleweed (*A.albus*). Other species are grown as ornamentals, some of which are perennials. Some of the many other so-called "Pigweeds", such as Lambs Quarters, are members of the *Chenopodiaceae* family.

PLANTING

Propagation: by seed. Dark seeds are sometimes removed before planting as these tend to produce very big, weedy plants. The seed is very small and is normally sown direct, by broadcasting, or sometimes in nurseries for transplanting. About 2 kg/ha of seed is required. Germination is affected by light and with some species only takes place in the dark.

Pollination: mainly by wind. Both self and cross-pollination occurs—a number of hybrids are found, formed naturally by cross-pollination between species.

Soil: similar requirements to maize ie best on fertile soils that have adequate P & N.

Spacing: similar to maize; wider spacing for the larger varieties.

GROWTH CONDITIONS

In general, Amaranths grow in similar conditions to maize, though there are species and varieties that can often be grown at higher altitudes and in colder areas.

Propagation: by seed, normally planted directly into the field (though they can be transplanted without any problem).

Day length: sensitive to photoperiod in that the timing of flowering depends on the day length. Most are short day plants, so they tend to bolt early if planted when the days are shortening. Different varieties have different responses, which can be a problem if they are introduced to one country or latitude from another.

Growth period: for leaf varieties, several cuttings of leaves can be taken before the plants set seed. Modern varieties mature quickly, in about 10–12 weeks.

Temperature: Amaranths are mainly a warm season crop.

Rainfall: soil must be moist for sowing; often irrigated in dry spells.

Pests: leaf eating caterpillars and beetles may become a problem, which can be picked off or controlled with Malathion or Sevin (carbaryl).

Diseases: not normally a problem, though Cercospora leaf spot and White rust may occur.

YIELD

Leaves can be continuously harvested for several months, yielding about 10 MT/ha. Seed yields of over 1 MT/ha have been reported in ideal conditions, though normal yields are only about 200–250 kg/ha, in theory, as they are rarely monocropped.

UTILISATION

- **Grain**–can be considered as a pseudo-cereal. The pale coloured grain is the best in appearance, flavour and its ability to "pop". It is usually parched and milled, and the dough made into pancakes. It can also be cooked for gruel, or popped and made into confections, or powdered and made into a drink. It contains about 14–18% protein and also high levels of the essential amino-acids lysine and methionine that are usually deficient in plant protein. The seed also contains tocotrienols, a form of Vitamin E which are believed to lower cholesterol levels in humans. It is also highly digestible, and therefore appropriate for people recovering from illness or famine. In Mexico it is popped and mixed with a sugar solution called alegria (happiness) and is also used to make a traditional Mexican drink atole.
- **Leaves and plant thinnings** are made into spinach type vegetable dishes, providing about 5% protein, especially from the Smooth Pigweed *Amaranthus hybridus*. Tea can also be made with the leaves, which is said to have astringent properties and to cure dysentery, diarrhoea and ulcers.
- **Other Amaranth species**—several are grown to be used as ornamentals, especially in the tropics, such as the Globe Amaranth *Gomphrena globosa* and *Amaranthus paniculatus*.

LIMITATIONS

- Amaranth plants can grow too vigorously and become a weed; in a survey of 15 States in North America, Pigweeds were reported to be one of the five weeds that caused the greatest yield loss in maize fields. The Alligator Weed (*Alternanthera philoxeroides*) is another example of an Amaranth that can become a serious nuisance.
- The seedhead shatters readily, which not only wastes food but also can create weed problems for the following crop.
- Little agronomic data is available—for example the response to day length of the different varieties is not well understood. India is one of the countries most involved with research. Some research is also conducted in the USA, China, Mexico and Nepal.
- Shortage of good quality seed, especially of named, improved varieties with well described characteristics.
- The pollen of the Noxious Pigweed *Amaranthus retroflexus* can cause an allergic reaction.

Bambara Groundnut
*Voandzeia subterranea (*Syn. *Vigna subterranea)*

Baffin Pea, Bambara Bean, Congo Goober, Earth Pea, Earth Nut, Ground Bean, Jugo or Juga Bean, Kaffir Pea, Madagascar Groundnut, Njugo Bean, Stone Groundnut, Underground nut, Aboboi, Akyii, Djokomaie, Epi roro, Gertere, Gobbo Voandzou, Haricot Pistache, Haricot de behanzin, Pois Arachide, Pois Bambara, Pois de Terre (French); Erderbse (German); Guisante de Tierra, Maní de Bambara, Maní de Africano (Spanish); Feijão Jugo (Portuguese); Oviélon (Angola); Ful Abungawi (Sudan), Guerte (Arabic); Njama (Malawi); Nela-kadalai (Malaysia); Kachang Bogor (India); Njugo Bean (South Africa)

Bambara Groundnuts originated in West Africa and have been grown in tropical Africa for many centuries. Today they are mainly grown in areas of tropical Africa with low rainfall and poor soils. They are also grown in central and South America, India, the Philippines, Malaysia, Indonesia and northern Australia. Zambia is the biggest producer. They are mainly grown on a small scale for local production and do not enter into world trade. They are occasionally found growing wild in West Africa.

There are many types of this annual legume: open (spreading), compact (bunched) and intermediate (semi-bunch). They have trailing stems 10–15 cm long, almost submerged. After the pale yellow or pink flowers have been fertilised (most varieties are self-pollinated, though cross-pollinated types also exist), the ovary grows down into the soil, like groundnuts. One or two centimetres below the soil surface pods develop, which are hard and wrinkled, 1–3 cm long, either round or oval. The pods contain one or sometimes two hard round seeds 7–15 mm in diameter of various colours (white, black, red and also often speckled or patterned), usually with a white hilum that may be edged by a black or brown eye.

The mature seed formed in these subterranean pods contains a nicely balanced composition of 14–24% protein, 4.5–6.5% fat and 50–60% carbohydrate (mainly starch). The calorific content is quoted as being between 367 and 412 per 100 mg edible portion. This well-balanced food is also quite easy to prepare and tastes good, though prolonged boiling is required.

The plant grows in hot, arid regions on poor soils where other legumes or high protein sources such as groundnuts do not grow well. This species has been neglected as a crop until recently when their potential as a source of protein in marginal areas is starting to be exploited.

The plant looks very similar to the **Groundbean** or **Kersting's Groundnut** (*Kerstingiella geocarpa*), which has grains which are usually white, brown, black or speckled and which look very similar to the seed of the Haricot or Common/Field Bean (*Phaseolus vulgaris*). The Groundbean (*K. geocarpa)* has broader leaves than the Bambara Groundnut, and the plant is less robust, though it can grow in even more arid areas than the Bambara Groundnut.

Online information on the Bambara Groundnut is available from the BAMNET website, and elsewhere.

PLANTING

Propagation: by seed; normally sown direct ie not transplanted.

Soil: Bambara groundnuts tolerate very poor and sandy soils that are unsuitable for groundnuts. The optimum pH is 5–6.5. Ideally the seed is inoculated if it is planted for the first time into a particular area, unless it is known that Cowpea rhizobia are present—inoculation of seed is discussed in **1Fe**, page 54. The flower stalks cannot penetrate a hard soil crust. The plant cannot tolerate waterlogging.

Seed rate: the average is 30–60 kg/ha, depending on seed size, to produce about 150,000 plants per hectare. The maximum is about 190 kg/ha, for closely spaced plants grown in a pure stand. There are about 1300–2000 seeds per kg.

Spacing: either in single rows about 45 cm apart with 5–15 cm between plants, or in double rows 8 cm apart on flat ridges about 90 cm apart.

Depth: 5–7.5 cm

Intercropping: sometimes with pearl millet, root crops or other legumes.

Seed treatment: Thiram or an equivalent is recommended for shelled seed. Sometimes the whole pod is planted.

GROWTH CONDITIONS

Growth period: 90–120 days for bunch types, 120–150 days for spreading types. The entire plant is pulled up and dried in the sun and wind, the pods being attached with tough, wiry stalks.

Temperature: the optimum is 20–28°C. Bambara Groundnuts prefer hot sunny climates and need a frost-free period of 100–120 days.

Rainfall: Bambara Groundnuts are one of the most drought resistant of the legumes. The optimum rainfall is about 900–1200 mm/a., though they can grow with 600–750 mm/a. They also tolerate heavy rainfall unless this occurs when the plants are mature.

Altitude: 0–1600 m

Rotation: often planted as the first crop in a rotation, followed by cassava, or after groundnuts or other legumes when their yields become very low.

Pests: very few, though root-knot nematodes, leaf-hoppers, crickets and rodents can sometimes cause some damage.

Diseases: also very few, though wilt, leaf spot and leaf virus can sometimes reduce yields, especially in humid conditions.

YIELD

Normal yields of Bambara Groundnuts are about 550–850 kg/ha, but in ideal conditions 3–4 MT/ha are possible.

The FAO estimate for the average global yield in 2004 was 779 kg/ha, the highest national average being recorded in Burkina Faso (1 MT/ha) and the lowest average in Mali (400 kg/ha).

UTILISATION
- **Grain**—eaten in various ways, either ripe (mature), pounded into flour and mixed with oil or butter to make a kind of porridge, or roasted in oil; or green (immature), eaten fresh or boiled or grilled. In Zimbabwe and Ghana the grain is canned on a commercial scale. The mature grain can be fed to animals after it has been soaked in water.
- **Whole green pods**—can be washed and boiled, and eaten either as a vegetable or used in soups. Easier and quicker to prepare than the mature grain.
- **Roasted, ground meal**—can be used as a coffee substitute.
- **Haulm**—a good animal food, containing about 16% protein.

LIMITATIONS
- Limited availability of tested, improved varieties of Bambara Groundnuts, and indeed of any seed at all.
- Many of the available varieties are only adapted to a small geographical area, so new varieties introduced to an area should be thoroughly tested first.
- During storage the grain is susceptible to both insect and fungal attack.
- The oil content is low, about 6% (4.5–6.5%).
- The mature dry grain is indigestible and hard and must be boiled for a long time or ground into a powder before it is edible.

Buffalo Gourd
Cucurbita foetidissima

Coyote Gourd, Fetid Gourd, Missouri Gourd, Prairie Gourd, Stinking Gourd, Wild Gourd, Wild Pumpkin, Mock Orange. Calabazilla (Amarga), Chilicote (Spanish)

North American Indians living in hot dry regions have used the Buffalo Gourd for thousands of years, as a food source, for washing clothes, cleaning animal skins, as ritualistic rattles and others (see "Utilisation").

The potential of this plant to produce oil and protein in very marginal areas has been recognised in recent years. Buffalo Gourds can produce similar yields to soybeans or groundnuts, but they do so in extremely hot and arid regions where conventional crops would die. The plants can survive with virtually no cultivation or effort by man.

The flowers are bright yellow, up to about 10 cm wide, pollinated by insects. Each plant produces about 50 hard-shelled, spherical fruits, either mainly yellow or mainly green with yellow stripes or markings, which are inedible but contain pulp and white, flat seeds. The seeds are about 12 mm long and 7 mm wide and contain 30–35% protein and up to 34% oil; seed yields of 2500 kg/ha are easily obtained.

Buffalo Gourds are also a form of root crop; the tubers can grow to an enormous size, up to 150 kg, which may reach down 5 m or more in search of water. The roots contain about 56% starch. The plant can therefore be viewed as a dual-purpose crop, being both a root crop and an oilseed crop.

The Buffalo Gourd is a member of the *Cucurbitaceae* family (the gourds, squashes and melons), and is a vigorous perennial that can survive for 40 years or more. It grows wild in Mexico and in the southwest of North America.

The plant can produce either very high or very low yields, without apparent reason—very little research work has so far been done on this potentially useful food crop. Further details online at: http://medplant.nmsu.edu

PLANTING
Propagation: in three ways: sexually, by seed; vegetatively from nodal roots, or vegetatively by fixing the long stems (vines) to the soil and watering them, rapidly forming new roots and plants. Seed takes about 2 weeks to germinate.
Soil: should be dry and well drained. Buffalo Gourds grow well in sandy and infertile soils, and can be useful as a soil binder to reduce erosion by wind and water.

GROWTH CONDITIONS
Growth period: seed can be harvested annually for up to 40 years.
Temperature: frost sensitive. The plants need long periods of hot, dry weather.
Rainfall: extremely drought resistant; once the plants are established they can survive with virtually no rain.
Pests: resistant to most pests, including cucumber beetles and squash bugs.

YIELD
Each fruit contains about 12 g of seed; on the basis of 60 fruits per plant, one hectare of plants can produce 2.5 tons of seed, containing 30–35% protein and up to 34% oil.

The root is enormous and can weigh up to 30 kg after just two growing seasons. Older plants can produce roots weighing up to 150 kg, of which about 70% is water.

UTILISATION
Buffalo Gourd is a dual-purpose crop, being both a root crop and an oilseed crop, but it also has many other uses:
- The seed can be crushed to obtain the edible polyunsaturated oil, used both as food and in industry and cosmetics. Seeds are also effective as a vermifuge (agent that causes the evacuation of intestinal worms).
- The pulp from undried fruit is used as cattle food.
- Traditionally, North American Indians used the seeds for food and soapy extracts of the fruit pulp and vine for washing clothes and cleaning animal hides.
- The Navajo used the dried gourds as ritualistic rattles.
- The roots can be used as laundry soap and shampoo. The roots can also be dried and used as firewood, in Afghanistan for example.
- The crushed leaves are used as an insecticide.
- The plant is used as an ornamental for its colourful fruits.

LIMITATIONS

- Buffalo Gourd plants require long periods of warm dry weather for optimum growth.
- they are sensitive to frost and intolerant of wet, poorly drained soil.
- there is wide variation in yield between plants, often without apparent reason.
- in common with most other plant protein, the protein contained in Buffalo Gourds is low in lysine and the sulphur containing amino-acids.
- the meal, or presscake, has a high phytic acid content and may also contain excessive saponins or other toxic substances.
- the plant is rather unattractive and some leaves have a stale, nauseating smell (hence "*foetidissima*").
- there is a shortage of research material on the plant and its agricultural requirements.

Leucaena

Leucaena leucocephala (formerly classified as *L. glauca*)

Horse Tamarind, Leadtree, White Popinac
Ipil-ipil, Lepile, Bayani (Philippines); Lamtoro (Indonesia); Guaje, Yaje, Vaxin (Central and South America); Koa-haole (Hawaii); Hediondilla (Puerto Rico); Tangatan (Guam)

The name "leucaena" may be pronounced as either "loo-see-na", "loo-kee-na", "loo-kay-na" or "loo-kuy-na". It is a productive, persistent and palatable leguminous tree or bush which originated in Central America and is now grown in almost all of the tropical and subtropical countries of the world.

It is not a human food crop, but is included in this handbook because of its significant potential as a multi-purpose crop.

Under optimum conditions Leucaena can grow incredibly fast, and for this reason as well as its multiple uses it has been called the "**Miracle Tree**". However it will only grow well in areas with appropriate soils, climate and altitude; the appropriate type and variety is also needed.

There are three main types of Leucaena:
- **Hawaiian**—1–5 m tall with very many branches. Starts to flowers at 4–6 months, then continues all year round. Very prolific (many seeds), and may become an aggressive weed. Main uses: erosion control, firewood, charcoal, shade crop and grazing (mainly cattle).
- **Salvador**—up to 20 m tall, with few branches on the lower part of the trunk. Also called "Arboreal" or "Guatemala" type. Plant breeders have selected many varieties of this type, known as "Hawaiian Giants" or by the letter K and a number, eg K8, K72, K132, etc. Main uses: timber and wood products, industrial fuel.

- **Peru**—up to 15 m tall, with many branches even low down on the trunk. A quite recent discovery, so only relatively few improved varieties are available. Mainly used for grazing.

PLANTING

Seed treatment: the seed has a covering (testa) which is hard, waxy and almost impermeable to water, so to ensure good germination the seed should be treated first. There are many methods; for small quantities the seed can be scarified by cutting a small "nick" in the side of the seed (not the base) with a knife or nail-clippers, or by rubbing it with sandpaper. For larger quantities, either immerse seed in 80°C water for 3 minutes or immerse the seed, in a mesh bag, in an equal volume of boiling water. Remove heat at the same time as the seed is immersed, remove bag after 3 minutes. In both cases the seed should be rapidly cooled in cold running water, then dried completely on a concrete or other hard floor, or in hessian bags.

Inoculation: a suitable strain of Rhizobia should be used if available. Also the seed should be lime-pelleted if sown in soils below pH 5.5, or if sown in contact with superphosphate.

Propagation: by seed, but Leucaena can also be propagated with cuttings or grafts. The plants coppice readily, producing what are known as "ratoons" or "pollards".

Soil: not well suited to acidic soils less than pH 5, nor poorly drained soils. Leucaena needs calcium and grows well on deep, calcareous or clay soils. Very salt tolerant, but sensitive to aluminium. Needs zinc, molybdenum and other nutrients (see "Limitations", below).

Seed rate: 0.5–5 kg/ha depending on type, row spacing, soil type, etc. 20–30,000 seeds per kg.

Spacing: the Salvador types are often grown very widely spaced to allow maximum growth.

For grazing, 1.5–5 m between rows, giving 75,000–100,000 plants/ha.

For timber, about 2 m × 2 m, giving 2500 plants/ha.

Depth: 2.5–5 cm

Intercropping: for grazing, grasses are usually planted, such as green panic, setaria, Rhodes grass or kikuyu grass. For soil enrichment, erosion control, etc. Leucaena can be intercropped with almost any crop.

GROWTH CONDITIONS

Growth period: a perennial. Seedlings grow slowly—less than 30 cm in the first six weeks.

Rainfall: optimum is 600–1700 mm per year, but can sometimes survive with 250 mm per year.

Temperature: grows best in full sun in hot places. Loses leaves (it "defoliates") with even light frosts, but soon recovers.

Altitude: best below 500 m; Leucaena does grow at high altitudes, but with less vigour. Latitude seems to make a difference to which altitudes are suitable.

Pests: rarely a problem. Seed Weevils attack young pods and eat the developing seed. Termites may attack young seedlings. Wild game, goats, rats, etc are also fond of eating young seedlings, which may have to be fenced.

Diseases: seedlings sometimes die ("damping off"), but diseases are rarely a problem.

Toxicity: Leucaena contains a toxic amino-acid, **mimosine**; this is most concentrated in young, growing plant tissue—shoot tips, for example, have 10%, while very old leaves have 1%. This may be a problem for grazing animals if Leucaena makes up more than about 30% of their diet for some weeks.

Symptoms: loss of hair or wool, goitres, loss of appetite, weight loss, excessive saliva, alopecia, growth retardation, cataracts and infertility in animals.

YIELD

Grazing: 2–20 MT/ha dry matter, equivalent to 430–4300 kg/ha of protein.

Timber: Salvador types can grow 4 m tall in six months, 9 m in two years and 17 m (25–35 cm diameter) in six years.

UTILISATION

- **Animal forage**.
- **Wood**—for construction, pulp and paper.
- **Fuel**—good firewood and charcoal (high calorie content). The rotation period is about seven years in the subtropics.
- **Soil improvement**—leaves which fall to the ground are equivalent to manure in Nitrogen content, so Leucaena greatly benefits other plants which are inter-cropped with it. Sometimes the leaves and young branches are cut off, then carried away and incorporated into other fields. The root system is aggressive, with a long tap root; this breaks up the topsoil and so improves water retention and reduces erosion, while nutrients are brought up from deep down where they would have remained beyond reach of other crops.
- **Reforestation** and **windbreaks**.
- **Young pods and seeds** (40–45% protein) are eaten by humans.
- **Shade crop**, for crops such as coffee and cacao.
- **Support crop**, for climbing crops such as Lima beans.
- In shifting cultivation, Leucaena can be **planted after the final crop**, which reduces the fallow period so that the next crop can be planted earlier.

LIMITATIONS

- Mimosine toxicity—see "**Toxicity**", above.
- Leucaena normally only grows well below about 500 m above sea level.
- The plants need a fertile, well drained, non-acidic soil with low aluminium levels and high levels of phosphorus, calcium, sulphur, molybdenum and zinc.
- The seedlings grow very slowly and so are often smothered by weeds.
- Despite this, the plants can grow so aggressively that they become a weed.
- Tsetse flies can breed in the plants, especially the more bushy Hawaiian types.

Lupin
Lupinus species

Lupine, {Texas} Bluebonnet (*Lupinus subcarnosus)* Lupin (French); Wolfsbohne (German); Lupino (Spanish & Italian)[1]; Tremoço{seed or plant}, Tremoceiro{plant} (Portuguese); Turmas (India); Turmus (Arabia)

[1] In the Andes *L. mutabilis* (Syn. *L. taurus*), the Pearl Lupin, has been cultivated for at least 1500 years and is known as Tarwi. Other names used for the Pearl Lupin include Taura, Tarin, Tarhui, Altramuz, Choco and Ullu.

Lupins have great potential as a high protein source in temperate and cool subtropical regions, as food for both humans and animals. Most species also have a high oil content, especially the Pearl Lupin which contains 12–24%. They are also widely grown to be used as green manure, a topic discussed in **Section 1Hd**, page 69.

Lupins are in the genus *Lupinus* of the pea family *Leguminoseae* (or *Fabaceae*). The name "lupin" comes from the Latin word for wolf, in the mistaken belief that the plants depleted or "wolfed" nutrients from the soil.

A great deal of plant breeding effort has been devoted to developing varieties of lupins with a low alkaloid content, to produce "sweet" lupins. The so-called "bitter" lupins contain 0.3–3% of alkaloid and are generally toxic to both humans and animals if eaten in the raw state. Before consumption the grain must be laboriously prepared, often by soaking in water and then cooking thoroughly.

The cultivated grain lupins are annuals. The seed normally germinates rapidly to produce vigorous fast growing seedlings. When the first flower head is formed, the main stem develops lateral branches which also form flowers; these in their turn produce more lateral branches, the process continuing indefinitely, producing flowers, pods, seeds and leaves whose numbers increase in geometrical progression.

Lupins have a strong taproot that penetrates deeply into the soil, and a well-developed root system. They are capable of producing 50 MT/ha of vegetation, containing 1.75 MT/ha of protein.

FAO estimated a worldwide production of 930,000 MT for the year 2004.

There are more than 300 species of Lupin, which can make their identification rather difficult. The Pearl Lupin and the White Lupin have been cultivated for thousands of years, in the Andes and the Mediterranean respectively. Today there are about seven species that are important in food production, as follows:

White (Egyptian) Lupin—*L. albus* (Syn. *L. termis*). Grown in the Mediterranean, Upper Nile, Madeira, Canaries and sometimes in central and southeast Europe, Georgia, Ethiopia, South Africa, Australia, southeast North America and South America. Flowers are white, tinged with blue or violet, and are not scented. Pods up to 13 cm long, large off-white seeds. 30–150 cm tall.

Pearl Lupin (Tarwi) —*L. mutabilis* (Syn. *L. taurus*). See above. Mainly found in the Andes. Determinate and indeterminate forms are found. Flowers are normally blue, with white or sometimes yellow marks, which turn violet and then brown before dying. The grain contains 12–24% oil and up to 50% protein, comparable to soybeans, but also contains bitter tasting toxic alkaloids. 90–180 cm tall.

Blue (Narrow-leaved) Lupin—*L. angustifolius*. Grown commercially in northern Europe, New Zealand, southeastern North America, France, South Africa and southwest Australia. Flowers light to dark blue, tinged with purple, or sometimes pink or white. 20–150 cm tall.

Yellow Lupin—*L. luteus*. Grown to a limited extent on sandy soils in northern Europe, South Africa, Australia, Spain, Portugal and the Mediterranean coast. Flowers are a bright golden yellow. 20–80 cm tall.

Sand Plain or **West Australian Blue Lupin**—*L. cosentinii*. Grown in northwest Africa, southwest Spain, south Portugal and Australia. Flowers are bright blue, with a yellow spot on the standard. Very vigorous growth. 20–120 cm tall.

Egyptian Lupin—*L. termis*. Grown in Sudan on flooded land which is too hard or saline for other crops. The grain is mildly bitter and poisonous when raw, but is edible after soaking and boiling.

The entire global production of lupin seed constitutes less than 2% of the total reported production of grain legumes, and they are normally grown for human food only in subsistence farming situations. The major producers are Russia, Poland, Australia, South Africa and Italy.

PLANTING

Soil: lupins generally prefer well drained, acid to neutral soil (classified as "tolerant" to soil acidity), without free lime, though each species has slightly different needs:

– *L. albus*—the most susceptible to waterlogging, though it tolerates some salinity.
– *L. mutabilis*—tolerates sandy and acid soils.
– *L. angustifolius*—best on moderately acid/neutral soils, and needs plenty of P and K.
– *L. luteus*—better adapted to infertile soils, with pH below 6.5–7.
– *L. cosentinii*—susceptible to Molybdenum deficiency, fairly tolerant of infertile soils.

All species of Lupins need plenty of phosphate and sulphur; about 200 kg/ha of 22% superphosphate is often recommended, with an additional 50 kg/ha of potash in soils where potassium levels are low.

Lupins are an excellent green manure crop, and can fix 400 kg/ha or more of Nitrogen. This crop could be much more widely grown in various crop rotation systems, both to increase the Nitrogen level and the organic matter content of soils.

Seed rate: some examples have been quoted, in kg/ha:
L. angustifolius—67–90 (North America), 67–78 (South Africa).
L. albus—179 (North America), 106 (South Africa).
L. luteus—45–67 (North America), 67–78 (South Africa).
An average seed rate of about 90 kg/ha can be used as a general guide. Approx. 3–4,000 seeds/kg.

Spacing: 10–20 cm between plants, 15–20 cm between rows. The seed is often broadcast, especially when sown as a green manure crop.

Rotation: should not be grown on the same land more than once in four years.

Depth: 2.5 cm, or even less in retentive soils with plenty of moisture.
Inoculation: recommended unless a well nodulated lupin crop was grown the previous year, but note "**rotation**" above ie only grow lupins in a 3–4 year rotation with other crops.

GROWTH CONDITIONS

Day length: most varieties are long-day, though this aspect is not well researched.
Growth period: about 105–180 days for most improved varieties. Some older types need six months or more in which to mature.
Temperature: lupins are semi-hardy and generally need a five month period with average temperatures of 15–25°C. The optimum for growth is 18–24°C. There is variation between spp.:
– *L. albus* and *L. mutabilis*—fairly frost tolerant.
– *L. angustifolius*—best in cool conditions, and can withstand minus 6°C in the vegetative state.
– *L. luteus*—only tolerates light frosts.
– *L. cosentinii*—susceptible to frost and needs heat for good growth.
Rainfall: lupins should have 400–1000 mm of evenly distributed water, as they suffer during any period of extended drought. *L. cosentinii* is the most drought resistant, and *L. mutabilis* also has some drought resistance.
Altitude: in Kenya lupins grow between 1500 and 2400 m. In the Andes *L. mutabilis* grows at 1800–4000 m.
Pests: Aphids—especially on sweet varieties of *L. luteus*, they attack at the bud stage; they also transmit virus diseases. Can be controlled with sprays (eg Bidrin) though reinfestation can occur.
– **Budworm (Earworm)**—the larvae enter pods and eat the seeds. Controlled by early planting, using early varieties and sprays (eg Dipterex or carbaryl compounds). *L. luteus* is the most susceptible to attack, and *L. angustifolius* the least susceptible. Other pests which can create problems are: Thrips, Red-legged Earth Mites, Lucerne Fleas, Root Weevils, Lupin Maggots, White-fringed Beetles, Grasshoppers, Root-knot nematodes, slugs and snails.
Diseases: these can be serious and cause large loss of yield. Lupinosis can be big trouble:
– **Lupinosis**—caused by *Phomopsis leptostromiformis* fungus which can produce a toxin that damages the liver of ruminants which eat the infected seed (or stems); sometimes fatal. The danger increases in the days following heavy rain. Controlled with resistant varieties and/or fungicides such as benomyl just before pod formation.
– **Brown Leaf Spot**—especially in cool, humid conditions. *L. albus* is very susceptible. The disease is seed-borne, and also survives on crop residues, which should be removed and destroyed. No effective control methods are available.
– **Anthracnose** and **Mildew**—especially in warm, humid conditions. Some varieties have some resistance.
– **Grey Leaf Spot**—especially with *L. angustifolius*. Neither *L. albus* nor *L. luteus* are infected.

– **Root Rots**—several kinds can cause problems.
– **Virus Diseases**—there are also several, including BYMV (Bean Yellow Mosaic Virus), which also infects beans and peas. Symptoms: a mosaic mottling, leaflets are distorted and the plants have a bunched, distorted growth. With *L. angustifolius* the growing point bends and blackens, and the plant dies if it is infected before flowering. Controlled with clean seed, early planting and aphid control.

YIELD

The yield potential of lupin grain is more than 5 MT/ha, though in many areas the average yields are no more than 500–600 kg/ha, due mainly to excessive flower drop. Yields in South Africa have been reported as varying from 290 to 1700 kg/ha, the difference often being attributed to planting date.

The FAO estimated average yield world-wide in 2004 was 1.2 MT/ha, varying from the highest national average of 2.7 MT in Chile to the lowest national average in Syria of 480 kg/ha.

With good growing conditions, yields of 1.5–2 MT/ha should be obtained.

UTILISATION

- **Seed/grain** of lupins can be used for either animal or human food: for humans the grain is soaked in water for some time, rinsed and then boiled. The most commonly used species is the White (Egyptian) Lupin *L. albus*. Pearl Lupin is also grown and eaten, mainly by the rural population of the Andes. Lupins can also be used as a substitute for coffee or eggs, or as a source of asparagine for the production of tuberculin. For animals, the concentrate is a protein source in cattle, sheep, pig and poultry food, or as a substitute for groundnut cake, fishmeal or soybean meal.
- **Flour** is used to fortify bread, or to substitute for soybean flour in meat products, noodles, pasta, bakery products, etc.
- **Whole plant** is used as a green manure crop, or for animal fodder, either grazed or for hay or silage. In Peru lupins are planted around pea and bean fields as a bitter hedge plant to discourage animals from grazing. In New Zealand and elsewhere lupins are used to enable trees to become established in very sandy regions, for erosion control, etc; the sand is first stabilised with Marram grass and other grasses, and lupins are then planted into the grass to enable the tree seedlings to survive.
- **Stubble**, the lower part of the stems left in the field after harvest, can be suitable for animal grazing, although whenever the plant stems of lupins are eaten there is always the possibility that lupinosis will develop.

LIMITATIONS

- **Alkaloids** are often present in the grain of older, "unimproved" and bitter varieties of lupin. These can be removed, but even after the grain has been steeped in water for some days and then cooked there is always some uncertainty as to the amount of insoluble alkaloids that remain.

- **Yield** of grain is often rather low, often due to poor seed set and other fertility problems. If beehives are placed in the lupin fields, pollination is usually greatly improved. Production of green matter, for green manure or animal food, is however often good.
- **Market prices** are often much lower for lupins than for cereals or other legumes.
- **Lupinosis**—see page 280, under "**Diseases**".
- **Shattering**, or "dehiscence", can be a problem with indeterminate varieties, especially *L. mutabilis* the Pearl Lupin (Tarwi). Even if the pods do not lose their seed from shattering, they tend to ripen over a long period so that the lower pods are mature while the upper pods are still green. This can be a problem for mechanised harvesting, but for subsistence farmers it may be an advantage as the supply of food continues for a long time.
- **Cross-pollination** occurs with many species, so that maintenance of pure strains of seed is difficult for seed producers. The rate of cross-pollination is about 20–25% for *L. luteus* Pearl Lupin and 10% for *L. albus* White (Egyptian) Lupin.
- **Adaptation** of many varieties is not very wide; they can be very specific in their soil and temperature requirements.
- **Diseases** are sometimes a serious problem; this fact may exclude lupins from being grown in certain areas, especially humid ones.

Tepary Bean
Phaseolus acutifolius var. *acutifolius* (Syn. *Phaseolus acutifolius* var. *latifolius*)

Pavi, Pawi, Rice Haricot Bean, Teparies, Texan/Texas Bean (North America) Tepary Bohne (Germany); Escomite (Spanish); Dinawa (Africa—also used for Cowpeas); Frijol Trigo (Chile); Garbancillo Bolando (Mexico); Haricot Riz (Algeria); Haricot Sudan (Senegal), Yori Mui

The Tepary Bean occurs wild in Arizona and northwestern Mexico, where it was cultivated by the Aztecs 5000 years ago. It is still grown to some extent in these areas, and also in some hot, dry parts of Africa.

It can be a useful crop if a rapid food supply is needed in areas with low rainfall and high temperatures, where the tepary bean can be grown as a catch crop. It will often produce some yield where other legumes would fail, and provides food for both humans and animals.

The plant is an annual, with pointed trifoliate leaves. Wild types have vines up to 10m long to enable the plant to climb desert shrubs. Cultivated types or mainly bush type (or semi-viny) about 30 cm tall. Several named varieties are available.

Dried beans contain approximately 9.5% water, 22–25% protein, 1–1.4% fat, 57–66% carbohydrate, 3.4–4.5% fibre and 4.2% ash.

The beans are almost round or oblong, about 8 × 6 mm, average weight 0.15 g, not glossy, white, yellow, brown, or deep violet, either entirely coloured or flecked. The fresh seeds absorb water easily, though the testa hardens during storage.

When grown as animal fodder, a reasonable crop of tepary beans can produce between five and ten MT/ha dry weight of hay.

PLANTING
Propagation: by seed, when soil temperature reaches about 70°C. Seed takes about 9 days to germinate. The flowers are thought to be self-pollinated. Inoculation: similar procedures and rhizobia to lima bean inoculation.

Soil: must be well drained. Not heavy clays. Moderately tolerant of saline soil, and tolerant of alkaline soils.

Seed rate: 11–17 kg/ha when sown in rows, 28–34 kg/ha when broadcast and 65–70 kg/ha when grown as a forage crop. 6000–7000 seeds per kg (approx. 0.15 g each).

Spacing: 60–90 cm between rows, 7.5–25 cm between plants. Sometimes 3 or 4 seeds are sown on mounds about 45 cm high. Can be grown in a 10 cm square grid.

Depth: 1.0–10 cm—varies according to soil type, moisture and the variety of bean.

GROWTH CONDITIONS
Day length: short-day (some day-neutral varieties exist).

Growth period: 60 days for early varieties in the tropics, 70–90 days for most varieties, up to about 120 days in cooler regions.

Temperature: adapted to hot, dry conditions with bright sunshine. Intolerant of frost. Night temperatures should not be lower than about 8°C.

Rainfall: very drought resistant, and by nature of its rapid growth tepary beans often "escape" from long drought periods. 500–600 mm per year or less is enough. However, moist soil is needed for germination and early growth. Tolerates heavy bursts of rain, but not much more than about 1000 mm per year. Grows well under irrigation, best done three times before flowering. Not suited to the wet tropics.

Altitude: in Mexico and Arizona, the homeland of the tepary bean, it is grown in the middle altitudes. When grown on the coast of Algeria it was observed that the growth period was extended.

Pests: tepary beans are fairly resistant to insect attack. The Black Bean Aphid can be a problem, but is easily controlled with nicotine sprays. In storage it can be attacked by the Rice Weevil. It seems to be much more resistant to the other common storage pest the Bean Weevil.

Diseases: rarely a problem. Rhizoctonia Root Rot may occur in wetter conditions.

YIELD
Seed yields of tepary beans are modest, as you would expect, since the crop is normally only planted in unfavourable conditions. Not only that, it produces its seed in record time, less than two months in some cases.

In Uganda yields of 450–770 kg/ha have been reported, while in North America dryland crops gave 500–780 kg/ha and irrigated crops 900–1680 kg/ha. Yield and production statistics for this crop are not available from FAO.

Fodder yields are around 5–10 MT/ha of dry hay.

UTILISATION

- The main use for the tepary bean is for the dried seed, which is similar to the haricot bean (*Phaseolus vulgaris*) in many ways, though it is generally less palatable. In Uganda the seed is normally boiled, then ground and added to soup. In Mexico the seed is sometimes soaked in water to produce a gelatinous extract used in soup preparation.
- The tepary bean is a fast growing catch crop (page 62) which can also be used for green manure (page 69), as a cover crop (page 63) and for animal fodder.
- After threshing the pods and haulm—which contain about 10% protein—can be fed to animals.

LIMITATIONS

- Tepary beans are very labour intensive at harvest time. The pods dehisce readily.
- The seed becomes very hard during storage and needs to be cooked for a long time, and unfortunately fuel is often in short supply in the very areas where the tepary bean would grow well.
- Food prepared from tepary bean often has a strong flavour, and may also have a bad smell ie less palatable than haricot beans.
- The plants need hot, bright sunshine with no frost nor temperatures at night below about 8°C.

Winged Bean (Four-angled Bean)
Psophocarpus tetragonolobus

Asparagus Pea or Bean, Dragon Bean, Four-cornered Bean, Goa Bean, Mauritius Bean, Manila Bean[1], Princess Pea, Wing Bean, Winged Pea; Haricot Dragon, Pois Ailé, Pois Carré[2] (French); Goabohne (German); Sesquidilla, Judía Careta (Spanish); Fava de Cavalo[3] (Portuguese); Too-a-poo, Tua Pu (Thailand); Amali, Batong-baimbing, Burma Haricot, Calamismis, Cigarillas, Sigarilya, Garbanso, Pallang, Parupa-gulung, Sabidokong, Segidilla, Sererella (Philippines)

[1]Also used for Marrowfat Peas (*Pisum sativum*) and sometimes for Grass Pea/ Chickling Pea/ Vetch (*Lathyrus sativus*).
[2]Also used for Bambara Groundnut (*Voandzeia subterranea*).
[3]Also used for small horse bean (*Vicia faba minor*).

The winged bean should not be confused with *Lotus tetragonolobus* (Syn. *Tetragonolobus purpureus*), also known as the winged pea or asparagus pea and which also has four-sided and four-winged pods. This plant grows wild in the Mediterranean and is occasionally grown in temperate regions for its young pods, eaten as a vegetable.

The winged bean is a fast growing perennial legume. It has particular value in the wet tropics where it provides not only oil and protein in its seed, but also has protein-rich tubers and edible pods and foliage. This species has excellent potential to provide protein in hot and humid conditions that are unfavourable to other leguminous crops. The crop also benefits subsequent crops planted since the roots have very many large root nodules, sometimes more than 1cm in diameter, which fix large quantities of Nitrogen.

Although winged beans are currently grown mainly as a subsistence or market garden crop in small areas, they have great potential as a source of protein and soil Nitrogen. They could be grown in very large areas which are too hot and wet for soybeans, although they are not a very suitable crop for mechanised agriculture.

All parts of the plant are edible: the seeds, pods, flowers, tubers and leaves can all be eaten, by both humans and animals. An edible oil is extracted from the seed, and the tubers have an exceptionally high protein content of 8–25% and up to 30% starch. Cassava, by contrast, has about 0.7% protein and sweet potatoes 1.2–1.5%.

The available varieties of winged bean are very variable, in shape, size and colour of flowers, pods and seeds. Some varieties produce both seed and tubers, others produce only seed.

The seed is about 1 cm long, smooth and shining and either white, yellow, brown or black in colour. It is a climbing perennial, but is normally treated as an annual. The stem can grow four metres high if it is supported.

The grain has excellent nutritive and cooking qualities, similar to soybeans, containing about 33% (29–37%) protein, 15–18% oil, 5% fibre and 32% carbohydrate. It is rich in lysine and is therefore a good supplement to a cereal based diet, which is deficient in lysine.

The tubers are also valuable, containing (@ 9% moisture content) 20–25% protein, 1% fat, 5.4% fibre and 56% carbohydrate.

The pods (per 100 mg of edible portion, @ 92% m.c.) contain 42 mg Calcium, 570 I.U. Vitamin A potency and useful amounts of other nutrients.

Nowadays winged beans are grown throughout the humid tropics, particularly in Papua New Guinea, southeast Asia and Sri Lanka. In Africa they are widely grown in Nigeria and Ghana, and they are also grown in the West Indies and southern Florida.

PLANTING

Propagation: winged bean seed is planted directly into the field. The seed coats are sometimes nicked with a file or sandpaper to improve germination rates and speed.

Soil: almost any soil which is not too saline or waterlogged, including soils with low Nitrogen. In clay soils however the tubers are less well developed.

Seed rate: about 4–6 kg/ha. 100 seeds weigh about 30–40 g ie 3000–4000 seeds per kg.

Spacing: very variable. For pods and seed the plants are normally grown up stakes, while for tubers they are left unstaked, and sometimes treated as perennials.

Two or three seeds are normally planted on hills, at spacings of between 60 × 60 cm and 4 × 4 m. Average is about 1.3 m × 0.6 m for seed and pod crops. For tuber crops, spacing is about 60 × 10 cm, without supports.

The plants prefer to be in full sunlight.

Depth: 2.5–6 cm.

Inoculation: winged beans are not normally inoculated as nodulation occurs naturally very readily, apparently with the Cowpea group of Rhizobia.

Intercropping: very common, with many different crops such as sweet potatoes, sugar cane, taro, bananas, green vegetables and other legumes.

GROWTH CONDITIONS

Day length: short-day. In long days, ie outside the tropics, older varieties tend to produce excessive vegetative growth and few flowers. Some recently developed varieties are day neutral, and so will set seed in higher latitudes.

Growth period: plants of some varieties can survive and produce for five years or more.

– pods—50–90 days for first pods, continuing for several weeks or months.

– seed—180–270 days. Some modern varieties mature in 110–120 days.

– tubers—120–240 days.

Temperature: winged beans require hot weather, with at least 180 frost-free days from planting seed to harvesting mature pods. After a frost the plants of some varieties will recover due to the large starchy tubers, provided these have not been frosted.

In Asia, winged beans are mainly grown between 20°N and 10°S.

Rainfall: if well distributed, 1500 mm per year is enough. Optimum is about 2500 mm or more per year. The crop is often irrigated, and does not tolerate long dry periods.

Altitude: 0–2100 m

Pests: rarely a problem, especially because winged beans are normally cultivated in mixtures in small areas. Bean Fly, caterpillars, leafminers, grasshoppers, spider mites and root-knot nematodes may cause some damage.

Diseases: also rarely a problem, though False Rust can be serious, in Papua New Guinea for example. Crown Rot and Leaf Spot also may infect winged beans.

YIELD

Little information is available, so the following may not be completely realistic.

Pods—harvested continuously over a long period. In trials in Malaysia more than 35 MT/ha were reported.

Seed (grain)—trial results:

Ghana 820–1380 kg/ha; IITA, Nigeria 2400 kg/ha; Malaysia 4580 kg/ha.

Tubers—2.4–11 MT/ha, increased when the flowers are removed, and when grown on hills (mounds) or when using varieties bred to produce tubers rather than pods and grain.

UTILISATION
- **Pods**—the winged bean's main use, they can be eaten raw, or sliced and boiled like haricot (French) beans, and used in soups and curries. They are pale green, 6-9" long and one inch wide when mature. The pods are square, with the four corners tapering into the thin wings. Protein content is about 2%.
- **Seeds**—nutritionally superior to groundnuts (see Introduction); sometimes roasted and eaten like groundnuts. They are very similar in composition to soybeans—and are more palatable—and could be used similarly in high protein foods, soap and cooking oil. In Indonesia, tempeh and tofu are made from the mature seeds. The half ripe seeds can be eaten raw, fried or steamed. **Flour** made from the grain is suitable as a milk substitute in treating kwashiorkor.
- **Presscake**, after oil extraction, is suitable for both human and animal food.
- **Tubers**—best eaten when they are about as thick as a thumb, either raw (peeled) or boiled like potatoes or roasted. They should be air dried for a few days, then peeled before cooking. To promote tuber development, the flowers are sometimes pinched off. The tubers are considered to be a delicacy in Myanmar. Exceptionally high protein content of about 25%.
- **Foliage**—the leaves and flowers can be eaten raw, or steamed and added to soups and curries. The flowers can be fried in oil, tasting like mushrooms. Both the stems and leaves make valuable animal food, with a protein content of about 6%.
- **Green manure** and **cover crop**—very useful for both purposes, especially due to its efficient Nitrogen fixation. In Myanmar, sugarcane was reported to have yielded up to 50% more when grown after a winged bean crop.

LIMITATIONS
- Shortage of seed of winged beans, especially of improved varieties with proven adaptability to different soils, climates and day length.
- Need to support the plants for the production of seed or pods. High labour requirement as a result, and there is also a need to supply stakes and other support material.
- Indeterminate growth habit ie long periods during which the pods and seeds are maturing. As a result the winged bean is not yet a suitable crop for large scale commercial planting. For subsistence farmers this is not a problem, in fact it is a useful quality as food is provided over a period of several months.
- The mature, dry seed (grain) must be cooked before eating, though pods and immature seed can be eaten raw without any ill effects.
- There is not enough practical, well researched agronomic information on the winged bean.

SECTION 3

3A. NAMING & CLASSIFICATION OF FOOD CROPS

This list of cultivated plant names is far from exhaustive, but it is included here in an attempt to minimise some of the misunderstandings that can arise when trying to identify the names of food crops. Confusion often arises with incorrect identification and naming of plant species, especially with legumes, and also when more than one language is involved.

Additional local names for the most common food crops are outlined in **Section 2, "Crop Descriptions—*Description and Characteristics of the Main Food Crops*"**.

Comprehensive details on plant taxonomy are available online at the USDA's website GRIN—the Germplasm Resources Information Network.

The following types of food plants are *not* included in the list below:
sugar crops, nut trees, oil palms, coffee, tea, beverages, stimulants, herbs, edible seaweeds, mushrooms or fungi, or wild food plants.

A. CEREALS AND PSEUDOCEREALS
Adlay *Coix lachryma-jobi*. Job's Tears
African Rice *Oryza glaberrima*. Arroz de Guinea (Spanish), Riz de Casamance (French)
Amaranths *Amaranthus* spp.—see **2G**
Barley *Hordeum vulgare*—see **2A**
Barnyard Millet *Echinochloa crus-galli (*Syn. *Panicum crus-galli)*
Bread Wheat *Triticum aestivum (*Syn. *T. sativum, T. vulgare)*. Wheat—see **2A**
Buckwheat *Fagopyrum esculentum (*Syn. *F. sagittatum)*—see **2A**
Cheena *Panicum miliaceum*. Common Millet, Proso Millet, Panic Millet etc—**2A**
Chicken Corn *Sorghum drummondii*. Shatter Cane, Sudan Grass
Durra (Sorghum) *Sorghum durra*
Durum Wheat *Triticum durum*. Macaroni Wheat—see **2A**
Einkorn (Wheat) *Triticum monococcum*—cultivated Einkorn, *T. boeoticum*—wild Einkorn
Emmer (Wheat) *Triticum dicoccum* and *T. dicoccoides*—wild Emmer
English Wheat *Triticum turgidum*. Rivet, Cone
Feterita *Sorghum caudatum*
Fonio *Digitaria exilis (D. iburua*—white-seeded form*)*. Acha, Fundi, Hungry Rice
Guinea Corn *Sorghum guineense*

Haraka Millet *Paspalum scrobiculatum (*Syn. *P. commersonii, P. polystachyum).* Kodo(a) Millet, Ditch Millet, Scrobic Millet, Ricegrass
Hegari—see **2A**, **"Sorghum"**
Hungry Rice *Digitaria exilis (D. iburua*—white-seeded form*).* Fonio, Fundi, Acha
Japanese Barnyard Millet *Echinochloa frumentacea (*Syn. *Panicum frumentaceum)*
Job's Tears *Coix lacryma-jobi.* Adlay (Millet). Larmes de Job (French); Hiobsträne (German); Juzudama (Japanese); Lágrimas de Job (Spanish), Lágrimas de San Pedro (Spanish)
Jungle Rice *Echinochloa colona.* Shama Millet
Kafir Corn *Sorghum caffrorum.* Sorgo Kafir (Spanish)
Kaniwa *Chenopodium pallidicaule* Canihua, Kuimi (Bolivia), Millmi (Bolivia), Ccañihua (Peru)
Kaoliang *Sorghum nervosum.* Chinese Sorghum, Brown-seeded Sorghum.
Kiwacha *Amaranthus caudatus*
Kodo (Koda) Millet *Paspalum scrobiculatum*
Maize *Zea mays*—see **2A**
Mands Forage Plant *Pennisetum typhoides.* Pearl or Bulrush Millet—see **2A**
Millets—see **2A** for local and botanical names of the various millet species.
Milo *Sorghum subglabrescens.* Sorghum—see **2A**
Oats *Avena sativa.* Yellow or White Oats—see **2A**
Perennial Buckwheat *Fagopyrum cymosum*
Quinoa *Chenopodium quinoa* Quingua. Petit riz, Riz du Pérou (French), Reismelde (German); Arroz del Perú, Quinua (Spanish).
Red Oats *Avena byzantina* Avoine Byzantine (French), Aveia Amarela (Portuguese), Avena Roja (Spanish), Mittelmeerhafer (German)
Rice *Oryza sativa*—see **2A**
Rye *Secale cereale*—see **2A**
Sanwa Millet *Echinochloa frumentacea (*Syn. *Panicum frumentaceum)*—see **2A**
Shallu *Sorghum roxburghii.* Indian Sorghum, Popping Sorghum. Shaaru (Japan)
Sorghum *Sorghum bicolor (*Syn. *S. vulgare)*—see **2A**
Spelt *Triticum spelta.* Espelta (Portuguese)
Tartary Buckwheat *Fagopyrum taricum* Green Buckwheat, India Wheat.
Ku Chiao Mai (Chinese), Ku Qiao (Chinese); Sarrasin de Tartarie (French); Tatarischer Buchweizen (German); Alforfón de Tartaria (Spanish)
Teff *Eragrostis tef (*formerly *E. abyssinica)*—see **2A**
Teosinte *Euchlaena mexicana* (Syn. *Zea mays* ssp *mexicana)*
Central Plateau Teosinte, Chalco Teosinte, Durango Teosinte, Mexican Teosinte, Nobogame Teosinte, Rayana Grass, Maíz Silvestre (Spanish)
Triticale *Triticosecale—**Triticum X Secale**—*see **2A**, **"Rye"**
Wheat *Triticum aestivum (*Syn. *T. sativum, T. vulgare)*—see **2A**
White Durra *Sorghum cernuum.* Jerusalem Durra, Egyptian Millet, Indian Grains.
Wild Rice *Zizania aquatica.* American Wild Rice, Canda Rice, Indian Rice, Water Oats—see **2A**

B. LEGUMES

Adzuki Bean *Vigna angularis (Phaseolus angularis)*. Atsuki, Azuki; Haricot Adzuki, Haricot à Feuilles Angulaires (French)

African Locust Bean *Parkia filicoidea*

Alfalfa (Lucerne) *Medicago sativa*. Rashqa (Dari)

American Groundnut *Apios americana* American Potato Bean, Ground Bean, wild Bean. Gland de terre (French); Erdbirne (German); Apio Tuberoso (Spanish)

Angola Pea *Cajanus cajan (*Syn. *C. indicus)*. Pigeon Pea—see **2B**

Arhar (Arhair) *Cajanus cajan (*Syn. *C. indicus)*. Pigeon Pea—see **2B**

Asparagus Bean/Asparagus Pea *Vigna sesquipedalis* (Syn. *V. sinensis* var. *sesquipedalis, Dolichus sesquipedalis*). Cowpea—see **2B**. A.k.a. Winged Bean—**2G**

Bambara Groundnut *Voandzeia subterranea* (Syn. *Vigna subterranea)*—see **2G**

Bengal Gram *Cicer arietinum*. Chickpea—see **2B**

Black-eye(d) Pea (Bean) *Vigna unguiculata*. Cowpea—see **2B**

Black Bean *Phaseolus vulgaris*. Haricot (French) Bean—see **2B**

Black Gram *Phaseolus mungo* (Syn. *Vigna mungo)* Mash, Urad, Urd, Woolly Pyrol. Mashang (Pashtu)

Blue Vetchling *Lathyrus sativus*. Grass Pea—see **2B**

Bodi Bean *Vigna sesquipedalis (*Syn. *V. sinensis* var. *sesquipedalis, Dolichus sesquipedalis)*

Bonavist(a) Bean *Lablab purpureus (*Syn. *L. niger, L. vulgaris, Dolichos lablab)*. Lablab, Hyacinth Bean, Egyptian Kidney Bean, Indian Butter Bean

Broad Bean *Vicia faba*—see **2B**

Butter Bean, Burma Bean *Phaseolus lunatus (*Syn. *P. limensis, P. inamoenus)*. Lima Bean—see **2B**

Cajan Pea *Cajanus cajan (*Syn. *C. indicus)*. Pigeon Pea—see **2B**

Chickling Pea (Chickling Vetch) *Lathyrus sativus*. Grass Pea—see **2GB**

Chickpea *Cicer arietinum*—see **2B**

China Pea (Bean)—see **2B**, "**Cowpea**"

Chufa *Cyperus esculentus*. Earth Almond, Tigernut, Water Grass, Yellow Nut Sedge/Grass. Amande de Terre, Chouta, Souchet Comestible (French); Erdmandel (German); Cebollín, Juncia Avellanada (Spanish)

Cluster Bean *Cyamopsis tetragonoloba*. Guar, Siam Bean (India); Cyamopse à Quatre Ailes (French); Mgwaru (Swahili)

Common Bean *Phaseolus vulgaris*. Haricot (French) Bean—see **2B**

Congo Goober *Voandzeia subterranea* (Syn. *Vigna subterranea)*. Bambara Groundnut—see **2G**

Congo Pea (Bean) *Cajanus cajan (*Syn. *C. indicus)*. Pigeon Pea—see **2B**

Cowpea *Vigna unguiculata* and other *Vigna* spp.—see **2B**

Crowder *Vigna* spp. Cowpea—see **2B**

Curry Bean *Phaseolus lunatus (*Syn. *P. limensis, P. inamoenus)*. Lima Bean—**2B**

Dragon Bean *Psophocarpus tetragonolobus*. Winged Bean, Four-angled Bean—**2G**

Earth Almond *Cyperus esculentus.* Chufa, Earth Almond, Tigernut, Water Grass, Yellow Nut Sedge, Yellow Nut Grass. Amande de Terre (French), Choufa (French), Souchet Comestible (French); Erdmandel (German), Chufa (Portuguese; Cebollín (Spanish), Juncia Avellanada (Spanish)

Earth Nut *Arachis hypogea.* Groundnut—see **2B**

Earth Pea(Nut) *Voandzeia subterranea* (Syn. *Vigna subterranea*). Bambara Groundnut—see **2G**

Earth-nut Pea *Lathyrus tuberosa.* Earth Chestnut, Earthnut Pea, Groundnut Peavine, Tuberous Vetch. Gesse Tubéreuse (French); Knollenplatterbse (German); Arveja Tuberosa (Spanish)

Field Bean *Phaseolus vulgaris* or *Vicia faba.* Haricot Bean or Broad Bean—**2B**

Field Pea *Pisum sativum* (Syn. *P. arvense*)—see **2B**

Four-angled (-cornered) Bean *Psophocarpus tetragonolobus.* Winged Bean—**2G**

French Bean *Phaseolus vulgaris.* Haricot Bean, Frijoles etc—see **2B**

Goa Bean *Psophocarpus tetragonolobus.* Winged Bean—see **2G**

Goober *Arachis hypogea.* Groundnut, Peanut—see **2B**

Golden Gram *Vigna radiata (Syn. Phaseolus aureus).* Mung Bean, Yellow Gram—**2B**

Grass Pea *Lathryus sativus.* Vetch—see **2G**

Green Bean *Phaseolus vulgaris.* Haricot (French) Bean—see **2B**

Green Gram *Phaseolus aureus* (Syn. *Vigna radiata*). Oorud Bean, Mung Bean—**2B**

Ground Bean *Voandzeia subterranea* (Syn. *Vigna subterranea*). Bambara Groundnut—see **2G**, and **Kersting's Groundnut** *Kerstingiella geocarpa*

Groundnut *Arachis hypogea*—see **2B**

Guar *Cyamopsis tetragonolobus.* Cluster Bean—see above

Haricot Bean *Phaseolus vulgaris.* French Bean—see **2B**

Hausa (Groundnut) *Kerstingiella geocarpa (Syn.Voandzeia geocarpa).* Kersting's Groundnut, Geocarpa (Groundnut), Ground Bean—also for Bambara Groundnut; Lentille de Terre, Fève de Kandela (French)

Horse Bean *Vicia faba.* Broad (Field) Bean—see **2B**. Also Jack Bean, see below

Horse Gram *Macrotyloma uniflorus (Syn. Dolichos uniflorus, D.biflorus).* Kulthi (Bean), Madras Gram, Horse Grain, Chickpea; Grain de Cheval (French)

Horse Tamarind *Leucaena leucocephala.* Leucaena, Leadtree, White Popinac—**2G**

Hyacinth Bean *Lablab purpureus (Syn. L. niger, L. vulgaris, Dolichos lablab).* Lablab, Bonavist(a) Bean, Seim Bean, Egyptian Bean, Indian (Butter) Bean; Ataque, Dolic (d'Egypte), Dolic du Soudan (French); Feijão Cutelinho, Feijão da Índia, Cumandatiá, Labelabe (Portuguese)

Indian Potato *Apios americana.* American Groundnut, Potato Bean

Indian Vetch *Lathyrus sativus.* Grass Pea—see **2B**

Jack Bean *Canavalia ensiformis.* Horse Bean, Common Jack Bean; Feijão Espada, Feijão de Cobra, Feijão de Porco, Feijão Holandês (Portuguese)

Jícama *Pachyrrhizus erosus.* Yam Bean (edible tubers and young pods)

Jerusalem Pea *Vigna radiata (Syn. Phaseolus aureus).* Mung Bean—see **2B**

Jugo(a) Bean *Voandzeia subterranea* (Syn. *Vigna subterranea*). Bambara Groundnut—see **2G**

Kaffir Pea (Bean) *Vigna* sp. Cowpea—see **2B**. Also name for Jugo Bean, above
Kerstings Groundnut *Kerstingiella geocarpa (*Syn. *Voandzeia geocarpa)*. Geocarpa (Groundnut), Ground Bean—also for Bambara Groundnut; Lentille de Terre, Fève de Kandela (French)
Kidney Bean/Red Kidney Bean *Phaseolus vulgaris*. Haricot (French) Bean—**2B**
Kudzu Vine *Pueraria lobata (*Syn. *P. thunbergiana, P. montana)*. Fan Kot
Lablab (Bean) *Lablab purpureus*—see Hyacinth Bean.
Lakh (Lakhori) *Lathyrus sativus*. Vetch, Grass Pea—see **2B**
Lathyrus Pea *Lathyrus sativus*. Vetch, Grass Pea—see **2B**
Lentil *Lens culinaris (*Syn. *L. esculenta, Ervum lens)*—see **2B**
Leucaena *Leucaena leucocephala (*formerly *L. glauca)*—see **2G**
Lima Bean *Phaseolus lunatus (*Syn. *P. limensis, P. inamoenus)*—see **2B**
Locust Bean *Parkia filicoidea (*Syn. *P. bussei)*. African Locust Bean, Nitta Tree
Long Bean *Vigna* spp. Cowpea—see **2B**
Lubia *Vigna* spp. Cowpea—see **2B**
Lupin *Lupinus* spp.—see **2G**
Madagascar Bean *Phaseolus lunatus (*Syn. *P. limensis, P. inamoenus)*. Lima Bean—see **2B**
Madagascar Groundnut *Voandzeia subterranea*. Bambara Groundnut—see **2G**
Manila Bean *Lathyrus sativus* Grass Pea (**2B**) or *Psophocarpus tetragonolobus* Winged Bean (**2G**) or *Pisum sativum* Marrowfat Peas
Marama Bean *Tylosema esculentum*. Gemsbok Bean, Tamami Berry
Marble Pea *Vigna* spp. Cowpea—see **2B**
Mash *Phaseolus mungo (*Syn. *Vigna mungo)*. Black Gram, Urd, Woolly Pyrol
Mat (Moth) Bean *Vigna aconitifolius (*Syn. *Phaseolus trilobus, P. aconitifolius)*. Moth Bean, Dew Bean Gram, Math, Mout Bean
Mauritius Bean *Psophocarpus tetragonolobus*. Winged Bean—see **2G**
Monkeynut *Arachis hypogea*. Groundnut—see **2B**
Mung Bean *Vigna radiata (*Syn. *Phaseolus aureus)*—see **2B**
Navy Bean *Phaseolus vulgaris*. Haricot (French) Bean—see **2B**
No-eye Pea *Cajanus cajan (*Syn. *C. indicus)*. Pigeon Pea—see **2B**
Nut Sedge *Cyperus esculentus*. Yellow Nutgrass. See Chufa, "Root Crops"
Pea *Pisum sativum* (Syn. *P. arvense)*. Field Pea, Garden Pea, English Pea—see **2B**
Peanut *Arachis hypogea*. Groundnut—see **2B**
Pearl Lupin *Lupinus mutabilis (*Syn. *L. taurus)*. Tarwi, Taura, Choco, Ullu—**2G**
Pigeon Pea *Cajanus cajan (*Syn. *C. indicus)*—see **2B**
Pillepesara *Phaseolus trilobus* (perennial). Simbí (Spanish)
Pindar *Arachis hypogea*. Groundnut—see **2B**
Pinto Bean *Phaseolus vulgaris*. Haricot (French) Bean—see **2B**
Pole Bean—name given to Haricot Bean, Broad Bean and Hyacinth Bean—see **2B**
Potato Bean *Apios americana*. American Groundnut, Indian Potato, Wild Bean
Potato Limas *Phaseolus lunatus (*Syn. *P. limensis, P. inamoenus)*. Lima Bean—**2B**
Princess Pea *Psophocarpus tetragonolobus*. Winged Bean—see **2G**
Rangoon Bean *Phaseolus lunatus* (Syn. *P. limensis, P inamoenus)*. Lima Bean—**2B**

Red Dahl *Lens culinaris (*Syn. *L. esculenta, Ervum lens).* Lentil, Split Pea—see **2B**

Red Gram *Cajanus cajan (*Syn. *C. indicus).* Pigeon Pea—see **2B**

Rice Bean *Vigna umbellata (*Syn. *Phaseolus calcaratus).* Climbing Mountain Bean, Mambi Bean, Oriental Bean, Haricot Riz (French), Reisbohne (German), Feijão Arroz (Portuguese), Frijol Mambe/Rojo/de Arroz (Spanish)

Rice Haricot Bean *Phaseolus acutifolius* var. *latifolius.* Tepary Bean—see **2G**

Runner Bean *Phaseolus coccineus (P. multiflorus).* Pole Bean, Multiflora Bean, Scarlet Runner Bean; Haricot d'Espagne (French); Feijão de Espanha, Feijão Flor (Portuguese)

Salad Bean *Phaseolus vulgaris.* Haricot (French) Bean—see **2B**

Sieva Bean *Phaseolus lunatus (*Syn. *P. limensis, P. inamoenus).* Lima Bean—**2B**

Snake Bean *Vigna sesquipedalis (*Syn. *V. sinensis* var. *sesquipedalis, Dolichus sesquipedalis)* Cowpea—see **2B**

Snap Bean *Phaseolus vulgaris.* Haricot (French) Bean—see **2B**

Southern Pea (Bean) *Vigna* spp. Cowpea—see **2B**

Soybean *Glycine max (*Syn. *G. soja, G. hispida, Soja max)*—see **2B**

Split Pea *Lens culinaris (*Syn. *L. esculenta, Ervum lens).* Lentil, Red Dahl—see **2B**

String Bean *Phaseolus vulgaris.* Haricot (French) Bean—see **2B**

Sugar Bean *Phaseolus lunatus (*Syn. *P. limensis, P. inamoenus).* Lima Bean—**2B**

Sword Bean *Canavalia gladiata.* Pearson Bean, Wonder Bean; Pois Sabre [Rouge] (French)

Tamarind *Tamarindus indica* Indian Tamarind, Kilytree. Tamarin, Tamarindier, Tamarinier (French); Tamarinde, Tamarindenbaum (German); Tamarindeiro (Portuguese); Tamarindo (Spanish)

Tarwi *Lupinus mutabilis (*Syn. *L. taurus).* Pearl Lupin—see **2G**

Tepary Bean *Phaseolus acutifolius* var. *latifolius*—see **2G**

Texan Bean *Phaseolus acutifolius* var. *latifolius.* Tepary Bean—see **2G**

Tick (Tic) Bean *Vicia faba* var. *minor.* Broad Bean, Windsor Bean etc—see **2B**

Tonka Bean *Dipterex* spp. Feijão Baru, Feijão Côco (Portuguese)

Tonkin Bean *Vigna* spp. Cowpea—see **2B**

Towe *Phaseolus lunatus (*Syn. *P. limensis, P. inamoenus).* Lima Bean—see **2B**

Tur *Cajanus cajan (*Syn. *C. indicus).* Pigeon Pea—see **2B**

Urd *Phaseolus mungo* Syn. *Vigna mungo.* Black Gram, Mash, Woolly Pyrol; Haricot Mungo, Ambérique (French)

Velvet Bean *Mucuna pruriens* var. *utilis.* Mauritius Bean—also for Winged Bean; Dolique de Floride, Haricot Velouté, Pois Mascate (French)

Vetch (Grass Pea) *Lathryus sativus*—see **2G**

White Popinac *Leucaena leucocephala.* Leucaena, Leadtree, Horse Tamarind—**2G**

Wild Bean *Apios americana.* American Groundnut, Indian Potato, Potato Bean

Windsor Bean *Vicia faba* var. *minor.* Broad Bean, Tick Bean etc—see **2B**

Winged (Four-angled) Bean *Psophocarpus tetragonolobus*—see **2G**

Yam Bean *Pachyrrhizus erosus* or *P. tuberosus.* Potato Bean, Jícama

Yardlong Bean *Vigna sesquipedalis.* Cowpea—see **2B**

Yellow Dahl *Cajanus cajan (*Syn. *C. indicus).* Pigeon Pea—see **2B**

Yellow Gram *Vigna radiata (*Syn. *Phaseolus aureus).* Mung Bean, Golden Gram—**2B**

C. OILSEED CROPS

African Oil Palm *Elaeis guineensis*—see Oil Palm
American Oil Palm *Corozo oleifera.* Dendêzeiro do Pará, Caiaué (Portuguese)
Babacu Palm *Orbignya martiana, O. oleifera*
Castor *Ricinus communis*—see **2C**
Coconut *Cocos nucifera.* Coco(tier) (French); Côco (Portuguese)
Cohune Palm *Orbignya cohune*
Field Mustard *Brassica campestris*
Flax(seed) *Linum usitatissimum.* Linseed—see **2C**
Goat Nut *Simmondsia chinensis*—see Jojoba
Indian Colza *Brassica campestris* var. *sarson*
Indian Mustard *Brassica juncea.* Leaf Mustard. Sharsham (Pashtu)
Indian Rape *Brassica campestris* var. *toria*
Jojoba *Simmondsia chinensis.* Goat Nut, Deer nut, Pignut, Wild Hazel, Coffeeberry, Quinine Nut, Gray Box Bush
Linseed *Linum usitatissimum*—see **2C**
Mahua *Madhuca longifolia*
Mole Bean *Ricinus communis.* Castor (Bean)—see **2C**
Mu-tree *Aleurites montana.* Tung, China Wood Oil; Noix d'abrasin (French)
Niger Seed *Guizotia abyssinica*—see **2C**
Oil Flax *Linum usitatissimum.* Linseed, Flax(seed)—see **2C**
Oil Palm *Elaeis guineensis.* African Oil Palm; Palmier à Huile (French); Dendê, Dendêzeiro, Coqueiro de Dendê (Portuguese)
Oilseed Rape *Brassica napus.* Swede rape
Olive *Olea europaea*
Palma Christi *Ricinus communis.* Castor (Bean)—see **2C**
Ramtil *Guizotia abyssinica.* Niger Seed—see **2C**
Rape *Brassica napus.* Colza (French, Spanish & Portuguese)
Safflower *Carthamus tinctorius.* Açafrão, Açafrão Bastardo (Portuguese); Kusum, Kurdi (Hindi); Suf (Ethiopia)
Sesame *Sesamum indicum (*Syn. *S. orientale)*—see **2C**
Sunflower *Helianthus annuus*—see **2C**
Tung *Aleurites montana & A. fordii.* China Wood Oil, Mu-tree; Noix d'abrasin (Fr.)
Turnip Rape *Brassica campestris (*Syn. *B. rapa).* Sarson, Toria

D. ROOT CROPS

Aerial Yam *Dioscorea bulbifera.* Air Potato, Bulbil-bearing Yam, Potato Yam
African Bitter Yam *Dioscorea dumetorum*
African Yam Bean *Sphenostylis stenocarpa.* Girigiri
Air Potato *Dioscorea bulbifera.* Bulbil-bearing Yam, Potato Yam
Aja *Dioscorea trifida.* Cush-cush Yam, Yampi, Indian Yam, Mapuey
American Bitter Yam *Dioscorea dumetorum.* Cluster Yam, Forest Yam
American Groundnut *Apios americana*

Angelica *Angelica archangelica*

Anu *Tropaeolum tuberosum*. Tuberous Nasturtium, Anyu, Quecha, Apina-mama, Mashua, Isanu, Ysano, Cubio

Ape *Alocasia macrorrhiza*. Giant Taro, Ta'amu—see **2D, "Taro"**

Arracacha *Arracacia xanthorrhiza (Syn. A. esculenta)*

Arrowhead *Sagittaria sinensis*. Tzi Koo (Chinese)

Arrowroot *Maranta arundinacea*. Bermudan or West Indian Arrowroot; Araruta (Portuguese)

Asiatic Bitter Yam *Dioscorea hispida*

Beetroot *Beta vulgaris* ssp. *vulgaris* Beterraba (Portuguese**)**

Brazilian Arrowroot *Manihot esculenta* (Syn. *M. utilissima, M. aipi, M. dulcis, M. palmata)*. Manioc, Mandioc, Cassava, Yucca—see **2D**

Bulbil-bearing Yam *Dioscorea bulbifera*. Air Potato, Aerial Yam, Potato Yam

Burdock *Arctium lappa*

Bush Yam *Dioscorea persimalis*

Camas *Camassia esculenta*

Capucine *Tropaeolum tuberosum* Mastouche tubéreuse (France), Mayua (Peru)

Carrot *Daucus carota*—see **2E**

Cassava *Manihot esculenta* (Syn. *M. utilissima, M. aipi, M. dulcis, M. palmata)*—**2D**

Celeriac *Apium graveolens* var. *rapaceum*

Celery *Apium graveolens*

Chinese Artichoke *Stachys affinis* A species of Woundwort

Chinese (Cinnamon Vine) Yam *Dioscorea opposita* (Syn. D. *batatas)*

Chinese (Lesser) Yam *Dioscorea esculenta*. Pana Yam

Chufa *Cyperus esculentus*. Tiger Nut, Earth Almond, Nutty Sedge, Yellow Nut Grass

Cluster Yam *Dioscorea dumetorum*. American Bitter Yam, Cluster Yam, Forest Yam

Cocoyam *Xanthosoma* spp.—see **2D, "Taro"**

Coleus *Coleus parviflorus* (Syn. *C. tuberosus)*

Cush-cush Yam *Dioscorea trifida*. Aja, Yampi, Indian Yam, Mapuey

Dasheen *Colocasia esculenta* var. *esculenta*—see **2D, "Taro"**

Earth Nut Pea *Lathyrus tuberosa*—see also under "Legumes"

East India Arrowroot *Curcuma angustifolia* and other *Curcuma* spp.

Eddoe *Colocasia esculenta* var. *antiquorum*—see **2D, "Taro"**

Edible Canna *Canna edulis*. Queensland or Purple Arrowroot; Acira (S. America)

Eight-month Yam *Dioscorea rotundata*. White Guinea Yam

Elephant Ear *Colocasia esculenta* var. *antiquorum*—see **2D, "Taro"**

Elephant's Foot *Dioscorea elephantipes*. Hottentot Bread—see **2D, "Taro"**

Elephant Yam *Amorphophallus campanulatus* (Syn. *A. rivieri)*

Florence Fennel *Foeniculum vulgare* var. *dulce*. Florentine Fennel

Forest Yam *Dioscorea dumetorum*. American Bitter Yam, Cluster Yam

Giant Alocasia *Alocasia indica* (Syn. *A. macrorrhiza)*

Ginger *Zingiber officinale*

Greater Yam *Dioscorea alata.* Water Yam, Winged Yam, Ten-month Yam, Asiatic Yam, White Yam—see **2D, "Yam"**

Hausa Potato *Plectranthus esculentus* (Syn. *Coleus dazo, C. esculentus*)

Horse-radish *Armoracia rusticana*

Indian Arrowroot *Curcuma angustifolia*

Indian Yam *Dioscorea trifida.* Cush-cush Yam, Aja, Yampi, Mapuey

Irish Potato *Solanum tuberosum*—see **2D**

Jerusalem Artichoke *Helianthus tuberosus*

Jícama *Pachyrrhizus erosus.* Yam Bean, Potato Bean, Mexican Water Chestnut

Hamburg Parsley *Petroselinum crispum*

Lesser Yam *Dioscorea esculenta.* Chinese Yam, Pana Yam

Livingstone Potato *Plectranthus esculentus (Coleus esculentus)*

Mangel *Beta vulgaris* ssp. *vulgaris.* Mangel-wurzel

Manioc (Mandioc) *Manihot esculenta* (Syn. *M. utilissima, M. aipi, M. dulcis, M. palmata).* Cassava—see **2D**

Mapuey *Dioscorea trifida.* Cush-cush Yam, Aja, Yampi, Indian Yam

Mashua *Tropaeolum tuberosum.* Tuberous Nasturtium, Anyu, Quecha, Apinamama, Isanu, Ysano, Cubio

Oca *Oxalis tuberosa.* Iribia/Ibia(Colombia), Cuiba (Venezuela), New Zealand Yam (New Zealand), Papa extranjera (Mexico)

Parsnip *Pastinaca sativa.* Cherivia or Pastinaga (Portuguese**)**

Potato *Solanum tuberosum.* Irish Potato—see **2D**

Potato Bean *Apois americana.* Wild Bean

Potato Yam *Dioscorea bulbifera.* Air Potato, Aerial Yam, Bulbil-bearing Yam

Radish *Raphanus sativus.* Rabanete (Portuguese)

Rampion *Campanula rapunculus*

Rutabaga *Brassica napus*

Saffron Crocus *Crocus sativus*

Salsify *Tragopogon porrifolius.* Oyster Plant, Vegetable Oyster; Cercefi (Portuguese)

Scolymus *Scolymus hispanicus.* Spanish Salsify, Spanish Oyster Plant

Scorzonera *Scorzonera hispanica.* Black Salsify

Skirret *Sium sisarum.* Chervis (French); Zuckerwurzel (German); Sísaro (Portuguese); Escaravía (Spanish)

Swedes *Brassica napobrassica*

Sweet Potato *Ipomoea batatas*—see **2D**

Tacca (Otaheite) Arrowroot *Tacca pinnatifida*

Tannia *Xanthosoma* spp.—see **2D, "Taro"**

Tarragon *Artemisia dracunculus* Little Dragon, Mugwort. Herbe au Dragon (Fr.)

Taro *Colocasia* spp.—see **2D**

Tiger Nut *Cyperus esculentus*—see Chufa

Turnip *Brassica napus.* Nabo (Portuguese); Shalgham (Dari)

Turnip rooted Chervil *Chaerophyllum bulbosum*

Ullucu *Ullucus tuberosus.* Papa Lisa (Peru), Melloca (Ecuador)

Wapato *Sagittaria latifolia*. Arrowhead, Duck Potato
Water-Chestnut—see **2E, "Vegetables"**, below
White Guinea Yam *Dioscorea rotundata*. Eight-month Yam
Wild Yam *Dioscorea sansibarensis*
Yacon *Smallanthus sonchifolius (*Syn. *Polymnia sonchifolius)*. Aricoma (Peru);
Poire de Terre (French); Jíquima, Jiquimilla, Llacon, Yacón (Spanish)
Yams *Dioscorea* spp.—see **2D**
Yam Bean *Pachyrrhizus erosus*. Yam Bean (edible tubers and pods) and
P. tuberosus. Yam Bean, Potato Bean, Jícama (edible tubers)
Yautia *Xanthosoma* spp. Tannia, Tanier, (New) Cocoyam—see **2D, "Taro"**
Yellow Guinea Yam *Dioscorea cayensis*. Twelve-month Yam, Cut-and-Come-
Again—see **2D**
Ysano *Tropaeolum tuberosum*. Tuberous Nasturtium, Anyu, Quecha, Apina-mama,
Isanu, Cubio

E. VEGETABLES

Alexanders *Smyrnium olusatrum*. Wild Celery
Amsoi—see Indian Mustard
Angled Loofah *Luffa acutangula*. Chinese Okra
Arrowhead *Sagittaria sinensis* (Syn. *S. trifolia* var. *edulis*). Tzi Koo (Chinese)
Asparagus *Asparagus officinalis*. Espargo, Aspargo (Portuguese); Marchoba
(Pashtu)
Babbington's Leek *Allium babbingtonii*
Balsam Apple *Momordica balsamina*
Balsam Pear *Momordica charantia*. Bitter Gourd/Cucumber, Carilla Gourd
Basella *Basella alba.* Ceylon Spinach, Malabar Spinach
Bayam *Amaranthus tricolor*. Hinn Choy, Calaloo
Beef Steak Plant *Perilla frutescens* var. *crispa*. Perilla, Shiso (Japan)
Beet *Beta vulgaris*. Chard, Seakale Beet, Swiss Chard, Perpetual Spinach;
Acelga Vermelha (Portuguese)
Beetroot *Beta vulgaris* ssp. *vulgaris*. Raiz de Beteraba (Portuguese)
Bindi *Abelmoschus esculentus (*Syn. *Hibiscus esculentus)*. Okra, Okro, Lady's
Finger, Gumbo, Gombo—see **2E**
Bitter Gourd *Momordica charantia*. Balsam Apple/Pear, Bitter Cucumber/Melon,
Carilla Gourd, Concombre Africain, Margose, Momordique (French), Balsambirne,
Bittergurke (German), Karela (India), Balsamo, Balsamito, Cundeamor (Spanish).
Black Mustard *Brassica nigra*
Black Salsify *Scorzonera hispanica*. Scorzonera. Zankai (Pashtu)
Bladder Campion *Silene vulgaris*
Bottle Gourd *Lagenaria siceraria (*Syn. *L. vulgaris)*. Calabash Gourd, White-
flowered Gourd. Abóbora Cabaça, Abóbora-de-Romeiro (Portuguese)
Borecole *Brassica oleraceae* var. *acephala*. Collard, Curly Kale
Broccoli *Brassica oleracea* var. *botrytis cauliflora* (or var. *italica*). Brócolis or
Brócolos (Portuguese); Sheen Gulpi (Pashtu)

Brussels Sprouts *Brassica oleracea* var. *gemmifera.* Couve-de-Bruxelas (Portuguese)

Buckshorn Plantain *Plantago coronopus*

Buffalo Gourd *Cucurbita foetidissima*—see **2G**

Bush Squash *Cucurbita pepo* var. *melopepo*

Brussels Sprouts *Brassica oleracea* var. *gemmifera*

Cabbage *Brassica oleracea* var. *capitata.* Head Cabbage, Common Cabbage—**2E**

Calabash Gourd *Lagenaria siceraria (*Syn. *L. vulgaris).* Bottle Gourd, White-flowered Gourd

Calabazilla *Cucurbita foetidissima.* Buffalo Gourd—see **2G**

Calabrese *Brassica oleracea* var. *italica.* Romanesco, (Green) Sprouting Broccoli

Cardoon *Cynara cardunculus* {Related to the Globe Artichoke (qv).} (Globe) Artichoke, Artichoke Thistle, Scotch Thistle. Cardon d'Espagne, Artichaut Commun *C. scolymus* (French); Gemüseartischocke, Kardone, Artischocke *[C. scolymus]* (German); Cardo (Portuguese & Spanish), Alcachofra (Portuguese); Cardo de comer, Alcachofa *[C. scolymus], A*lcaucil *[C. scolymus]* (Spanish)

Carrot *Daucus carota*—see **2E**

Cauliflower *Brassica oleracea* var. *botrytis.* Couve-flor (Portuguese); Gulpee (Pashtu)

Cayenne *Capsicum* spp.—see **2E, "Peppers"**

Chaya *Cnidoscolus aconitifolius* ssp. *Aconitifolius*

Chayote (Choyote) *Sechium edule.* Cho-cho, Choko, Christophine, Vegetable Pear, Chou-chou (French), Chuchu (Portuguese), Chocho, Pipinela, Tallote (Spanish), Chinchayote, Camochayote, Ichintla (South America)

Cherry Tomato *Lycopersicum cerasiforme*

Chilicote *Cucurbita foetidissima.* Buffalo Gourd—see **2G**

Celery *Apium graveolens* var. *dulce (*Syn. *A. dulce)* Aipo, Celeri (Portuguese); Silari (Pashtu)

Celtuce *Lactuca sativa* var. *augustana.* Asparagus Lettuce, Chinese Stem Lettuce, Woo Chu, Woh Sun

Chicory *Cichorium intybus.* Chicória (Portuguese)

Chinese Artichoke *Stachys affinis*

Chinese Broccoli *Brassica oleraceae* var. *alboglabra.* Chinese Kale, Gai Lohn

Chinese Cabbage *Brassica chinensis* var. *pekinensis.* Chinese Leaf (Leaves)

Chinese Chives *Allium tuberosum (*not *A. odoratum).* Nira, Kau, Tsoi

Chinese Leaves *Brassica rapa* ssp. *pekinensis*

Chinese Mustard *Brassica juncea.* (Chinese) Mustard Greens (in Snow), Brown Mustard, Kai Tsoi

Chinese Small Onion *Allium cepa* var. *aggregatum (*Syn. *Allium fistulosum)*

Chinese Spinach *Amaranthus* spp.—see **2G, "Amaranths"**

Chinese Water-chestnut *Eleocharis dulcis* (Syn. *E. tuberosa).* Matai

Chives *Allium schroenoprasum.* Cebolinho, Ceboletas de França (Portuguese)

Chopsuy greens *Chrysanthemum carinatum.* Shungiku

Choyote (Chayote) *Sechium edule.* Christophine etc, see below

Christophine *Sechium edule*. Choyote (Chayote) etc, see below

Ciboule *Allium cepa* var. *aggregatum*—see Welsh Onion

Collard *Brassica oleraceae* var. *acephala*. Borecole, Curly Kale

Corn Salad *Valerianella eriocarpa (V. locusta, V.olitoria)*. Lambs Lettuce, Nusslisalad (Switzerland), Mache (France)

Cranberry Gourd *Abobora tenuifolia*. Abóbora-do-Campo (Portuguese)

Cress *Lepidium sativum*. Garden Cress. Taratizak (Pashtu)

Cucumber *Cucumis sativus*. Pepino (Spanish & Portuguese); Concombre (French); Badrang (Dari)

Cucurbits *Cucurbita* spp.—see **2E**

Curly Kale *Brassica oleraceae* var. *acephala*. Collard, Borecole

Currant Tomato *Lycopersicum pimpinellifolium*. Tomatoes groseille à grappes (French)

Cushaw *Cucurbita argrosperma* syn *C. mixta*. Ayote (German and Spanish)

Dishcloth Gourd *Luffa* spp. Luffa, Loofah, Sponge Gourd, Vegetable Sponge

Edible Snake Gourd *Tricosanthes anguina*

Egg Plant *Solanum melongena*. Aubergine, Brinjal, Melongene; Aubergine (French); Berinjela, Brinjela, Jiloeiro, Jiló (Portuguese); Torbanjan (Pashtu)

Egusi Melon *Cucumeropsis edulis* and *C. manii*

Endive *Cichorium endivia*. Chicorée (in French, Witloof Chicory is called "Endive")

Epazote *Chenopodium ambrosoides*

Fennel *Foeniculum vulgare (*Syn. *F. oficinale)*

Fig-leaf Gourd *Cucurbita ficifolia*. Malabar Gourd; Abóbora Chila (Portuguese)

Garland Chrysanthemum *Chrysanthemum coronarium*. Tangho, Shungiku

Garlic *Allium sativum*. Alho (Portuguese); Ooga, Ozha (Pashtu)

Garlic Chives *Allium tuberosum*

Gboma Eggplant *Solanum macrocarpon*

Gherkin *Cucumis anguria*

Glasswort *Salicornia europaea*. Sea Asparagus, Marsh Samphire

Globe (French) Artichoke *Cynara cardunculus* var. *scolymus*—see Cardoon.

Good King Henry *Chenopodium bonus-henricus*. Mercury, Lincolnshire Asparagus

Gourds *Cucurbita* spp.

Great-headed Garlic *Allium ampeloprasum*

Guinea Pepper *Xylopia aethiopica*

Gumbo (Gombo) *Abelmoschus esculentus (*Syn. *Hibiscus esculentus)*. Okra, Lady's Finger, Bindi—see **2E**

Hamburg Parsley *Petroselinum crispum*

Herb Patience *Rumex patientia*

Horse Radish *Armoracia rusticana (Cochlearia armoracia)*

Huazontli *Chenopodium berlandieri*

Ice Plant *Mesembryanthemum cristallinum*. Brakslaai (Africaans)

Inca Wheat *Amaranthus* spp.—see **2G**

Indian Mustard *Brassica juncea* Amsoi

Indian Spinach *Basella alba (*Syn. *B. rubra)*

Italian Dandelion *Cichorium intybus*
Jamaican Sorrel *Hibiscus sabdariffa*
Jamberry *Physalis ixocarpa.* Tomatillo
Japanese Bunching Onion *Allium cepa* var. *aggregatum (*Syn. *A. fistulosum)*—see Welsh Onion
Jesuit's Nut *Trapa natans.* Water Caltrops
Joseph's Coat *Amaranthus* spp.—see **2G, "Amaranths"**
Kale *Brassica oleracea* var. *acephala.* Borecole, Collard
Kohlrabi *Brassica oleracea* var. *gongylodes (B. caulorapa).* Couve Rábano (Portuguese)
Komatsuna *Brassica rapa komatsuna*
Korila *Cyclanthera pedata.* Caygua (Haiti), Achoccha (Peru)
Kurrat *Allium ampeloprasum* var. *kurrat*
Love-lies-bleeding *Amaranthus* spp.—see **2G**
Lamb's Lettuce *Valerianella eriocarpa (V. locusta).* Corn Salad, Nusslisalad (Switzerland), Mache (France)
Lamb's Quarters *Chenopodium album.* Pigweed
Land Cress *Barberea verna*
Leaf Beet *Beta cicla.* Acelga Brava (Portuguese)
Leek *Allium porrum (*Syn. *A. ampeloprasum* var. *porrum).* Alho-poró, Alho-porro (Portuguese); Gandana (Dari)
Lettuce *Lactuca sativa.* Laitue (French); Lechuga (Spanish); Alface (Portuguese); Kahoo (Pashtu)
Ling *Trapa bicornis* Water-chestnut
Loofah *Luffa* spp. Luffa, Dishcloth Gourd, Sponge Gourd, Vegetable Sponge
Lotus *Nelumbo nucifera*
Madeira Vine *Boussingaulia cordifolia (Anredera cordifolia)*
Malabar Gourd *Cucurbita ficifolia.* Fig-leaf Gourd; Abóbora Chila (Portuguese)
Malabar Spinach *Basella rubra* var. *alba.* Ceylon Spinach
Mallow *Malva verticillata (M. crispa)*
Marrow *Cucurbita* spp.— see **2E**
Milk Thistle *Silybum marianum*
Miner's Lettuce *Montia perfoliata (Claytonia perfoliata).* Winter Purslane
Missouri Gourd *Cucurbita foetidissima.* Buffalo Gourd—see **2G**
Mizuna *Brassica rapa* ssp. *nipposinica* var. *laciniata* (or var. *japonica*)
Mock Orange *Cucurbita foetidissima.* Buffalo Gourd—see **2G**
Mountain Spinach—*Atriplex hortensis.* Orache, Orach
Multiplier Onion *Allium cepa* var. *aggregatum.* Potato Onion—see **2E, "Onion"**
Mustard *Sinapis alba (Brassica juncea).* Sharsham (Pashtu and Dari)
Naples Onion *Allium neapolitanum.* Cebolinho Branco (Portuguese)
New Zealand Spinach *Tetragonia tetragonioides (*Syn. *T. expansa)* Tetragone Cornu (French); Espinafre da Nova Zelândia, Beldroega do Sul, B. de Fólha Grande (Portuguese)

New Zealand Tree Tomato *Cyphomandra betacea*
Okra *Abelmoschus esculentus (*Syn. *Hibiscus esculentus).* Lady's Finger—see **2E**
Onion *Allium cepa*—see **2E**
Orache *Atriplex hortensis.* Mountain Spinach
Oyster Plant *Tragopogon porrifolius.* Salsify
Pak Choi *Brassica rapa* ssp. *chinensis*
Peppers *Capsicum* spp. and *Piper nigrum* (White or Black Pepper)—see **2E**
Pepper Tree *Schinus molle*
Perilla *Perilla frutescens* var. *crispa.* Beef Steak Plant, Shiso (Japan)
Pe-tsai *Brassica pekinensis.* Chinese Leaf
Pokeweed *Phytolacca americana*
Pumpkin *Cucurbita* spp.—see **2E**
Purslane *Portulaca oleracea*
Purslane Pink *Montia sibirica*
Radish *Raphanus sativus.* Rabanete (Portuguese); Muli (Dari)
Rakkyo *Allium chinense (*Syn. *A. bakeri).* Ch'iao t'ou (China)
Rape *Brassica napus*
Red Russian Kale *Brassica napus.* Canadian Broccoli
Rhubarb *Rheum rhabarbarum.* Ruibarbo (Portuguese); Pakhai (Pashtu)
Rocambole *Allium sativum* and *A. scorodoprasum*
Rocket *Eruca sativa (*Syn. *E. vesicaria).* Arugula, Roquette, Rucola; Eruca
(Portuguese)
Roka *Eruca sativa.* Roquette
Romanesco *Brassica oleracea* var. *italica.* Calabrese, Sprouting Broccoli
Roselle *Hibiscus sabdariffa.* Red Sorrel
Rutabaga *Brassica napobrassica (*Syn. *B. napus* var. *napobrassica).* Swede
Salad Burnet *Sanguisorba minor*
Salsify *Tragopogon porrifolius.* Oyster Plant, Vegetable Oyster
Scorzonera *Scorzonera hispanica.* Black Salsify
Scurvy Grass *Cochlearia officinalis*
Sea Beet *Beta vulgaris* var. *maritima*
Sea Kale *Crambe maritima.* Chou Marin (French)
Seakale Beet *Beta vulgaris* var. *cicla.* Chard, Swiss Chard. Acelga (Portuguese)
Shallots—see **2E**, "Onion"
Shepherd's Purse *Capsella bursa-pastoris* var. *auriculata*
Siberian Kale *Brassica napus*
Singhara Nut *Trapa bispinosa.* Water-chestnut
Skirret *Skium sisarumi* Zuckerwurzel (German), Sisaro (Italian), Chirivia (Spanish)
Snake Gourd *Trichosanthes cucumerina.* Serpent Gourd, Chicinda
Sorrel *Rumex acetosa.* Azêda (Portuguese). French Sorrel—*R. scutatus*
Spinach *Spinacia oleracea.* Epinard (French); Espinafre (Portuguese); Palak (Dari)
Spinach Beet *Beta vulgaris*
Sponge Gourd *Luffa* spp. Loofah, Luffa, Dishcloth Gourd, Vegetable Sponge

Spring Onion *Allium fistulosum*
Sprouting Broccoli *Brassica oleracea* var. *italica*. Calabrese, Romanesco
Squash (Winter and Summer). *Cucurbita* spp.—see **2E**
Sunberry *Solanum intrusum (*Syn. S. *Nigrum* var. *guineense)*. Garden Huckleberry
Swede *Brassica napobrassica (*Syn. *B. napus* var. *napobrassica)*. Rutabaga
Sweet Cicely *Myrrhis odorata*
Swiss Chard *Beta vulgaris* var. *cicla*. Chard, Seakale Beet. Acelga (Portuguese)
Tamarillo *Cyphomandra betacea*. Tree Tomato
Thousandhead Kale *Brassica oleracea*
Tomatillo *Physalis ixocarpa*. Jamberry, Sugar Cherry
Tomato *Lycopersicon esculentum (*Syn. *L. lycopersicum, Solanum lycopersicum)*—**2E**
Tree Onion *Allium cepa* var. *proliferum*. Walking Onion—see **2E**, **"Onion"**
Tree Tomato *Cyphomandra betacea*. Tamarillo
Trick Madame *Sedum reflexum*. Stonecrop Houseleek, Trip Madame
(Turks) Turban squash *Cucurbita maxima* var. *turbaniformis*
Turnip *Brassica rapa (B. campestris* var. *rapifera)*. Nabo (Portuguese); Tipar
(Pashtu)
Vegetable Sponge *Luffa* spp. Loofah, Luffa, Dishcloth Gourd, Sponge Gourd
Walking Onion *Allium cepa* var. *proliferum*. Tree Onion—see **2E**, **"Onion"**
Wasabi *Wasabia japonica (*Syn. *Eutrema japonica)*
Water-chestnuts *Trapa bicornis* (Ling), *T. bispinosa* (Singhara Nut) and *T. natans*
(Water Caltrops, Jesuit's Nut)
Watercress *Nasturtium officinale (*Syn. *Rorippa nasturtium-aquaticum)*. Agrião
(Portuguese)
Water Spinach *Ipomoea aquatica*. Kancon, Green Engstai
Wax Gourd *Benincasa hispida*. Ash Pumpkin, Chinese Preserving Melon,
Chinese Watermelon, Chinese Fuzzy Gourd; Tung Kwa, Mo Kwa, Cham Kwa,
Fa Kwa, Tsit Kwa
Welsh Onion *Allium cepa* var. *aggregatum (*Syn. *Allium fistulosum)*. Japanese
Bunching Onion, Ciboule, Scallions (also used for Shallots)
White-flowered Gourd *Lagenaria siceraria (*Syn. *L. vulgaris)*. Bottle Gourd,
Calabash Gourd
White Mustard *Sinapis alba (*Syn. *Brassica hirta)*
Wild Celery *Smyrnium olusatrum*. Alexanders
Wild Pumpkin *Cucurbita foetidissima*. Buffalo Gourd—see **2G**
Winter Cress *Barbarea vulgaris*. American Cress, Land Cress, Upland Broad Leaf
Cress, Herbe de Sainte Barbe
Zucchini *Cucurbita pepo*. Squash, Courgette, Vegetable Marrow; Courgette
(French)

F. FRUITS AND NUTS
African Breadfruit *Treculia africana*
African Locust Bean *Parkia filicoidea*

Akee *Blighia sapida*. Achee, Ackee (Apple), Akee Tree, Vegetable Brains
Alligator Pear *Persea americana*. Avocado—see **2F**
Almond *Prunus dulcis*. Badam (Dari)
Alpine Strawberry *Fragaria vesca*. Sow-teat SB, Wild SB, Woodland SB
Angled Loofah *Luffa acutangula*
Apple *Malus pumila* (Syn. *Pyrus malu*). Pomme (French); Maçã/Macieira (Portuguese); Saib (Dari)
Apple Serviceberry *Amelanchier* X *grandiflora*
Apricot *Prunus armeniaca* (Syn. *Armeniaca vulgaris*). Zard Ahlu, Mandata (Pashtu & Dari)
Argus Pheasant Tree *Dracontomelon mangiferum*
Atemoya *Annona cherimola* X *A. squamosa* (Cherimoya X Sugar Apple/Sweet Sop)
Australian (Moreton Bay) Chestnut *Castanospermum australe*
Australian Desert Lime *Eremocitrus glauca*
Australian Wild Limes *Microcitrus*
Autumn Olive *Elaeagnus umbellata* Japanese Silverberry, Oleaster, Elaeagnus
Avocado (Pear) *Persea americana*—see **2F**
Banana *Musa* spp.—see **2F**
Banana Passion Fruit *Passiflora mollissima* (Syn. *Tacsonia mollissima*)—pink flowers, and *Passiflora antioquiensis* (Syn. *P. van-volxemii*)—red flowers
Barbados Cherry *Malpighia glabra* (Syn. *M. punicifolia*)
Beet Berry *Chenopodium capitatum*. Strawberry Spinach
Bilimbi *Averrhoa bilimbi*. Cucumber Tree, Tree-sorrel
Blackberry *Rubus ulmifolius*
Blackcurrant *Ribes nigrum*
Black Haw *Crateagus douglasii*
Black Persimmon (Black Sapote) *Diospyros digyna* (Syn. *D.ebenaster*) Chocolate Pudding Fruit. Barbicoa (French); Ebenholzbaum (German); Zapote Negro(Spanish)
Borassus Palm *Borassus flabellifer*. Palmyra
Box Blueberry *Vaccinium ovatum*. Evergreen Huckleberry.
Boysenberry—Loganberry X Blackberry X red Raspberry hybrid
Brazil Nut *Bertholletia excelsa*. Pavory or Pava Nut; Neuz de Para (Spanish); Castanha-do-Pará (Portuguese)
Brazilian Guava *Psidium guineense* (Syn.*P.mole*). Wild Guava, Araca, Guayabillo, Guisaro, Guayaba Agria / Llanera / Coyote
Bread Fruit *Artocarpus altilis* (Syn. *A. communis, A .incisa*). Bread Nut; Arbe à pain, Fruit à pain (French); Fruta-de-pan (Spanish); Fruta-pão (Brazil); Uto (Fijian)
Bullock's Heart *Annona reticulata*. Custard Apple, Ramphal, Sweet Sop; Anone, Coeur de Boeuf (French); Anona (colorado), Corazon (Spanish); Coração de Boi (Brazil); Araticumape (Portugal)
Butternut *Juglans cinerea*. White Walnut, Demon Walnut, Oilnut

Cainito Star Apple *Chrysophyllum cainito.* Star Apple; Caimito, Cainito/i, Abiú-do-Pará, Maçã-Estrelada (Portuguese)

Calamondin *Citrofortunella microcarpa (Citrus reticulata X Fortunella* spp.*)*

Cape Gooseberry *Physalis peruviana (P. pruinosa)* Goldenberry, Husk or Ground Cherry

Caper *Capparis spinosa* Caper, *Cappero*, Alcaperro, Caper Berry, Caper Bud, Caperbush, Caper Fruit, *Kápari*, Smooth Caper, Spiny Caper, *Tapèra*

Carambola *Averrhoa carambola.* Star Fruit, Coromandel Gooseberry, Averrhoa; Carambole (French)

Carob *Ceratonia siliqua.* St John's Bread, Locust Bean. Caroube, Caroubier (French); Johannisbrotbraum (German); Alfarrobeira (Portuguese); Algarrobo, Caroba (Spanish)

Cás *Psidium friedrichsthalianum* Costa Rican Guava

Cascade Oregon Grape *Mahonia nervosa* Cascade Barberry, Cascade (Dull) Oregon Grape

Cashew *Anacardium occidentale.* Acajú, Cashew Nut; Cachou (French); Merci (Spanish, in Venezuela); {Castanha de} Caju (Portuguese)

Cattley *Psidium littorale (P. cattleianum).* Chiku, Sapodilla, Strawberry Guava

Ceriman *Monstera deliciosa (*Syn *M. lennea, Pholodendron pertusum)* Swiss-cheese plant, Windowleaf, Mexican Breadfruit, Split-leaf Philodendron

Champedak *Artocarpus integer (*Syn. *A. champeden).* Lemasa

Checkerberry *Gaultheria procumbens.* (Creeping) Wintergreen, Mountain-tea Teaberry. Gaultherie, Petit Thé des Bois (French); Wintergrün (German)

Cherimoya *Annona cherimolia.* Anone, Cherimolier (French); Cherimolia, Anona do Chile, Querimólia (Portuguese)

Cherry *Prunus* spp., including *P. avium* (Sweet Cherry) and *P. cerasus* (Sour Cherry)

Cherry Plum *Prunus cerasifera.* Myrobalan Plum

Cherry Prinsepia *Prinsepia sinensis.* Dong bei rui he (Chinese)

Chiku *Manilkara achras (*Syn. *M. zapotilla, Achras zapota)* Sapodilla, Sapota, Naseberry; Sapotille (French); Sapoti (Portuguese); Nispero (Spanish, in Columbia); Chika (Malaya)

Chinese Gooseberry *Actinidia deliciosa (*Syn. *A. chinensis).* Kiwi Fruit

Chinese Jujube *Zizyphus jujuba.* Chinese Date

Chinese Mulberry *Cudrania tricus pidata* Che, Cudrang, Silkworm Tree

Chinese Watermelon *Benincasa hispida.* Wax Gourd, Ash Pumpkin, Chinese Preserving Melon, Chinese Fuzzy Gourd; Tung Kwa, Mo Kwa, Cham Kwa

Citrange *Citrus sinensis X Poncirus trifoliata*

Citron *Citrus medica.* Cédrat (French); Cidra (Portuguese)

Citrus Fruits—see **2F:**

Citrus limon—lemon (*citron*, in French)

Citrus aurantifolia—lime (*limette*, or *citron vert*, in French)

Citrus sinensis—sweet orange

Citrus aurantium—sour (Seville) orange

Citrus paradisi—grapefruit
Citrus reticulata*—tangerine, mandarin, satsuma (& Rangpur lime) [* Also classified as *C. nobilis*]
Citrus maxima**—pomelo (shaddock) [** Syn *C. grandis, C. decumana, C. paradisi, C. aurantium)*]
Citrus medica—citron (*cedrat*, in French)
Poncirus trifoliata—trifoliate orange
- Citrus Hybrids: *C. reticulata* X *C. paradisi* = tangelo, *C. reticulata* X *C. sinensis* = tangor, *C. sinensis* X *P. trifoliata* = citrange
Clementine cv. of tangerine *Citrus reticulata,* or *C. reticulata* X *C. sinensis*
Cloudberry *Rubus chamaemorus*
Coast Gooseberry *Ribes divaricatum*
Coconut *Cocos nucifera*
Cornelian Cherry *Cornus mas* Sorbet. Cornouiller Mâle (French); Kornelkirsche (German); Cornejo Común, Cornejo Macho (Spanish)
Creeping Barberry *Mahonia repens.* Creeping Oregon Grape
Cuachilote *Parmentiera aculeata* (Syn. *P. edulis* Cuajilote (Spanish, Guatemala)
Custard Apple—at least 3 species are called "Custard Apple": 1. Papaw (Pawpaw) *Asimina triloba,* 2. Bullock's Heart (Ramphal, Sweet Sop) *Annona reticulata,* and 3. Sugar Apple (Sweet Sop, Anona) *Annona squamosa*
Damson *Prunus damascena*
Date Palm *Phoenix dactylifera* Date. Dattier, Palmier Dattier (French); Dattelpalme (German); Tamareira (Portuguese); Palmera Datilera (Spanish)
Dewberry *Rubus caesius.* European Dewberry. Ronce Bleuâtre (French); Acker-Brombeere, Kratzbeere (German); Ou zhou mu mei (Chinese)
Duku *Lansium domesticum.* Langsat; Lansium (French); Arbol de Lanze (Portuguese); Lansone, Ayer (Malaya)
Durian *Durio zibethinus.* Durione (French); Durianbaum, Stinkfrucht (German)
Egg Fruit *Lucuma bifera, L. nervosa,* and *L. salicifolia*
Elderberry *Sambucus canadensis*
Ensete *Ensete* spp.—see **2F, "Banana"**
Feijoa *Feijoa sellowiana.* see Pineapple Guava
Fig *Ficus carica.* Inzar (Pashtu)
Gooseberry *Ribes grossularia* (Syn. *R. uva-crispa*)
Granadilla *Passiflora quadrangularis.* Barbadine (French); Maracujá melão (Brazil)
Grapes *Vitis vinifera.* Angoor (Pashtu)
Grapefruit *Citrus paradisi.* Toranja {fruit}, {Toranjeira-tree} (Portuguese)
Greengage *Prunus italica*
Green Sapote *Calocarpum viride*
Ground Cherry *Physalis pruinosa.* Husk Tomato, Strawberry Tomato, Dwarf Cape Gooseberry
Guama *Inga laurina*—see Sacky sack
Guaraná *Paullinia cupana*
Guava *Psidium guajava*—see **2F**

Guisaro *Psidium guineense (*Syn. *P. molle)* Brazilian/Wild Guava
Heartnut *Juglans sieboldiana cordiformis*
Highbush Blueberry *Vaccinium corymbosum*
Hog Plum *Spondias mombin*
Hoogly *Citrus reticulata* X *C. paradisi.* Ugli
Huckleberry *Solanum burbankii*
Husk Cherry *Physalis pruinosa* Ground Cherry, Cape Gooseberry
Ilama *Annona diversifolia*
Indian Wood Apple *Feronia limonia*
Jaboticaba *Myrciaria cauliflora (*Syn. *Eugenia cauliflora)*
Jackfruit *Artocarpus heterophyllus (*Syn. *A. integra, A. integrifolia).* Jakfruit; Jack (French); Jaca (Spanish & Portuguese); Buah Nangka (Malaya); Langka
Jambolan *Eugenia cuminii*
Java (Rose) Apple *Eugenia javanica (*Syn. *Syzygium javanica).* Jambosa, Wax Jambo, Macopa, Java Apple, Semarang, Jambu Ayer Rhio
Indian Jujube *Zizyphus mauritania*
Italian Pine *Pinus pinea*
Japanese Medlar *Eriobotrya japonica.* Loquat
Japanese Persimmon *Diospyros khaki.* Persimmmon, Oriental Persimmon, Date Plum, Sharon Fruit
Japanese Raisin Tree *Hovenia dulcis*
Japanese Wineberry *Rubus phoenicolasius*
Juneberry *Amelanchier canadensis*
Kapok *Ceiba pentandra* Silk-cottontree. Capoc, fromager, kapokier (French), Kapokbaum (German), Árbol capoc, Ceiba (Spanish), Pochote
Kiwi Fruit *Actinidia* spp. Chinese Gooseberry, Strawberry Peach
Korean Pine *Pinus koraiesis*
Kumquat *Fortunella polyandra, F. margarita, F. japonica* and other spp.
Langsat (or **Duku**) *Lansium domesticum.* Lansium (French); Arbol de Lanze (Portuguese); Lansone, Ayer (Malaya)
Lemon *Citrus limon.* Citron (French); Limão (Portuguese)
Lime *Citrus aurantifolia.* Limette (French); Limão doce/Limeira (Portuguese)
Litchi *Litchi chinensis (*Syn. *Nephelium litchi).* Lychee, Litchee
Loganberry *Rubus loganobaccus*
Locust Bean *Ceratonia siliqua* See Carob
Loquat *Eriobotrya japonica.* Japanese Medlar, Japanese PLum
Lychee *Litchi chinensis (*Syn. *Nephelium litchi).* Litchi, Litchee
Mabolo *Diospyros discolor.* Macassar ebony, Velvet Apple
Macadamia Nut *Macadamia ternifolia (*Syn. *M. integrifolia, M. tetraphylla).* Queensland Nut
Mahua *Madhuca longifolia (*Syn. *M. indica).* Moatree, Mowra-buttertree
Malay Rose Apple *Eugenia malaccensis (*Syn. *Syzygium malaccensis).* Pomme de Tahiti, Jambossier Rouge, Jamalae (French); Pomoreja de Malaca, Jambe de Malacea (Spanish); Ohia (Hawaiian); Jambu Merah (Malaya)

Mammey Apple *Mammea americana.* Tropical Apricot. Abricotier (French)
Mammey Sapote *Calocarpum sapota (*Syn. *C. mammosum)*
Mandarin *Citrus reticulata (*Syn. *C. nobilis).* Satsuma, Tangerine; Mexerica (Portuguese)
Mango *Mangifera indica*—see **2F**
Mangosteen *Garcinia mangostana.* Mangoustan (French); Mangostan (Spanish); Mangostão (Portuguese)
Medlar *Mespilus germanica (*Syn. *Mimusops elengi, Vangueria infausta)*
Melon *Cucumis melo.* Melão (Portuguese)
Mountain Papaya *Carica candamarcensis (*Syn. *C.pubescens, Vasconcellea cundinamarcensis,V. cundamarcensis, V. pubescens).* Papayer de Montagne (French); Bergpapaya (German); Chamburú, Chamburo , Chiluacán, Papaya de Tierra Fría (Spanish)
Mountain Soursop *Annona montana*
Mulberry *Morus alba* (White Mulberry), *Morus nigra* (Black Mulberry). Amora/Amoreira (Portuguese); Tut (Pashtu, Persian/Dari)
Naranjilla/o *Solanum quitoense*
Natal Plum *Carissa grandiflora (*Syn. *C. macrocarpa).* Amatungulu. Carisse (French); Amatúngula, Cereza de Natal (Spanish)
Nectarine *Prunus persica* var. *nectarina.* Nectarina (Portuguese)
Orange *Citrus sinensis* (Sweet Orange), *C. aurantium* (Sour Orange), *C. trifoliata* (Trifoliate Orange)—see **2F, "Citrus"**
Oregon Hollygrape *Mahonia aquifolium.* Holly Barberry, Tall Oregon Grape
Osoberry *Osmaronia cerasiformis.* Indian Peach
Oval Kumquat *Fortunella margarita*
Oval-leafed Blueberry *Vaccinium ovalifolium*
Pacific Dewberry *Rubus ursinus.* California Blackberry, California Dewberry, Pacific Blackberry
Pacific Madrone *Arbutus menziesii.* Madroña, Madroño (Spanish)
Palmyra Palm *Borassus flabellifer.* Borassus Palm
Papaya *Carica papaya.* Pawpaw, Papaw—see **2F**
Passion Fruit *Passiflora edulis, P. quadrangularis* and other *Passiflora* spp. (Purple) Granadilla, Simitoo, Sweet Cup; Pomme Liane (French); Maracujá (Portuguese)
Pawpaw (Papaw) *Asimina triloba.* Custard Apple
Peach *Prunus persica (*Syn. *P. vulgaris).* Shaftahlu (Pashtu & Dari); Pêssego/ Pessegueiro (Portuguese)
Pear *Pyrus communis.* Nak (Pashtu and Dari); Pêra/Pereira (Portuguese)
Pepino *Solanum muricatum*
Pineapple *Ananus comosus.* Ananas (French); Abacaxi (Brazil); Ananás (Portuguese); Pina (Spanish)
Pineapple Guava *Feijoa sellowiana (*Syn *Acca sellowiana).* Feijoa, Guavasteen; Goiaba do Campo, Goiaberra Serrana (Portuguese), Guayaba Chilena (Spanish).
Pitahaya *Celenicereus* sp.

Pitanga Cherry *Eugenia uniflora*
Plantains—see **2F, "Bananas"**
Plum(European) *Prunus domestica*. Ameixa (Portuguese); Olocha (Dari)
Pomegranate *Punica granatum*. Romã (zeiro) (Portuguese); Anar (Pashtu)
Pomelo *Citrus maxima***—Pomelo (shaddock), Pummelo, [** Syn *C. grandis, C. decumana, C. paradisi, C. aurantium*]
Prickly Pear *Opuntia ficus-indica* Barbary Fig, Indian Fig , Mission Cactus, Tuna cactus. Figuier d'Inde/ de Barbarie (French); Feigenkaktus (German); Figo-da-Índia/Espanha/Barbária, Palma de Gado (Portuguese); Chumba, Higuera, Nopal Pelón, Tuna (Spanish)
Pulasan *Nephelium mutabile* (Close relation of the Rambutan)
Pummelo—see Pomelo
Quince *Cydonia vulgaris*
Rambutan *Nephelium lappaceum*. Hairy Lychee; Ramboutan, Litchi Chevula (French); Ramustan (Spanish)
Raspberry *Rubus idaeus*. Framboesa/Framboeseira (Portuguese)
Redcurrant *Ribes sativum, R. petraeum* and *R. rubrum*
Red Huckleberry *Vaccinium parvifolium*
Rhubarb *Rheum rhaponticum*. Ruibarbo (Portuguese)
Rose Apple *Eugenia jambes (*Syn. *Syzygium jambes)*. Pomarosa (Spanish); Jambu Mawar (Malayan)
Round Kumquat *Fortunella japonica*
Russet buffalo-berry *Shepherdia canadensis*
Sacky sack *Inga laurina* Guama, Jackysac, Spanish Oak, Sweet Pea. Pois Doux (Blanc) (French); Cujinicuil, Paternillo, Palal, Jina, Guavo, Guamo Rosario, Guabo, Guamo, Guamá (Spanish)
Salmon Berry *Rubus spectabilis (*Thimbleberry *Rubus parviflorus* is also sometimes called Salmon Berry*)*
Sapodilla *Manilkara zapotilla (*Syn. *M. achras, Achras zapota)*. Sapota, Naseberry, Chiku; Sapotille (French); Sapoti (Portuguese); Nispero (Spanish, in Columbia); Chika (Malaya); Breiapfelbaum, Kaugummibaum, Sapodillbaum (German)
Sapote *Calocarpum mamosum*. {Taxonomy very confused with Sapodilla}
Sapucaia Nut *Lecythis* spp.
Saskatoon Berry *Amelanchier alnifolia* June Berry, Pacific Service Berry, Service Berry, Western Service Berry, Western Shadbush
Shaddock *Citrus maxima***—Pomelo, Pummelo, [** Syn *C. grandis, C. decumana, C. paradisi, C. aurantium*]
Shallon *Gaultheria shallon*. Salal
Siberian Pea Tree *Caragana arborescens*
Siberian Kiwi *Actindia arguta* Hardy Kiwi
Smooth Loofah *Luffa cylindrica*
Soncoya *Annona purpurea* Atier, Tête de Négre (French); Anona Rosada,Catagüire, Manirote, Soncoya, Turagua, Toreta (Spanish)
Sorbet *Cornus mas*. See Cornelian Cherry

Soursop *Annona muricata (A. montana.* Mountain Soursop) Guanabana, Sappadille (French); Graviola, Jaca do Pará (Brazil); Guanábana (Spanish); Anona, Araticum (Portuguese); Durian Belanda (Malaya)

Stagbush *Viburnum prunifolium.* Black Haw (also common name for *Crataegus douglasii*)

Star Apple *Chrysophyllum cainito.* Cainito Starapple, Golden Leaf Tree, Eslo, Nessarrjo, Tuko; Caimito, Cainito/i, Abiú-do-Pará, Maçã-Estrelada (Portuguese), Cauje, Maduraverde, Ablaca, Pipa, Sapotillo; Abiaba, Caimite, Caïnite (French)

Star Fruit *Averrhoa carambola.* Carambola. Sternfrucht (German)

Strawberry *Fragaria* spp. Morango (Portuguese)

Strawberry Guava *Psidium littorale (P. cattleianum).* Cattley

Sugar Apple *Annona squamosa.* Anona, Custard Apple, Sweetsop; Anón (Spanish); Ateira (tree), Ata (fruit), Fruta-do-Conde, Condêssa, Pinha (Portuguese)

Surinam Cherry *Eugenia uniflora (E. michelli).* Brazilian or Cayenne Cherry; Cerise de Cayenne (French); Ceraza de Cayena (Spanish); Pitanga (Portuguese)

Sweet Granadilla *Passiflora ligularis.* Granadina China. Cranix (Spanish)

Sweetsop *Annona squamosa.* Anona, Custard Apple, Sugar Apple; Anón (Spanish); Ateira (tree), Ata (fruit), Fruta-do-Conde, Condêssa, Pinha (Portuguese)

Tamarind *Tamarindus indica.* Tamarindo (Portuguese); Asam Jawa (Malaya)

Tangelo *Citrus reticulata* X *Citrus paradisi*

Tangerine *Citrus reticulata (*Syn. *C. nobilis).* Mandarin, Satsuma; Mexerica (Port.)

Tangor *C. nobilis (C. reticulata* X *C. sinensis).* King Orange/of Siam. Kunembo

Thimbleberry *Rubus parviflorus* Salmon Berry. Nutka Himbeere (German)

Thin Leaved Blueberry *Vaccinium membranaceum.* Black Huckleberry

Tomato Litchi *Solanum sisymbriifolium.* Dense Thorn Bitter Apple, Wild Tomato. Doringtamatie, Wildetamatie (Afrikaans)

Trifoliate Orange *Poncirus trifoliata*

Ugli *Citrus reticulata* X *C. paradisi.* "Hoo-Glee", Uniq Fruit, Unique Fruit.

Walnut *Juglans* spp. Black Walnut (*J. nigra*), English/Persian Walnut (*J. regia*). Charmarghz (Dari)

Water Lemon *Passiflora lauriflora.* Jamaica Honeysuckle, Sweet Cup, Bell Apple, Yellow Granadilla. Pomme de Liane/d'or (French); Parcha (Spanish)

Watermelon *Citrullus lanatus (*Syn. *C. vulgaris, Colocynthis citrullus)*—see **2F**

Watery Rose Apple *Syzygium aqueum (*Syn. *Eugenia aquea).* Bellfruit, Water Apple, Watery Rose Apple. Jambo Ayer (French); Wasserjambuse (German); Perita Costeña, Tambis (Spanish); Jambu Air (Malaya)

White-bark Raspberry *Rubus leucodermis.* Blackcap

Whitecurrant—same species as Redcurrant but without red pigment (anthocyanin)

White Sapote *Casimiroa edulis.* Casimiroa, Mexican Apple. Matasano (Spanish)

Wineberry *Rubus phoenicolasius (*Syn. *Aristotelia serrata, Rubus phoenicolasius)* Wine Raspberry, Japanese Wineberry. Duo Xian Xuan Gou Zi (Chinese); Jananische Weinbeere, Rotborstige Himbeere (German)

Yamaboshi *Cornus kousa* var. *chinensis.* Japanese Dogwood, Kousa Dogwood. Cornouiller Kousa (French); Japanischer Blumen (German)

3B. SEED PURCHASE PROCEDURES

The following notes attempt to give some guidance to individuals and organisations (NGOs for example) involved with buying seed in large quantities. Depending on the circumstances, it may not be necessary to include all of the following points in a seed purchase contract. However, the more details that are included in the contract, the lower is the danger of misunderstandings, delays and possible litigation arising between the seed supplier/seller and seed purchaser/buyer.

The Contract

Purchase / Contract Number.
Name, address, phone & fax numbers, e-mail address of the **buyer** (eg NGO) and the **seller** (vendor or supplier).
Goods. eg Maize Seed *(Zea mays)*
Currency. eg Somali Shillings
Unit. eg Metric Ton (MT or tonne)
Quantity. eg 25MT (2500 × 10 kg packed in 500 × 50 kg bags)
Description. eg Open pollinated maize seed, variety Katumani
Unit Price. eg 10,000
Total Cost. eg 250,000 Somali Shillings (two hundred and fifty thousand Somali shillings). This is a fixed price Contract and increase of taxes, duties or other costs will not influence the Contract in any manner whatsoever and the price agreed herein shall remain binding without variation.
Specifications. Year/season of harvest, min.% pure seed (99%), other crop seed (0.2% max.), infected/infested seed (0.02% max.), inert matter (2%), germination (X% min.), moisture content, seed treated/dressed (no, or what type of dressing). Percentages should follow the laws/standards of the country in question. *Absence of Striga weed seed (and certain seed borne diseases) should be guaranteed.*
Inspection. This Contract is subject to quality acceptance by the survey/inspection company mandated by the buyer (or government authority), and at the buyers cost, before loading at the seller's warehouse. Inspection regarding quality, quantity and purity will be carried out on a lot defined as a quantity of X MT of the seed. The seller will inform the buyer on the readiness of this quantity X days in advance (2–4 is reasonable) so that the buyer has time to organise the inspection. The seller will be responsible for the quantity and quality delivered according to the agreed specification. In case of delivery of seed that does not conform to the conditions stated above, the buyer has the right to take the appropriate legal steps according to the law of the country in question. The certificate of weight and quality will be established by the survey company mandated by the buyer. If samples are taken, they will be sealed in packets, which shall be signed by the survey company and the seller. One sample will remain with the seller and two samples will be delivered to the survey company for laboratory analysis. The seller should note that failure to meet the buyer's standards outlined in this Contract may affect the awarding of future contracts.

Packing Instructions. eg 5 × 10 kg new polypropylene bags packed in 50 kg Net, new polypropylene bags, all bags to be stitched at the mouth. The packing should be strong enough to withstand wear and tear during multiple handling and transportation. 2% extra bags to be provided by the seller (free of charge) for re-bagging purposes.
Marking Instructions. On "Master" eg 50 kg bags only:
Seller's name and/or logo, name of variety and class of seed, Purchase Order/Contract Number. Buyer's name and/or logo. No other markings should be on the bags.

Delivery Schedule
All goods to be delivered as follows:
Delivery date(s): …………………...: Loading date(s) …………………..
    Upon loading (the same day) of the goods the buyer should receive the following information from the seller: Truck Registration Number, number of bags loaded, time of departure, estimated time of arrival at the destination.
Payment Terms
    State with or without taxes in force in the country. Against Invoice and upon presentation of required documents: X shillings/USD within seven days of receipt of the entire supply of goods and the laboratory analysis report.
Penalty Clauses.
1. *Late Delivery.* Failure by the seller to deliver the consignment to …. by the date stated above (Delivery Schedule) will result in the seller being penalised as follows:
    – one day or part thereof late a 0.50% reduction in the cost of undelivered goods;
    – two days or part thereof late a 0.75% reduction in the cost of undelivered goods;
    – three days or part thereof late a 1.00% reduction in the cost of undelivered goods;
    – and for every day thereafter a further reduction of 1% per day in the cost of undelivered goods.
    The penalty clause will be imposed for late delivery of the goods unless *force majeure*, in which cases the issue shall be resolved between the seller and buyer, or by court of law.
2. *Germination Failure.* If the report from the survey/inspection company mandated by the buyer (or government authority) indicates that the seed fails the germination rate specified in the Contract, the following penalties will be applied:
    – 2% reduction on the amount of the failed lot if the germination is below X%, as stipulated in "Specifications" (eg 80–85%);
    – 4% reduction on the amount of the failed lot if the germination is below X%, as stipulated in "Specifications" (eg 75–80%);
    If the germination rate is below 75% the seed will be rejected and the seller will be held liable for all costs related to the transportation, warehousing, inspection and all other costs incurred. The rejected lot(s) of seed will be removed and replaced at the seller's expense.

3. *Weight Shortage.* If the report from the survey/inspection company mandated by the buyer (or government authority) indicates that the quantity delivered is less than the quantity contracted, the seller is obliged to supply the remaining quantity up to the total amount stipulated in the Contract.

Delivery Address: X warehouse in Y town, truck(s) to arrive on site before midday.
Consignee: X warehouse in Y town, Z organisation/NGO.
Delivery Terms: FCA, X warehouse in Y town.
Address for Invoice:
Documents Required:

* commercial Invoice, original plus three copies.
* bank Details for Payment Transfers.
* delivery Forms plus copies.
* certificate of Origin.
* germination Certificate.

Special Instructions:
Rejected Seed. The seller will be held responsible for all costs involved with the rejected lot(s) of seed. In the event of the buyer accepting seed which deviates in any way from the specifications stated in this Contract, the buyer reserves the right to renegotiate the price. In case of rejection the seller will have to collect all the rejected seed within two working days from the warehouse. Should the seller fail to do this, the buyer reserves the right to charge demurrage for the quantity remaining in the buyer's warehouse. No sorting or rectification is to take place in the buyer's warehouse, but must be done outside the buyer's premises.
Language of this Contract.
Applicable Law. X country law to apply.

Dispatch of Documents: One set to the buyer. Please confirm receipt of this order and keep the buyer informed. Acceptance of this Contract entails the waiving by the seller of its General Conditions of Sales.
This Contract will come into force after signing by both parties.

Authorised Signatures:
On behalf of the Buyer
Name(s) and Position(s)
On behalf of the Seller
Name(s) and Position(s)

This Contract is hereby signed on: /   /

3C. CONVERSION TABLES AND STATISTICS

The International System of Units (the SI System—*Système International d'Unités*) is in essence an expansion of the metric system, which was established in the 18th century. The SI System is gradually replacing the old metric system so as to incorporate more recent technological and scientific developments. The SI System uses symbols that should not be used in the plural, or be followed by a full-stop or be written in capitals (except when a capital letter is a symbol). It is based on 7 base units for 7 base quantities:length (m), mass (kg), time (s), electric current (A), thermodynamic temperature (K), amount of substance (mol) & luminous intensity (cd).

LENGTH
SI / Metric
1 centimetre (cm) = 0.394 inch
1 metre (m) = 39.370 inch = 1.094 yards
1 kilometre (km) = 0.621 mile
Imperial
1 inch (in.) = 2.540 cm
1 hand = 4 inches
1 foot (ft.) = 12 inches = 30.480 cm
1 yard (yd.) = 36 inches = 0.914 m
1 furlong = 220 yards
1 statute mile = 1760 yards = 1608.64 m
1 nautical mile = 2025.333 yards = 1851.154 m

AREA
SI / Metric
1 sq. centimetre (cm^2) = 0.155 $inch^2$
1 sq. metre (m^2) = 10.764 $foot^2$
1 are = 100 m^2
1 hectare (ha) =10,000 m^2 = 2.471 acres
1 sq. kilometre (m^2) = 0.386 $miles^2$
1 manzana = 0.700 ha
Imperial
1 sq. inch (in^2) = 6.452 cm^2
1 sq. foot ($foot^2$) = 0.093 m^2
1 sq. yard ($yard^2$) = 0.836 m^2
1 acre = 4840 $yard^2$ = 0.405 ha

WEIGHT
SI / Metric
1 gram (g) = 0.035 ounce
1 kilogram (kg) = 2.205 pounds
1 quintal (qu) = 100 kg
1 metric tonne (MT or tonne) = 1000 kg = 2204.6 pounds = 0.984 short ton

Imperial
1 ounce (oz.) = 28.350 g
1 pound (lb.) = 16.000 oz. = 0.454 kg
1 short hundredweight (Cwt.) = 100 pounds
1 long hundredweight (Cwt.) = 112 pounds (50.802 kg)
1 short ton = 907 kg = 2000 pounds
1 long ton = 1016 kg = 2240 pounds

VOLUME
SI / Metric
1 cubic centimetre (cc) =1 millilitre (ml) = 0.061 in^2
1 litre * = 10 ×10 × 10 cm^3 = 1000 cc
1 litre = 1.760 pints (Imperial) = 2.113 pints (US)
1 litre = 0.220 gallons (Imperial) = 0.264 gallons (US)
1 cubic metre (m^3) = 35.315 cubic feet = 1.308 cubic yards
1 m^3/ha = 14.292 feet^3/acre
{* *"Litre" is normally written in full, to avoid confusion with 1 (one).*}

Imperial
1 cubic inch (in.^{3}) = 16.387 cc
1 cubic foot (foot^3) = 0.028 m^3
1 foot^3/acre = 0.070 m^3/ha
1 cubic yard (yard^3) = 0.765 m^3
1 pint = 473.167 cc
1 gallon (Imperial) = 8 pints = 4.546 litres = 1.201 gallons (US)
1 gallon (US) = 3.785 litres = 0.833 gallons (Imperial)

       1 litre of water at its maximum density at 4.2°C weighs 1 kg

YIELD
Metric: 1 kg/ha = 0.892 lb/acre
Imperial:
   1 lb/acre = 1.181 kg/ha
   1bushel (60 lb.)/acre = 67.26 kg/ha

TEMPERATURE
0°Celsius (Centigrade) = 32°Fahrenheit

To convert Fahrenheit to Celsius:	To convert Celsius to Fahrenheit:
$\frac{5}{9}$ × (°F minus 32) = ×°C	$\frac{9}{5}$ × (°C plus 32) = ×°F

From Fahrenheit to Celsius: subtract 32, multiply by 5, divide by 9.
From Celsius to Fahrenheit: multiply by 9, divide by 5, add 32.

CALIBRATION

For field applications of fertiliser, sprays etc:

To convert: into

g/litre—parts/million (ppm): **multiply by 1000**
g/litre—ounces/Imperial gallon: **multiply by 0.160**
g/m^2—oz./$yard^2$: **divide by 33**
kg/ha—lb./acre: **multiply by 0.892**
oz./$yard^2$—g/m^2: **multiply by 33**
lb./acre—kg/ha: **multiply by 1.121**
miles/hour (MPH)—km/hour: **multiply by 1.6**

How to Determine Miles per Hour (MPH) using a tractor:

1. With two stakes, mark out a distance of 26.8 metres.
2. Select the gear and accelerator setting of the tractor to be used when spraying, or applying fertiliser etc.
3. From a running start, measure the time taken to cover the 26.8 metres, using a stopwatch.
4. Divide 60 by the time in seconds. This gives the speed in mph; to covert into km/hour, multiply by 1.6.

Converting Spray Broadcast Rate to Row Basis

Instructions for applying sprays, and sometimes fertilisers, are often only given for coverage of the entire surface area. But if the chemicals are to be applied in bands, ie along the plant row only, then the actual amount of solution to use per hectare can be calculated by using this formula:

Broadcast Rate (litres/ha) multiplied by
[the Band Width (cm) divided by the Row Spacing (cm)].

CONVERSION FORMULAE

Multiply	By	to calculate:
LENGTH		
Centimetres	0.394	Inches
Metres	39.370	Inches
Metres	3.281	Feet
Metres	1.094	Yards
Kilometres	0.621	Miles
Inches	2.540	Centimetres
Feet	30.380	Centimetres
Feet	0.305	Metres
Yards	0.914	Metres
Rod	5.029	Metres
Statute mile	1.609	Kilometres
Nautical mile	1.852	Kilometres
AREA		
Square metres	1.196	Square yards
Hectares	2.471	Acres
Square feet	144.000	Square inches
Square feet	0.093	Square metres
Square yards	0.836	Square metres
Acres	0.405	Hectares
Square miles	640.000	Acres
Square miles	258.999	Hectares
Square miles	2.590	Square kilometres
WEIGHT		
Grams	0.039	Ounces
Kilograms	2.205	Pounds
Ounces	28.350	Grams
Pounds	453.592	Grams
Pounds	0.454	Kilograms
VOLUME		
Cubic inch	16.387	Cubic centimetres
Cubic foot	0.028	Cubic metre
Cubic yard	0.765	Cubic metre
Acre-foot	1233.482	Cubic metre
Litres	0.035	Cubic feet
Gallons (US)	3.785	Litres
Gallons (Imperial)	4.546	Litres
Gallons (Imperial)	3787.879	Cubic centimetres
Bushel (US)	35.238	Litres
Bushel (Imperial)	36.368	Litres

3D. SOME ISSUES TO CONSIDER WHEN PLANNING OR ASSESSING AGRICULTURAL DEVELOPMENT OR REHABILITATION PROGRAMMES

This section is concerned with issues which aid and development agencies, and their staff working "in the field", agricultural students, health staff concerned with nutrition and others may wish to have a think about when they are attempting to understand the agricultural situation in an area of interest.

The agricultural situation can be viewed from many perspectives, and it is important that field workers have some understanding of at least some of these perspectives, or "issues", if they are to respond in a way which is aware of the predicaments and useful to the people they are attempting to assist. Field workers can usually only discover the answers to a few of the issues raised—time is always too short, there are often language and cultural bridges to cross, and the worker may not be sufficiently well motivated to get to the root of these fundamental issues. The following pages attempt to clarify some of the questions that can be asked in order to either assess, or intervene in, an agricultural system.

For any agricultural programme to succeed it is always important to spend plenty of time talking with the local farmers. After all, it is the farmers on the spot who have the best understanding of the problems in their area, and who hold the key to finding the best solutions.

One current approach is to hold sessions called "farmer led discussions"; the problems are discussed by the farmers themselves and the outsiders are just there to facilitate and to listen and take note. This is part of the process known as Participatory Rural Appraisal (PRA), a process of careful interviewing and observation to unlock the principal factors involved.

Below are listed some of the issues to consider when planning to either study, or intervene in, an agricultural system, under 14 headings:
Cropping Calendar, Climatic Data, Agricultural Practices, Land Area, Nutrition, Seed, Logistics, Soils, Altitude, Harvest, Markets, Livestock, Other Agencies/ NGOs and **Information Bank.**

The questions following each issue are in italics, followed by notes on action which may be taken, and other comments.

1. Cropping

1. What types of crops are normally grown in the area?
List crops in order of importance, including local names. Make sure that the species (and variety names, if available) are correctly identified.
2. When are these crops normally planted and harvested? Why are they planted at these times?
Give ranges, eg "October–November, 75% in November", "all year round, mainly in May–June".
See **1Ga, "Cropping Calendar"**, page 59. Rainfall is normally the most critical factor.

3. Which other crops are also sometimes grown which are acceptable to farmers and also adapted to local conditions, but are not commonly cultivated by farmers?
Discover the reasons for the under utilisation of these crops eg "Not widely known", "seed unavailable or too expensive", "the grain/root takes too long to cook (fuel in short supply)" etc. If possible, obtain small quantities of named, improved varieties of these crops so they can be grown and tested in the area (leader farmers).
4. Are all the crops harvested at approximately the same time?
Investigate the promotion or introduction of crops which can be harvested during the "hungry gap", or which grow quickly and so produce food (and/or income) earlier in the year. Fruit trees may be appropriate, to spread the harvest labour peak.

2. Climatic Data
1. What is the rainfall, temperature (and humidity) pattern?
Obtain any available meteorological data.
Is the data reliable? It is often incomplete or inaccurate, based on too few years' data.
2. Is rainfall the main limiting factor to crop production?
Investigate local irrigation methods and means to improve them, and possible use of more drought resistant crops and/or cultivars. See **Section 1N, "Irrigation"**, page 93.

3. Farming Practices
1. What is the purpose of the agricultural rehabilitation programme?
There are often a number of goals to achieve eg to reduce food aid, to increase self reliance—including production of a surplus for sale—to introduce improved crop varieties or tools etc.

Take into account the programme's negative aspects, such as possible depression of local market prices, fostering a dependence on handouts, and imbalance/unfairness created between beneficiaries and non-beneficiaries.

A wide scale distribution of small quantities of several crops and varieties may well have a greater impact *in the long term* than distribution of a larger weight per family of just one or two crops or varieties.
2. Is the shortage of seeds or tools a major problem?
If the shortage of tools is considered to be a problem, try to obtain samples of the exact types normally used. Photos can also be very useful when describing what is required to the Head Office or potential suppliers.

For both tools and seed, investigate local and imported availability, prices, delivery dates etc. Ensure that the need is widespread, ie applicable for the majority of beneficiaries.

The use of questionnaires, for example when many people are gathered for a distribution/vaccination programme, market day etc, can often provide valuable information. Local farmers (non-beneficiaries) should also be asked the same questions. The questionnaire should be relevant to future monitoring. Local staff and/or Ministry of Agriculture personnel who are known and trusted by the people are the key to obtaining useful and accurate data.

3. What are the other constraints to increasing food production?
Make notes of replies, such as "Security—crop X is more easily stolen than crop Y", "shortage of land, oxen, rain" etc.
4. Are there often problems with insects, diseases or birds? What can farmers do about this? Is what they do effective?
Note and describe the problems, and the months in which crops are damaged.
5. Is fertiliser or compost used, and if so to what extent?
Discover the types of fertilisers and quantities used, when and on which crops, and the cost. What materials are available for making compost and are these used? Discover farmers' attitudes to the use of compost, manure and fertiliser.
6. Is intercropping normally practised?
List approximate yields for each crop, and total yields/hectare (See **1Hb, "Mixed Cropping"**, page 66).
7. Is cooking/heating fuel a problem?
Investigate. If possible, encourage the use of crops that can be eaten raw or with short cooking times, and the possible establishment of fuelwood lots.
8. Is it normally the men or the women who work in the fields and make decisions? Who does what?
Make sure to ask questions to the right groups/gender, and in a way which does not influence the answer or give any offence. In many places women cannot talk directly to male "outsiders".
9. Do both young and old people work in the fields? If it is the old, where are the young people, and why are they not in the fields?
Older people may need help at peak work periods, so crops that require lower labour inputs may be appropriate. However older people are even less likely to adopt new practices than the young are.
    Large families (more than 4 or 5 average) may indicate a precarious social system, where security in old age is mainly provided from within the family.
10. Are the inputs such as seed, tools, agrochemicals and oxen available locally?
Investigate the local supply (quantity and prices) of the inputs required. The root of the problem is often the shortage of cash not the availability of inputs.
11. Are the beneficiaries farming in familiar surroundings or are they displaced?
If they are displaced/refugees, make sure that their skills are combined with the skills and knowledge of local farmers. Talk with the local farmers about their crops, problems etc. They have experimented for generations with the crops and farming systems which work best in their area.

4. Land Area
1. How much land on average is available for each family? Does this depend on family size, or on what? What is the average family size? Who is responsible for distribution of the land or the right to cultivate the land?
Ask the farmers to show you their land, as they very often do not know the size in hectares, acres etc. Even if they do know they will normally be reluctant to tell you, or they will underestimate the area. Ask local agricultural staff, such as the Ministry of

Agriculture. Take the time to probe this important, and often contentious, issue; it is normally difficult to accurately estimate the answer—see **Section 1Hh, "Land Area Measurement"**, page 74.

2. Are the farmers certain that they will be able to cultivate the same land in future years? ie Do they have security of tenure/ownership?

Security of tenure is often one of the biggest problems. If farmers cannot be certain to cultivate the land for more than one or two seasons they will be most interested in growing early maturing, reliable annuals. Aspects such as crop rotation, green manure, organic matter in the soil, tree planting etc will be of little or no relevance for them.

5. Nutrition

1. Does the "normal" food of the beneficiaries contain a balanced diet?

Investigate the diet, and if possible consult with nutritionists and/or medical staff.

2. Are there any vitamin deficiency problems? Do the symptoms occur all the year round, or only in certain months?

Consider the introduction or promotion of crops which could provide the appropriate vitamins, especially during the dry months before the rains.

Examples: dark green leaves for Vitamin A, tomatoes and fruit for Vitamin C, etc.

6. Seed

1. Which seed of different crops and varieties is most in demand? (These may not necessarily be the most commonly grown crops). Is good quality seed of adapted varieties available locally? Is it certified by any official agency?

If positive, estimate how much seed can be purchased locally without disrupting local market prices. If negative, investigate the feasibility of importing seed, including timing of arrival in country, import/customs and logistic constraints etc.

2. Are the varieties which are available proven to be adapted to the area?

Consult the Ministry of Agriculture, local seed suppliers, agencies and NGOs involved with agriculture etc.

3. What is the normally used seed rate?

Estimate for each crop grown locally. One good way to do this is to show farmers 5 kg, say, of seed and ask them how much land this would plant. Often the seed rate is a theoretical figure because the seed is planted as an intercrop in a mixed cropping system —see **Section 1Hb, "Mixed Cropping"**, page 66.

4. When will the seed be available, both local and imported seed?

Calculate the time available and needed for transport and distribution in advance of the planting season.

5. What are the prices, for both local and imported seed?

Compare prices (CIF to distribution points) between local and imported seed. Note any significant changes in local market prices, though big variations from month to month are common even in normal circumstances (normally lowest just after harvest).

More expensive suppliers are often preferable (justifiable) if their seed is guaranteed high quality and if timely delivery is assured.

6. Is the quality and germination of the seed acceptable?
Check if the seed has been, or can be, reliably and independently tested. If necessary, have the seed tested, by the best means available—see **3B, "Seed Purchase Procedures"**, pages 310–312 and **1Fa, "Germination"**, pages 46–52.
Very often "old" seed (12–18 months or more) is acceptable, depending on the species, storage conditions etc.
7. Is the seed treated, or does it require inoculation?
Investigate if this is desirable or not. If the seed is treated, make sure the beneficiaries are informed of any possible dangers, and if appropriate make a food distribution at the same time as the seed distribution ("seed protection")—see **1Fe, "Inoculation/Nitrogen, Fixation"**, pages 54–57 and **1Ff, "Seed Treatment"** page 57–59.
8. Is germination likely to be a problem?
Check the age (ie harvest date) of seed and find out how it has been stored and handled—see **1Fa, "Germination"**, pages 46–52.

7. Logistics
1. Is there sufficient logistic support for the planned programme?
Investigate the availability of trucks, tractors, planes etc for hire, in addition to any "in-house" transport. Sometimes it happens that imported seed or tools can be delivered closer (and therefore more rapidly and/or cheaply) to the distribution points than those purchased "locally" within the country (and vice versa).
2. Are all the tools and seed to be delivered at the same time?
If transport is limited, investigate the possibilities of spreading the logistic burden by distributing other types of seed with different planting dates.
3. Is the available storage large enough, and dry and secure?
Investigate. Seed should be stored as dry and cool as possible—see **1O, "Storage"**.

8. Soils
Soil samples must be taken carefully to get an accurate idea of the whole field or area. One easy way is to walk right across the field to be sampled in a "W" shape. The W is walked and several samples are taken at random points along it.
1. Are the soils mainly acid, alkaline or neutral?
If it is appropriate, have representative soil samples analysed. Check on the influence of the soil pH—see **1Cc, "pH Value"** and **2 A–G, "Crop Descriptions—soil"**.
2. Are the soils deficient in any of the essential elements?
If the soils are low in Nitrogen, consider introducing legumes such as *Leucaena* or *Acacia albida* trees, clovers, beans etc—see **1C, "Soil"**, pages 11–19, **2B, "Legumes"** pages 150–188 and **2G, "Underexploited Crops"**, pages 266–287.
3. Do the soils have harmful levels of any trace elements or salts?
If so, investigate the possibility of providing seed/cuttings etc of crops which tolerate these conditions—see **1Ca, "Saline Soils"**, pages 15–17 and **1Cd, "Trace Elements"**, pages 23–29.

9. Altitude

1. At what height(s) above sea level are the farmers growing their various crops?
Obtain the best available (contour) maps and/or use a pocket altimeter in the field.
Some crops are only adapted to certain altitudes, and some varieties are better adapted
to high altitudes than others.
2. What is the latitude?
Check on a map. Day-length sensitive crops such as soybeans may not grow well when
they are introduced from an area of different latitude—see **1Ej, "Day-length/
Photoperiodism"**, pages 42-43.

10. Harvest

1. What are the normal yields in the area for each crop?
Give both lowest and higher figures, such as "from less than 300 kg to about 900 kg",
with explanations for the differences. The by-products may be very valuable, such as
the stems of sorghum and maize used for fuel, animal food and building material—see
1Ib, "Yield", page 76.
*2. Will some of the harvested crop from the seed you provide be retained for sowing
in the next season?*
Consider distributing different crops, or varieties, or tools, in the second year of the
programme. Advise farmers to select seed or cuttings/roots from the best plants. The
distribution of small quantities of less well-known crops may be very valuable—see **2G,
"Underexploited Crops"**, pages 266–287.
3. What are the problems with post-harvest (storage) losses?
If there are serious problems, consider either providing seed of crops which store more
easily, such as Finger Millet, or assistance with improved storage facilities (eg seed
banks), or pesticides/traps/bounties etc. Because seed should be stored as dry and cool
as possible, in hot or humid places consider transportation of the seed to cooler and less
humid places until planting season—see **1O, "Storage"**, pages 100–103.

11. Markets

*1. What kinds and quantities of food, seed and animals are available in the local
markets?*
Observe quantities, prices and seasonal variations. If possible, maintain weekly records
of prices of all available staples and animals.
Market prices, and the way these rise and fall, can be a very sensitive indicator of the
food supply situation. Ideally this data should be collected by a local(known) person.
*2. Are there goods for sale in the markets which have been distributed by your
agency/NGO or others, or by government?*
Investigate the reasons for them being sold and not used, and in what quantity and at
what prices, and modify your programme if required.
3. How do farmers transport their produce to the market? Is this a problem for them?
Investigate the transport systems. Sometimes, assistance with transportation can be a
highly relevant form of support.

12. Livestock

1. How important are animals for the families' food production—for meat, milk and blood as well as for transport and field cultivations?

2. What are the main problems associated with livestock, such as diseases, water supply, grazing?

3. What part do animals play in the economy of the area?

Discuss with farmers, traders and local vets. Continually monitor market prices, and note which prices rise and which fall during periods of food shortage. Compare with prices in previous years when possible.

Sometimes providing assistance with, for example, vaccinations or providing oxen or ploughs may be a more appropriate response than providing seed or tools.

By-products such as dung can be valuable, as compost/mulch/manure and/or fuel and building.

13. Other Agencies and NGOs

1. Are there any other aid agencies, NGOs or government agencies engaged in the same or similar areas?

2. Which crop species, varieties, tools etc are they distributing, and why?

Discuss with the other players involved their plans, past experience, problems etc. Inform all the relevant government and other agencies of your plans. Try to avoid duplication, and learn from the experience and mistakes of others.

14. Information Bank

Is the information outlined above readily available for others to use?

Establish agricultural files and reference material, preferably in one place such as your agency's Head Office. Agricultural research stations can be a good source of up to date information and trends. Experienced local agriculturalists, college students etc. who can check and augment this information can also be invaluable in collecting, collating and providing data.

BIBLIOGRAPHY

General Agriculture and Horticulture

Agricultural Compendium. [Elsevier Science Publishers B.V] ISBN 0-444-42905-0

Tropical Crops—Monocotyledons. J.W. Purseglove [Longman] ISBN 0-582-46606-7

Tropical Crops—Dicotyledons. J.W. Purseglove [Longman] ISBN 0-582-46666-0

Fundamentals of Modern Agriculture. C.D. Blake (Editor) [Sydney University Press, 1967]

Principles of Field Crop Production. J. Martin and W. Leonard. [Macmillan Co., New York], Library of Congress Catalog Card No. 67-16360

(Freams) Elements of Agriculture. D.H. Robinson (Ed.) [J. Murray, London] ISBN 0-7195-2579-9

The New Oxford Book of Plants. [OUP 1997] ISBN 0-19-850567-1 (Paperback)

Primrose McConnell's The Agricultural Notebook. [Butterworth and Co., London] ISBN 0-408-00208-5

Blacks Agricultural Dictionary. D.B. Dalal-Clayton A. & C. Black Ltd, London ISBN 0-7136-2130-3

Farm Crops. G.Boatfield [Farming Press Ltd, Suffolk, England] ISBN 0-85236-129-7

The Plant Kingdom. H.C. Bold [Prentice-Hall] ISBN 0-13-680389-X.

The Self Sufficient Gardener. J. Seymour [Faber and Faber, London] ISBN 0-571-11212-9

East African Crops. J.D. Acland [Longman] ISBN 0-582-60301-3

Vegetable Growing in Zambia. P.J. Mathai [Zambia Seed Company Ltd]

Agricultural Science. G.H. Owen [Longman] ISBN 0-582-60611-X (Book I), 60612-8 (Book II) and 60613-6 (III)

Crop Production in Dry Regions, Volume II. I. Arnon [Leonard Hill, London] ISBN 0-249-44086-5

Crops of the Drier Regions of the Tropics. D. Gibbon & A. Pain [Longman] ISBN 0-582-77506-X

Tree and Field Crops of the Wetter Regions of the Tropics. Williams, Chew and Rajaratnam [Longman] ISBN 0-582-60319-6

Evolution of Crop Plants. N.W. Simmonds (Editor) [Longman] ISBN 0-582-44496-9

Agriculture Tropicale en milieu paysan africain. H. Dupries & P. de Leener [Terres et Vie]

Agricultural Botany. Gill and Vear [G. Duckworth & Co., London]

The Greening of Africa. P. Harrison [Paladin] ISBN 0-586-08642-0

Vegetables. The Garden Plant Series. R. Phillips and M. Rix [Macmillan, London] ISBN 0-333-62640-0

Encyclopaedia Britannica CD '99

Legumes
Crop and Product Digest. No.3 Food Legumes. D.E. Kay [Tropical Products Institute, London] ISBN 0-85954-085-5
Tropical Legumes: Resources for the Future. [National Academy of Sciences, Washington DC, USA] Library of Congress Catalog Card No.79-64185
Modern Soybean Production. Scott and Aldrich [S. And A. Publications, Illinois, USA] Library of Congress, Catalog CardNo. 73-113066
The Winged Bean—A High Protein Crop for the Tropics; Leucaena - Promising Forage and Tree Crop for the Tropics; Underexploited Tropical Plants with Promising Economic Value—three booklets from the National Academy of Sciences, Washington DC, USA
Technical Handbook on Symbiotic Nitrogen Fixation. [FAO] ISBN 92-5-101440-X
Les Inoculums de Legumineuses et Leurs Applications. [FAO] ISBN 92-5-201441-1

Fertiliser
Fertiliser and Plant Nutrition Guide. [FAO] ISBN 0-92-5-102160-0
Fertiliser Guide for the Tropics and Subtropics. Jan G. De Geus [Centre d'Etude de l'Azote, Zurich]
Hunger Signs in Crops. H.B. Sprague [McKay, New York]

Plant Diseases and Fungicides
Plant Pathology in Agriculture. D.W. Parry [Cambridge University Press, 1990]

Seed
Successful Seed Programs: a Planning and Management Guide. J.E. Douglas [Westview Press, USA] ISBN 0-89158-793-4
Seed Physiology. J. Bryant [E. Arnold, Australia] ISBN 0-7131-2898-4
Improved Seed Production. [FAO] ISBN 92-5-100243-6
World List of Seed Sources. [FAO] ISBN 92-5-001297-7
Agricultural and Horticultural Seeds. [FAO] ISBN 92-5-100475-7
An Introduction to Seed Technology. J.R. Thompson [L. Hill, London and Glasgow]
Seed Dressing (Treatment). K.A. Jeffs (Editor) [British Crop Protection Council] ISBN 0-948404-00-0
Crop Genetic Resources: Conservation and Evaluation. J. Holden and J. Williams (Editors) [IBPGR] ISBN 0-04-581018-4
Seeds and Their Uses. C. Duffus and C. Slaughter [John Wiley and Sons] ISBN 0-471-27798-3

Nutrition
Human Nutrition in Tropical Africa. [FAO] ISBN 92-5-100412-9
Tables of Representative Values of Foods Commonly used in Tropical Countries. B.S. Platt [Medical Research Council/HMSO, London] ISBN 0-11-450009-6

Insects

UK Pesticide Guide 2000 (The Green Book) [The British Crop Protection Council/ CABI] ISBN 0-85199-468-7
Pest Control and its Ecology H.F. van Emden [E. Arnold, London] ISBN 0-7131-2472-5
Formulation of Pesticides in Developing Countries. [UNIDO]
Poisonous Chemicals on the Farm. [HMSO] ISBN 0-11-883414-2
Produits Phytosanitaires Ciba-Geigy [Bale, Switzerland]
The Use of Plants and Minerals as Traditional Protectants of Stored Products. Golob and Webley [Tropical Products Institute, London] ISBN 0-85954-115-0

Weeds

The World's Worst Weeds. Holm, Plucknett, Pancho and Herberger [University Press of Hawaii] ISBN 0-8248-0295-0
Biological Control of Pests and Weeds. M.J. Samways [E. Arnold, London] ISBN 0-7131-2822-4

Tools

Tools for Agriculture. [Intermediate Publications Ltd, London] ISBN 0-946688-36-2

Storage

Processing and Storage of Foodgrains by Rural Families. [FAO Training Series] ISBN 92-5-101276-8
Prevention of Post-harvest Food Losses. [FAO Training Series] ISBN 92-5-102209-7

INDEX